BEYOND MAMMOTH CAVE

BEYOND
Mammoth Cave

A Tale of Obsession in the World's Longest Cave

James D. Borden and Roger W. Brucker

Southern Illinois University Press
Carbondale and Edwardsville

Printed in the United States of America

03 02 01 00 4 3 2 1

Frontispiece: A typical vertical shaft in Kentucky, formed by vertical flowing water seeking master base level. Photo by Art Palmer.

Library of Congress Cataloging-in-Publication Data
Borden, James D., 1957–
Beyond Mammoth Cave : a tale of obsession in the world's longest cave / James D. Borden and Roger W. Brucker.
p. cm.
1. Caving—Kentucky—Mammoth Cave. 2. Mammoth Cave (Ky.)—Discovery and exploration. I. Brucker, Roger W. II. Title.

GV200.655.K42 M253 2000
796.52′5′09769754—dc21
ISBN 0-8093-2345-1 (cloth : alk. paper)
ISBN 0-8093-2346-x (pbk. : alk. paper) 00-021610

The paper used in this publication meets the minimum requirements of American National Standard for Information Sciences—Permanence of Paper for Printed Library Materials, ANSI Z39.48-1992. ♾

Dedicated to the memory of

Bob Keller

Ralph Powell

Barbara Lipton

Roberta Swicegood

John Bridge

James Quinlan

Frank Reid

Fred Benington

Joe Kulesza

Omne ignotum pro magnifico.
(Anything little known is assumed to be wonderful.)

—Tacitus

Lose your dreams and you will lose your mind.

—The Rolling Stones

CONTENTS

PREFACE

This is a very personal story of obsessive cave exploration in the world's longest cave. It is told from two points of view by two central participants who are separated by more than a quarter of a century in age and outlook. The big cave sucked us in. For long periods of time, it was our lives, sometimes to the detriment of family, friends, and jobs. But we are satisfied. When this story began, Mammoth Cave was just over 144 miles long; when it ended, the world's longest cave was over 300 miles long—and was predicted to reach the 500-mile mark.

So who cares about these caves? We do. We have always been explorers. Both of us have vivid childhood memories of madly wanting to explore the cliffs and creeks around our homes.

I, Roger Brucker, taught myself layback climbing in 1933 when I was four years old. I climbed over a board fence to discover a world of giant rocks among whose dark recesses I crept. I had found the hidden place! Within minutes (probably), an adult head appeared over the top of the fence. I don't remember the words, but the message was clear: Don't you ever do that again!

Of course, I paid no attention at all. Each time I explored was more thrilling than the last. Those rocks were mine, my world. I also discovered that exploring involved sneaking off when nobody was looking and slipping back before anyone noticed I had gone.

I, Jim Borden, was four years old in 1961. Trails, mysterious and endless, stretched from my house under arching tree branches to distant hills. One day I spied a large tree house in a huge oak. It was wonderful beyond all my dreams. This big tree house was mine because it was discovered by me!

"Don't you ever go back there!" said my mom. "It's rotten. You'll fall and be killed." I went there all the time. My imagination fired into orbit from that tree house. And when a friend set a fire that burned the forest down, I learned that exploration is risky.

To explore Mammoth Cave, we did some neat things and some nasty things. Here we tell all. Other cavers, honest or crafty, may disagree with our version of this story, but we can only present the facts as we know them. So this book is not like its predecessor, *The Longest Cave* by Roger W. Brucker and Richard A. Watson. In 1973, they wrote in the third person; in this book, we write almost exclusively in the first person, indicating the speaker at the beginning of particular chapters as needed. They showed their manuscript to many of the participants in order to incorporate corrections and amplifications in their book; we showed our manuscript to only a few friends and enemies.

Some cavers have colossal egos and a limited appreciation of the contributions of others. Even we have been characterized this way! A few may try to promote themselves by downplaying others. And even we have been accused of this! We'll just have to let the readers decide. So let's get to the damn cave!

ACKNOWLEDGMENTS

We are grateful to the hundreds of cavers whose work we describe herein.

For extraordinary assistance, we thank the National Park Service, the Department of the Interior, and park superintendents Joseph Kulesza, Amos Hawkins, Robert Deskins, Frank Pridemore, and David A. Mihalic. As protectors and interpreters of most of Mammoth Cave, they made the exploration possible and have provided material and moral assistance. Their encouragement and personal help can never be repaid.

We thank the Cave Research Foundation (CRF) and its president, Patricia Kambesis, and the Central Kentucky Karst Coalition (CKKC) and its president, James Wells, for access to records and editorial assistance.

Landowners on whose properties the various cave entrances are located have our grateful thanks for letting us come and go. Thanks to John Logsdon, Jerry Roppel, Bill Downey, and Elroy and Marilyn Daleo.

We thank cartographer Patricia Kambesis for her preparation of maps that help make our story understandable and for her extraordinary effort in assembling hundreds of miles of cave surveys from multiple individuals and organizations. This compilation made drawing the maps possible.

We thank Linda Heslop for her creative illustrations, without which our story would be far from complete.

For the several black-and-white photographs, we are indebted to Art Palmer. We thank Bill Eidson, David Black, Ron Simmons, Paul and Lee Stevens, and Pete Lindsley for use of their color photographs.

If it weren't for the nagging, cajolery, encouragement, and editorial guidance of Richard "Red" Watson, this book would not exist. He has been a faithful friend to both of us through thick and thin. We thank him.

For editorial and critical assistance, we thank Rick Olson, Ron Bridgemon, R. Scott House, Karen Willmes, Peter Zabrok, Leonard B. Taylor, Patricia Porter, Stanley Sides, Richard Zopf, and Elizabeth Winkler. We gratefully acknowledge the professional editorial assistance of Carol Burns at Southern Illinois University Press as well as that of freelance editor Julie Bush.

Accounts used in this book are from various sources, but most of the information is from the memories and correspondence of the authors. Additional information was obtained from conversations with the participants.

Details of specific trips were obtained from the unpublished official trip report logs of the CKKC and the CRF. Those organizations throughout the period covered in this book required each party leader to prepare a detailed narrative report describing each trip. These files represent the most extensive contemporaneous record of who did what and when.

We also consulted the personal journals and memoirs of Pete Crecelius, Pete Lindsley, Lynn Brucker, and Darlene Anthony. Some of the information about Don Coons and Sheri Engler's secret work in Morrison Cave is from "In Morrison's Footsteps" by Don Coons and Sheri Engler, *NSS News* 38, no. 6 (1980): 127–32, 137, and from "In Morrison's Footsteps" by Don Coons and Sheri Engler, *Caving International* 12 (1981): 28–37. A summary of the Proctor Cave–Morrison Cave connection with Mammoth Cave appeared in "New Kentucky Junction" by Roger W. Brucker and Pete Lindsley, *NSS News* 37, no. 10 (1979): 231–36. A summary of the Mammoth Cave–Proctor Cave–Morrison Cave connection with Roppel Cave appeared in "The Joy of Connecting" by Roger W. Brucker, together with "The Roppel-Mammoth Connection" by Jim Borden and Pete Crecelius, in the *NSS News* 42, no. 2 (1984): 105–9.

Background information about organizational policy was obtained from the CRF and CKKC archives or from public information published by the National Park Service, Department of the Interior. Mike Dyas and Bill Mixon prepared several synopses and analyses of the policy, politics, and events of the central Kentucky cave scene. These were circulated privately to the participants and appeared in whole or in part in one or more NSS Grotto publications. These were helpful in refreshing our memories on several points.

Historical information was checked against two books: Robert K. Murray and Roger W. Brucker's *Trapped!* and Roger W. Brucker and Richard A. Watson's *The Longest Cave.* Some details of the discovery of Proctor Cave are from William Stump Forwood's 1870 book, *An Historical and Descrip-*

tive Narrative of the Mammoth Cave of Kentucky. Additional historical information is from reminiscences of John Bridge, Stanley Sides, and local residents obtained by oral interviews.

Some data used for the maps were provided by Joe Saunders, the Detroit Urban Grotto of the NSS, and Don Coons. We thank them for their contribution.

Our acknowledgment would not be complete without pointing out that none of the named organizations or individuals necessarily endorse the viewpoint that this book presents. This is not an "official account": it is our personal story of how we experienced this grand adventure.

AUTHORS' NOTE

Caves are fragile. Their features may be thousands or even millions of years in the making. Cave animals are rare. They live precarious lives on the knife edge of ecological balance. Both cave features and cave life can be destroyed through ignorance on the part of cavers who do not understand the fragility of underground relationships. Irresponsible people have taken stalactites that will never grow again. They have killed bats by entering and disturbing habitats at a time when food supplies were low. Caves are unique places for adventure and discovery, but before you enter a cave, we urge you first to learn about safe and careful caving. Please contact the National Speleological Society, 2813 Cave Avenue, Huntsville, Alabama, 35810-4431. They will guide you to a safe and fulfilling adventure in caving.

INTRODUCTION

Mammoth Cave National Park is located about one hundred miles south of Louisville, Kentucky, near Interstate 65 on top of an upland area known as the Mammoth Cave Plateau. This upland is part of a two-hundred-square-mile region of interior drainage called the Central Kentucky Karst. As in the limestone karst region of Yugoslavia (where the name "karst" originates), rainwater runs off the surface and plunges underground through sinking creeks and sinkholes. The water eventually drains through hidden passages into the Green River and Barren River.

Nearly three hundred feet of Mississippian limestone was laid down about three hundred million years ago. Perhaps two million years ago, water began to drain through the network of cracks in the rock to form the caves by erosion and solution. Enlarged cave passages consist of vertical shafts, which divert the surface water into the underground streams, and horizontal tubes and canyon passages, which carry water to the river-level spring outlets. The upper levels of these caves are abandoned drainage systems, whose waters have drained to successively lower levels of the cave as the Green River lowered its bedrock floor.

How are limestone caves made? Four things are necessary and sufficient in Kentucky: (1) a source of water over a long period of time (rainfall is about fifty inches per year); (2) rocks with cracks (the limestones have abundant joints, bedding planes, and faults); (3) rocks that can be removed (roughly two cubic miles of rock have been dissolved or transported away); and (4) a place for water to drain (the Green River and Barren River). Caves are created only if all four requirements are met.

A special set of circumstances accounts for the great lengths of the cave systems in this part of Kentucky. Covering the soluble, cracked limestone is a caprock layer of tough sandstone that resists weathering. Sometimes a thin, impermeable shale layer at its base prevents groundwater from penetrating to the limestone below. This combination of sandstone and shale acts as a cap or roof that diverts runoff water to the edges of the ridges. This concentrates the waterflow along a narrow front. Funneled underground at the edges of the caprock, the descending water dissolves vertical shafts that resemble the insides of grain storage silos.

Such aggressive, high-energy vertical flows of water underground are responsible for removing more than 90 percent of the missing rock from this karst region—a karst chainsaw ripping away the landscape. Water flowing horizontally through cave passages has removed less than 10 percent of the missing limestone.

Under the resistant sandstone caprock on the tops of ridges, miles of horizontal cave passages are protected for a time against the most aggressive rock removal.

These are the caves we explore.

All the caves in the Central Kentucky Karst are connected, but the connections may be hidden behind rocks, down tiny crawlways, down vertical shafts, or even beneath the surface of the water in the base-level streams such as Echo River in Mammoth Cave. People may or may not fit through the connections.

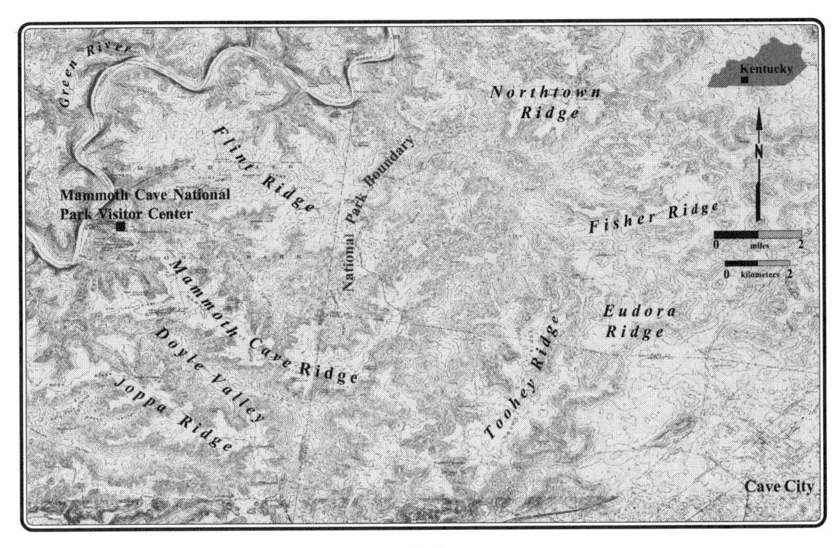

Mammoth Cave area

Green River, the master base level of the Mammoth Cave region

"Connecting caves" means that cavers find elusive natural connecting passageways rather than make their own by tunneling or blasting.

About 2500 B.C., aboriginal Americans began to explore Mammoth Cave, Salts Cave, and Lee Cave in what is now Mammoth Cave National Park and mined mirabilite salts, gypsum, and selenite crystals. Ancient people explored as far as two miles into the caves.

Prior to the mid-1950s, cave discoveries were usually fortuitous and infrequent results of exploration. Most old-timers just followed the cave. Sometimes they abandoned the search when the passages ended in breakdowns, piles of rocks produced by the collapse of the ceiling and walls that barred further progress.

During the fifties, a group of explorers—who founded the Cave Research Foundation in 1957—began to see distinctive passage patterns in the cave maps they plotted from their surveys. The patterns gave insights that were the key to connecting these caves. Principles discovered through years of CRF work were then applied with spectacular results to find more miles of cave passages than anywhere else in the world.

These explorers swept into mile after mile of new cave after their initial connection of Unknown Cave to Crystal Cave in 1955. In 1960, they found the connection between Colossal Cave and Salts Cave in Flint Ridge in

Lee Cave trunk passage, one of the many dry upper-level passages used by aboriginal Americans in search of mineral salts

Mammoth Cave National Park. The following year, CRF explorers found the connection between Floyd Collins' Crystal Cave–Unknown Cave and Colossal Cave–Salts Cave in Flint Ridge.

The stage was set for a series of trips in the summer and autumn of 1972, culminating in the discovery of a connection between the Flint Ridge Cave System and Mammoth Cave on 9 September 1972. The cave's combined passageway length was 144.4 miles, 72.6 miles longer than the second longest cave in the world, Hölloch Höhle in Switzerland.

John Wilcox, the leader of the CRF Flint Ridge–Mammoth Cave connection trip, built his breakthrough trip on the discoveries of many explorers, starting with the ancient aboriginal Americans on the Mammoth Cave end and Floyd Collins on the Flint Ridge end. (Collins, a farmer and cave explorer, began caving when he was only six years old; at the apex of his career, he discovered Floyd Collins' Crystal Cave in 1918. On 30 January 1925, he became trapped in Sand Cave. For more than two weeks, rescuers tried to reach him, but their struggle ended on 16 February when the rescue shaft broke through and Collins was found dead. The story was one of the most sensational media events of that time.)

The caves are always dark, so every explorer carries at least three separate sources of light. Most of the exploration in Mammoth Cave was lit by carbide lamps. This two-part brass lamp contains a lower chamber into which lumps of calcium carbide are placed. A top compartment is filled with water and adjusted with a valve to a rate of one drip per second. When the two halves of the lamp are screwed together, the water reacts with the calcium carbide to generate an acetylene gas jet. When ignited with a spark from the built-in flint lighter, the gas burns with a bright yellow-white flame. A polished reflector focuses the light ahead of the caver. Cavers can carry enough carbide in two baby bottles to last for thirty-six hours.

Electric headlamps are sometimes used in Mammoth Cave, but carbide is cheaper and more reliable. Cavers debate the merits of the two types of light, but veteran cavers know there's a place for both.

Other sources of light include flashlights, plumber's candles, matches, or spare carbide lamps. The reason cavers bring so many types of light is that if a caver gets lost in the dark, he or she could die.

Cavers wear the headlamp on a high-impact plastic helmet or hard hat that protects against ceiling scrapes and rockfalls. Rubber pads protect knees while crawling and permit friction for the knees while climbing.

A caver's pack contains food and supplies for about twenty-four hours of hard exploring. It resembles a large purse and holds everything that the old-timers tried to stuff into their pockets. Carbide, water bottle, flashlights, candles, survey gear, canned food, and other useful items fit into the pack.

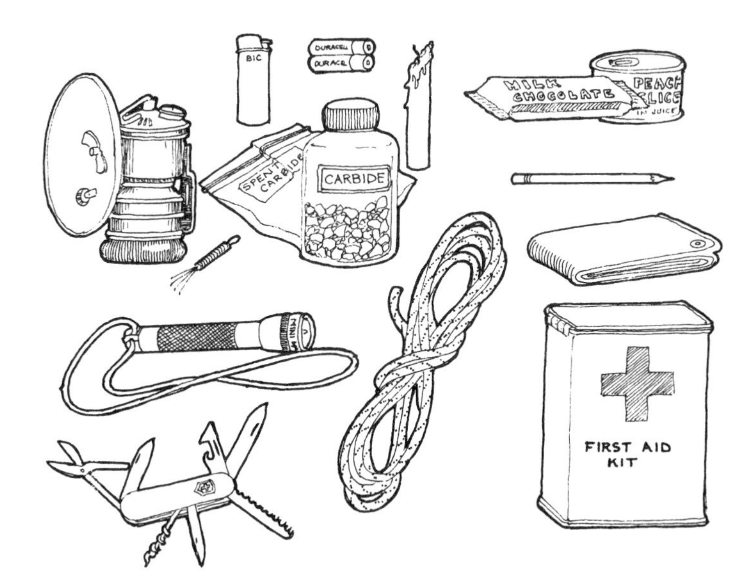

Contents of a cave pack

A variety of climbing hardware is used for descending and ascending vertical pitches in caves. Cavers use single-rope techniques of rappelling—a controlled friction slide down a sheathed, non-stretch nylon rope. They ascend the same rope using a variety of cam or ratchet devices that permit an inchworm-type of climbing movement upward.

For surveying cave passages, one uses, for example, a Suunto magnetic compass that displays the bearing from one survey station to the next with a precision of one-half degree or less. Survey teams measure the distance between survey stations with a fifty-foot tape and the vertical angle up or down with a Suunto clinometer. The compass reader calls the observed measurements out loud to a note taker, who writes the data in the survey notebook. The note taker also draws a sketch map of the passage. In short, a cave is mapped by describing a line through three-dimensional space, then by describing the cave surrounding the line. A survey party may return to the surface with a survey of a half-mile in easy walking passageway, or maybe only a few tens of feet in a tough, small, muddy crawlway.

Why survey caves? Cave maps drafted from the surveys are the prime tools of discovery. In the Mammoth Cave region, few serious cavers even think about cave exploring without doing a simultaneous survey. We made the big discoveries and connections that way, although many who do not survey still think our findings in Mammoth Cave were pure, enviable luck.

Getting lost in a cave is a horror imagined by all readers of Mark Twain's *Tom Sawyer.* The fear is exaggerated and unwarranted for most simple caves, because their plan is obvious. Serious cavers pay attention to the confusing

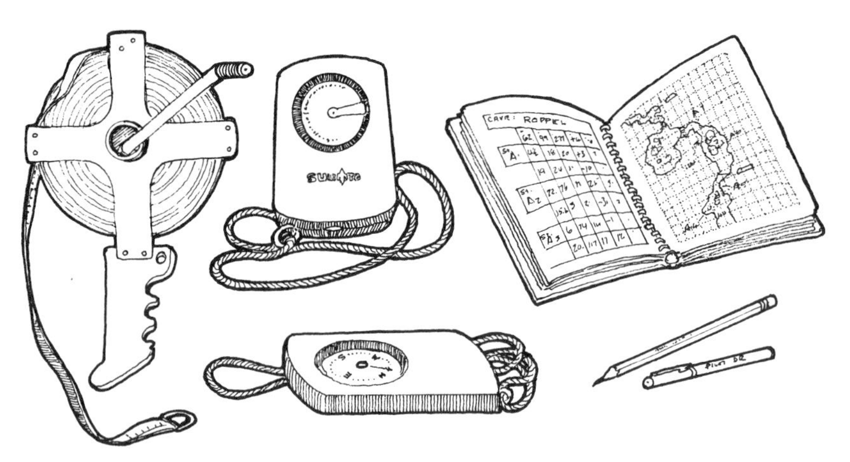

Tools used for cave surveying

junctions, as a visitor to a city would note the streets. But large caves can be so confusing that inexperienced spelunkers get lost. Cave maps, drafted from cave passage surveys, untangle this confusion, just as a road map shows the complex route for a long trip. Other dangers are falling rocks or cavers falling into holes. Death by drowning is a risk on the lowest levels of caves and is a relatively frequent occurrence in cave diving. Caves can kill or injure in many ways, even if explorers are careful.

Jim Borden taking the lead in stream passage

1

Blazing Glory and Blind Ambition

Roger Brucker Tells of the Thrill and Envy

"Whump! My wetsuit pack hit the sand floor seven feet below me. I—Roger Brucker—sat on the edge of the drop and braced on a foothold three feet below. Then I jumped down. I scooped up the bulky but lightweight wetsuit pack and headed south in a walking passage to join the others.

We carefully picked our way down an easy free-climb that connected a passage containing a B Survey with the Third A Survey in Cocklebur Avenue in Mammoth Cave. Six of us were laden with shoulder packs stuffed with cave gear: extra carbide and water, food, surveying compass and measuring tape, and flashlights. Each caver clutched an additional pack containing a wetsuit.

A pool of yellow light from our combined carbide lamps filled a room about fifteen feet wide, twenty-five feet long, and six feet high. Some of us stood and some sat as we all stripped off cave clothes and crammed ourselves into the sponge rubber wetsuits that scuba divers wear.

Sheri Engler's piercing shriek cut through the murmur of voices: "Don, help me with this goddamn zipper! You said you were going to fix it. If I get cold, I can't read compass!" Her husband, Don Coons, rolled his eyes toward the ceiling, stood up, and began to fumble with her wetsuit zipper.

Dave Weller talked nonstop about tearing apart an old building in Louisville. Jim Borden peered into the gloom ahead, where the water started and the ceiling dipped almost to meet it. Lynn Brucker, my wife of eight months, pulled her wetsuit on with no wasted motions, while I struggled into my patched, secondhand, much-used wetsuit.

Our purpose on this September day in 1983 was to find and survey a connection between Mammoth Cave and Roppel Cave, and our excitement about this "trip of a lifetime" was expressed in various ways. The "engineers," whether talking or listening, methodically prepared for their plunge into water. These included Dave, a self-taught, practical engineer and cave entrance constructor; Lynn, an electrical engineer by profession; and Jim, a systems analyst with IBM. The rest of us were "liberal arts"–oriented. Sheri had a degree in history, Don was a farmer, and I worked in an advertising agency. Our preparations were more spontaneous and chaotic than the others'. All of us were seasoned cave explorers who had devoted from five to thirty years of our lives to exploring the big caves of Kentucky.

What were our prospects for success? Certainly slim. Finding connections between these caves took years, often frustrating years of search, disappointment, and despair. There was no luck to it. Nobody "stumbles" onto a connection. A cave connection is the product of sequential discoveries, of building on new knowledge from mapping and from crawling on the backs of previous explorers. While the probability of finding a connection was low, there was always the eternal driver: possibility.

Our friendly, echoing chatter filled the passage as we plunged into the river to wade upstream. Wetsuit caving is warm and comfortable; cold cave water seeps in between suit and skin but quickly heats to body temperature. The noise of splashing water drowned out conversation except between adjacent pairs of cavers. Since I was leading the party, I could see deep into the unsilted, clear pools of water. I saw several white cave crayfish and a pale pink cave blindfish and reduced the pace so I wouldn't miss any of this fine glimpse of the cave ecosystem.

After three hundred feet, Jim Borden cruised by me in a crashing wave of foam at high speed. A natural leader, he had grown impatient at my scenic pace. Rather than negotiate with the old man, he had shifted into second gear and passed him. He did not relinquish the lead for the rest of the trip.

Our long hike in the underground river was a delightful succession of wading, swimming, and portaging across gravel bars for thousands of feet, all in a passage from twenty to thirty feet wide by twenty feet high. Jim and Dave had never been in this passage, but the rest of us had. It was the finest cave anyone could want, and our spirits were high. We knew we would have a productive trip, even if we didn't find a connection.

How many times had we been confident of finding a connection, only to meet a dead end? Truthfully, most of the time. Only a few perpetual enthusiasts like Jim and me thought in romantic terms of "sure" connections and "certain" discoveries. Much of our palaver had been bait to lure new cavers. Yet, secretly we knew we would find the connection between Roppel Cave and Mammoth Cave. Some of the old-timers with us, especially the engineers by temperament or vocation, regarded us as bullshitters.

What makes cave connections rare is collapse. Flowing underground water enlarges a crack into a passage. As the passage enlarges, its walls and ceilings become stressed. Rock slabs and arches fail, and the fallen rock closes off the passage. The worry is not that the roof will fall on the caver's head (which is what non-cavers think) but that the fallen rocks will terminate the passage and frustrate the caver to death.

At 12:30 P.M., our party of six reached the spot where the river upstream had formerly terminated in a sump. Dave and I brought up the rear. Don and Jim led the party in a rush under the low ledge that used to be underwater. They left us behind as Dave changed the carbide in his lamp. This was the first moment of quiet on the trip.

We then caught up with the others at the end of a long stretch of river wading, where a breakdown filled most of the passage and covered the river. We could move along the left wall of the passage near the ceiling. This was extremely wet cave. Water beads stood out on the walls and ceiling and fresh mud coated the floor. Shattered rocks jutted out from the right side. I knew this kind of passage must lie beneath a valley, where all the waterflow and rock weakening processes were maximized. We bumped some rocks and they fell, reminding us that while general collapse was unlikely, local collapse of a few pounds of this unstable rock could trap, injure, or kill us.

Breakdown

A piece of orange flagging tape marked the last survey station from the Mammoth Cave passages behind us. Jim and Don were now leading.

I heard a faint cry ahead of us.

Don shouted, "It's them!"

We all yelled simultaneously, then got better organized and shouted "Hello" in unison. The muffled shouts from the distant caver seemed closer now. My heart pounded!

Ding! My lamp struck a rock and went out, plunging me into darkness.

"Dave," I yelled, "I need a light." I could have opened my pack for a flashlight but thought Weller was close by. He reversed and came back to me to light my lamp with the flame of his.

In a few minutes we both moved forward a couple of hundred feet, then climbed up to the right where the traffic scuff marks led through a crack. The room we squeezed into was about four feet high, floored by large breakdown slabs extending the width of the fifteen-foot-wide passage. There was mud everywhere. In the small room we joined Jim, Lynn, Don, Sheri, and John Branstetter.

John, who had heard our shouting, was a member of a team of explorers who had entered Roppel Cave that same morning. Their objective was to search for a connection to Mammoth Cave from the Roppel end, following a river passage that certainly seemed to be the same river in both caves.

We shook hands with John, an unusual gesture for cave explorers in a cave, but it seemed a suitable greeting for our counterpart connection party. John said that the others in his party were some distance back, squeezing through a tight crawlway. One by one, Roberta Swicegood, Bill Walter, and Dave Black popped into the connection room, the site of our long-awaited success. We laughed, hugged each other, and shook hands again, then took photographs of the combined connection party.

John described how the Roppel party had arrived at the farthest point of penetration. They had concluded that the previous party had not penetrated far into the breakdown. As John and Roberta were placing new knee-print marks in the mud of a virgin crawl, he had yelled, "Hey!" but had heard nothing in return.

Don Coons built a cairn to mark the connection spot. I smoked a sign on a flat rock with my carbide lamp: "9–9–83–293," which indicated the date and the number of combined cave miles we had joined together by finding and traversing the connection route. Nobody wanted to leave.

The plan was that if a connection were made, the parties would survey the connection route and tie the survey into stations on either side, and each party would exit the other's entrance. A grand traverse, portal to portal, with a party crossover was an exciting prospect, but we were way-to-hell-and-gone in the cave and had plenty of survey work to do.

Jim Borden asked me, "How does it feel to actually be in on a connection?"

I smarted a little, thinking to myself, You squirt, I was exploring these caves before you were born! Back in 1954, three years before Jim existed, I was plotting how to connect Floyd Collins' Crystal Cave with Mammoth Cave in Mammoth Cave National Park. I had helped or led expeditions—that is, I set the expeditions up but didn't directly participate—when the connections of 1960 and 1961 were made. I helped with the connection of the Flint Ridge caves and Mammoth Cave in 1972 and Proctor Cave to Morrison Cave in 1979. Later in 1979, I had led a trip on which the connection between Mammoth Cave and Proctor-Morrison Cave was found. True, I had never been on a connection party, but wasn't that just a detail? A stinging little detail? Well, this was not the time for sarcasm.

I said to Jim, "I'm pleased to be invited." This was the first time I had been on an actual connection. It felt wonderful.

We started surveys about 2:30 P.M. from the connection room and tied them back to the respective caves. Our party left via the Roppel Entrance around midnight. John Branstetter's party emerged from the Ferguson Entrance about 2:30 A.M. Then both parties celebrated with champagne. The three-hundred-mile cave was 98 percent complete, a blaze of glory indeed.

Becoming president of the Cave Research Foundation had been my ambition since the summer of 1957, when Philip Smith, Burnell Ehman, David Huber, Jim Dyer, Dave Jones, Jack Lehrberger, and I formed the CRF on Smith's front porch in Springfield, Ohio. In November 1974, I did become

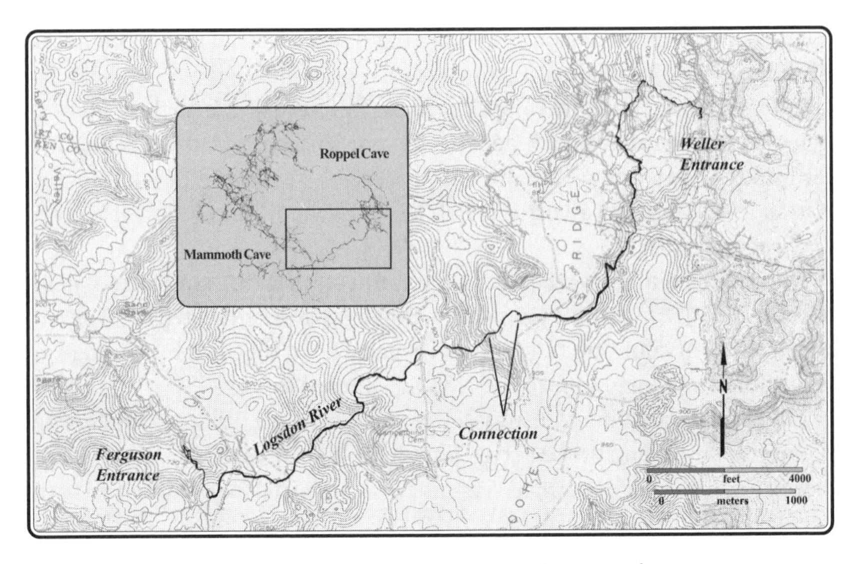

Connection route between Roppel Cave and Mammoth Cave

CRF president. Much of a rich lifetime of caving was behind me. At forty-four, I had been crawling in Flint Ridge caves since Jim Dyer had introduced us to the sport twenty-one years before. With Joe Lawrence Jr. I wrote a book, *The Caves Beyond,* about the National Speleological Society's 1954 expedition into Floyd Collins' Crystal Cave. I had participated in setting up connection trips in the Flint Ridge Cave System in Mammoth Cave National Park, Kentucky. Some of my discoveries were short but vital pieces of the trail that became the connection route to Mammoth Cave in 1972. I felt a romantic thrill when my friend John Wilcox and his connection party telephoned me at 4:00 A.M. on 10 September 1972 to tell this "grand old man" of caving that the big connection had been made.

While I also felt keen disappointment at not being on the Mammoth Cave connection trip, I knew I was too big to fit through the squeeze known as the Tight Spot that blocked the route for me. Were I Wilcox, I'd have left me at home, too.

Following the connection, I had two thoughts: First, this 1972 grand Kentucky junction was not going to be the last of the big connections. I could continue to help set up and make future cave connections. Second, I wasn't the premier cave connection strategist I thought I was. John Wilcox had succeeded with a better strategy, and we were as different in temperament as could be.

Two seventeenth-century French philosopher-scientists illustrated our two polarized approaches, René Descartes and Blaise Pascal. Descartes believed the world was governed by reason and that it would yield its mysteries to reasoned inquiries. Wilcox the engineer was like that: logical, systematic, thorough. I viewed him as a thinker with never an impulsive thought. He had methodically fitted together the Mammoth Cave connection route, lead by lead, eliminating each possibility one by one. Wilcox was successful.

Blaise Pascal, a contemporary of Descartes, changed his whole way of life upon receiving a revelation from God in his study one evening. He argued that reason must be abandoned altogether in favor of an important human factor, intuition. I fancied myself a Pascal. To me, the Gestalt psychology idea of the "Aha!" experience summed it up. I could imagine the cave in my mind in three dimensions and crank it forward or backward like a movie film. The sudden breakthrough of insight was what revelation meant to me. I thought like a shotgun, Wilcox like a high-powered rifle. Wilcox was a doer, I was a communicator. It wasn't that I saw myself as unsuccessful in relation to Wilcox; rather, I was not yet successful.

In the flurry of CRF activity following the 1972 connection, Red Watson and I plunged into the tasks of photographing the connection route, writing news releases, and planning a press conference. We set Wilcox to work making a map of the connection route.

Red and I collaborated on a book about everything leading up to the connection. Red and I had lived and crawled through the toughest places. Together, over the years, we had enjoyed each other's full involvement in exploring, mapping, strategizing, and writing. We were old friends. Gathering historical materials and seeking cooperation of the participants absorbed us completely.

John Wilcox's careful draftsmanship on the cave maps delighted me and set my imagination racing. I had been drawing cave maps for years, and I envied Wilcox's style. Almost twenty years earlier, I had railed against Bill Austin's idea of a cave map—simple straight lines connecting pin pricks on the vellum. You couldn't tell anything about the cave passages and their relationship to the land above from Bill's surveys, and he was educated as a civil engineer!

My artistic talent had eclipsed Bill as the cave mapper, and I had dragged in and coached like-minded cave mappers over the years. Those mappers—Micky Storts, Denny Burns, and now John Wilcox—had much more patience than I did. My concern for cartographic beauty and unity was supplanted by their more practical attention to accuracy and detail. Wilcox elevated cave cartography by orders of magnitude over earlier efforts.

Well, this was a big operation, requiring the very best talents of the very best people. Delegation and sharing were the keys. The old-timers had kept their precious secrets and never gave anything away, but we were smarter, giving it all away as a lure to promising newcomers—the way John Bridge and Denny Burns had first snared John Wilcox.

Through 1973, Red Watson and I worked hard writing the book. My talk that year to the National Speleological Society convention in Bloomington, Indiana, brought an emotional standing ovation from the audience of one thousand cavers in response to my account of the "final connection story."

By 1974, *The Longest Cave* was largely completed. Pat Crowther, a key figure in the connection story, spent many days at my house drafting maps for the book. Her mapping skill rivaled Wilcox's. Then in November, I was elected the fifth president of the CRF, succeeding Stan Sides.

My heart was with the continuing cave exploration in Kentucky, but my objectives as CRF president were to help strengthen the organization's western operations at Carlsbad Caverns National Park in New Mexico and to spearhead the largely political undertaking of protecting the Mammoth Cave National Park environment. It was noble missionary work.

In truth, I relished the chance to move the CRF forward in the direction of our founding vision. If the politics took me out of the cave, then it was time I paid my dues running interference for scientific projects, negotiating with National Park Service officials, and developing future CRF leadership by helping to launch the National Cave Management Symposium. Phil Smith, Red Watson, Joe Davidson, and Stan Sides had taken their presidential turn in the management arena so the rest of us could explore the caves. Now it was my turn.

Mammoth Cave continued to grow with a set of large discoveries: In 1973, Mystic River in Mammoth Cave yielded several miles of trunk passage. Carlos Way, a muddy walking canyon leading off from River Hall in Mammoth Cave, was developing into a five-mile complex of shafts and drains under Eaton Valley. One of the passages looked for a time as if it might give us yet another connection to the Flint Ridge caves. John Wilcox, still a spark plug of exploration, began running parties into East Bransford Avenue at the south end of Mammoth Cave. What was he up to?

In the late summer of 1974, Jim Borden had shown up at our field station on Flint Ridge. He was only seventeen, six feet, one inch tall, with a shock of dark brown hair and big glasses. His smooth, bony face was as thin as his frame.

A mouth breather, I thought. The girls won't discover him for years. Maybe he'll take the bait and get sucked into big caving like the rest of us.

Jim Borden wanted to go caving, all right, but the CRF's agreement with the National Park Service required all cavers to be at least eighteen. Denied participation on a "real cave trip," Jim was taken by expedition leader Stan Sides on a spellbinding tour of the formerly commercial parts of Floyd Collins' Crystal Cave in October. It was the old Flint Ridge con, the same covert recruitment tactic I had worked on Stan ten years earlier and that Jim Dyer had used to net me ten years before that. I argued with the CRF's personnel officer, seeking special dispensation from the age rule for Jim Borden, since he wouldn't be old enough until the next May.

Jim's conversations convinced me we had a promising caver. He could mentally project cave passages in three dimensions—another Pascal! But too often such raw talent had left the CRF in frustration over bureaucratic red tape.

And hadn't Joe Saunders brought him around? Guys like Joe had cave smarts and could recruit cavers. I liked Joe personally but thought of him as just another project caver with his own agenda; a risk taker, not a cave connector. He could steal away a kid like Jim Borden. And we needed an uninterrupted supply of enthusiastic, unspoiled talent.

I wrote Red Watson and asked him to send the Sincere Encouragement Letter to Jim. Red promptly complied:

> Dear Jim,
>
> I got a letter from Roger Brucker telling me to write a little encouraging note to you, because you have great potential, and we want to keep you interested. Now I don't want particularly to imply that you are a stupid son of a bitch, but it is true that young men have been joining the Marines at the age of 14 from time immemorial. You don't even seem to be capable of figuring out how to join CRF, an organization whose investigative forces, I can assure you, are considerably less than that of the U.S. Marines. Enclosed is a Joint Venture form with your name typed on it, with witness signatures of two of the most venerable persons in CRF. Now if you can't figure out how to fill out the DATE OF BIRTH blank so you can do a bit of caving with us on the expedition, I give up on you.
>
> Sincerely,
> Red
>
> (And quit shooting off your mouth about your age; it bores us people who are over forty to hear about the accomplishments of the young. Want to hear how good I was when I was only fifteen? I thought not.)

When his high school let out, Jim came back to Flint Ridge and went on several survey trips. In August 1975, Jim and I were assigned to two parties going to Carlos Way. He was full of good questions (meaning those I could answer) and seemed a quick study, but it concerned me that he was going on a wet trip without wool underwear or knee pads. Several times his party leaders bitched about his growing cold and running out of energy. In November, several party leaders wanted to boot Jim out of the CRF. It was time for my own tough letter to him: Shape up or ship out.

The CRF brand of big-time caving had earned the well-deserved reputation of elitist. It was conceived in rebellion against the slow, fumbling methods of old-time explorers. It was weaned on tough, adversarial relationships with the National Park Service. It had teethed on audacious visions of establishing speleology as Real Science. The CRF had recruited good minds and strong bodies. Collectively, we had learned how to harness the cave and the experience to Challenge the Promising Men and Women to Significant Accomplishment and had alienated some of the loudest mouths in American sport caving (proof of success!). Most of all, we had become an institution that could reproduce itself. Detractors never accused the CRF of being unsuccessful.

Whether Jim stayed or left was up to him. No matter; we had lots of good prospects in the pipe. Richard Zopf, who stumbled into John Wilcox's net when we needed strong cave connectors, was within a year leading survey and exploring trips. Tom Gracanin, a geology student from Ohio State, had become addicted to this kind of caving. Diana Daunt, a small woman with an apparently limitless capacity for getting what she wanted, led hard trips, drafted maps, and fueled the big caving engine with her drive. Most cave project managers would give their eyeteeth to land even one such promising caver; the CRF had scores of them! It had to. Attrition of cavers was high due to the physical punishment of caving, frustration from the persistence required, changes in life goals and values, aging, and finite tolerance for bullshit and arrogance. We had learned to tweak the personnel intake valve to keep the good people flowing in at a slightly faster rate than the good cavers drained away. From the pool of prospects, only the determined applied—thus demonstrating they had a key quality we needed.

Making big discoveries was a goal of my CRF administration. But what I really wanted was finally to be on a connection party myself. According to my wonderful intuition, the next connection had to be with Proctor Cave in nearby Joppa Ridge in Mammoth Cave National Park, and I intended to make it. Well, it didn't exactly work out that way. Embarrassing as it is, I owe the completion of my dreams to that damned twerp, Jim Borden. He set me up for my final connection trip.

Jim Borden encountering hundreds of feet of crawlways in Crump Spring Cave

2

Roots of a System

Jim Borden Seeks Involvement in the Longest Cave

"Shit!" I muttered. I finally had to stop in the low crawlway, panting, sweat dripping from my nose. What the hell was I doing here? I—Jim Borden—mindlessly gazed at the dark brown pebbles on the floor two inches from my nose. They were everywhere, laid down by fast-moving water sometime in the distant past. They were rounded and polished, probably carried from some eroding hilltop miles away. That hilltop might not even exist today. Now, the crawlway was more like a desert, long abandoned by any water. My mouth was dry from all the crawling.

Still panting, I raised my helmet off the rough floor and peered farther into the cave; I felt alone. So far, we had crawled better than a thousand feet in this passage, the first foot looking just as the last. For all I knew, it could continue a thousand feet more before relenting. "Right," I snickered to myself, "this thing will probably dump us into something even worse, half filled with water, no doubt."

Ahead, I could see the passage stretching nearly a hundred feet before the wide elliptical tube gently turned to the right and out of sight. The sounds of my heart beating, amplified in the close confines of the crawl, echoed through my head. Once in a while, I could hear the telltale scraping and groaning from a moving caver. Keith Ortiz was out of sight around the corner ahead. The illumination from his bobbing carbide light cast shadows that danced eerily off the ceiling. Somewhere behind was Joe Saunders, who completed our party of three cavers. I strained to listen, holding my breath. Nothing. Joe was either too far back or was also resting. I turned my head backward to look behind: nothing but blackness. "Well, he can't get lost," I mumbled.

Resigned, I sighed. "God, is this what caving in the Mammoth Cave region is like?"

It was the spring of 1974. I was in Crump Spring Cave, my introduction to caving in Kentucky. Five miles to the west lay Mammoth Cave, the longest cave in the world. Since 1969, cavers from Illinois and Wisconsin had been exploring Crump Spring Cave in hope that it would lead to another Mammoth Cave . . . well, at least to a big cave. So far, all they had to show for their efforts were ten miles of caving hell. A monumental effort indeed—almost heroic—but a Mammoth Cave it was not.

I dropped my helmet back onto the floor and fell asleep, the floor's coolness soothing me.

Six months earlier, I had listened with awe as Joe Saunders presented a slide program at the regular monthly meeting of the District of Columbia Grotto (D.C. Grotto), a chapter of the National Speleological Society (NSS). Cavers from all over the Washington area gathered at these meetings held at the local library near my home. After the end of the obligatory business session and before the program, people would look over some new maps or tell tales about the most recent discoveries. They swapped lies and planned future cave trips. I, like the others, savored the fellowship. This was why we came—the new as well as the experienced caver. Each month, I was learning all about the world of caves: where to go and who the real explorers were. I deliberately sought the company of these "real" cavers.

Before the program had begun, the chairman of the D.C. Grotto stood in front of the room and called the meeting to order. He watched as the clumps of chattering cavers began to break up and slowly take their seats. Cavers are generally unruly; it usually took cajoling to quiet them down. Stragglers, oblivious to the glares of the chairman, continued their animated conversation until a well-thrown paper wad hit the ring leader on the side of his head. Grumbling, the remaining few shuffled to their seats. As the chairman made his introduction, a plump, red-faced fellow took his place at the podium, the evening's featured speaker.

This was Joe Saunders. His beaming smile captivated me as he enthusiastically began his program, describing a cave dear to his heart in the hills of central Kentucky. He gestured a lot and raised and lowered his voice, a skilled storyteller. As the room darkened for the slide program, his story mesmerized me. With each slide and movement of his waving arms, Joe spun out vivid descriptions about exploring in Crump Spring Cave in Fisher Ridge, five miles east of Flint Ridge. Crump Spring certainly was not a large cave but was located in an area influenced by the same hydrologic conditions that formed nearby Mammoth Cave. "Big cave is surely there to be found," he repeated as slide after slide showed only the bottoms of some unknown caver's boots, heels scraping the ceiling.

"Yes, lots of crawling," Joe bantered, "but we do have some walking passage too!" On the screen flashed an underexposed photograph filled with fog. A shadowy figure could barely be seen.

I squinted at the scene. Yes, you could see someone standing on something. Was that Kentucky? Before I could study it further, the next slide flashed onto the screen. More boot soles.

As slide after slide flashed by, Joe's presentation never lost its animation. I was hypnotized.

From my perspective, it seemed that I had entered the caving scene relatively late. The golden age of exploration in the caves of West Virginia had been in the late 1950s and 1960s. For years I had listened to my uncle Bob's spellbinding tales of cutting-edge exploration in the caves of eastern West Virginia.

Uncle Bob had started me in caving in 1966 when I was nine years old. He outfitted me with a helmet and a brass carbide lamp hooked onto its front. The weight of the carbide lamp caused the helmet to slide over my eyes continually.

We began by exploring the hundred-year-old gold mines in the Maryland forests along the Potomac River. At the time, I did not know that I was exploring gold mines. To me, these were caves, and I crawled through the low tunnels as if I were the first ever to see them. The feeling of being a cave explorer was a high I had never experienced before.

I had been especially proud when, before entering the mine, I did not throw the lamp down when it burst into flames in my hands due to a missing gasket. I was frightened, yes, but I would not show fear to my uncle. I calmly placed the lamp on the rock beside me. I smiled, ignoring the pain from the burn.

The trip was short, probably no more than an hour or two. I yearned for more of this caving. My uncle saw the glimmer in my eye and smiled. He knew he had a caver.

The next month, Uncle Bob, my dad, and I crawled through the low and muddy passages of Molers Cave—my first "real" cave. Vastness! This cave in West Virginia, two hours from my home, seemed to go forever. At a small room, we came upon some other explorers. They talked about exploring the "Third Level" and "Fourth Level," crawling through hundreds of feet of mud and water. These were super-cavers! I envied these explorers and longed to be just like them.

Later, as I squinted in the bright sunshine outside the entrance, I thought to myself that Molers must be a giant cave. I was shocked to learn years later that it was barely three thousand feet long—a small cave by most standards.

I took the bait. The hook was set. Like many other male adolescents, I had an obsession. Mine was not with guns, sports, or other "normal" endeavors; mine was an exploration obsession. I read everything I could find about caving. My thirst was insatiable. To quench it, I had to find a cave that I could lose myself in. I sought after and found other cavers with a passion like my own.

When Uncle Bob caved in the 1960s, new cave was everywhere, waiting only for someone to look in the right place. On weekends, scores of cavers would fan out into the cave-rich counties. The friendly local folk were always happy to lead inquiring cavers to holes on their farms. Often they told age-old tales of bottomless pits or of caves miles long with a dog entering one place and emerging from hillsides far away. This was fuel for imagination. In one such cave, Uncle Bob and his party had crawled for hours in water to discover big cave. On a later trip, they found an obscure back entrance, verifying a local yarn. They squirmed through the small hole into a raging winter gale outside. Faced with the choice of crawling back through miles of mud or braving the numbing cold and snow, they stoically headed off into the woods, using a compass to lead them in the direction of their waiting cars. Frostbite, hypothermia, and wet clothes that froze into boards turned this already epic trip into near disaster.

Back at the car, unable to pull the frozen zippers on their wetsuits that they wore beneath overalls, they climbed into their cars and drove to the nearest restaurant. Denim-clad patrons sipping coffee after a tough day working in the snow were agape to see this muddy crew in black rubber suits stagger through the front door. Frozen zippers were not going to stop this hungry group from devouring a hot meal.

During this heyday, big cave systems were discovered everywhere. But in 1973, it seemed that the ripest fruit had already been plucked and big discoveries were much fewer. No, I concluded, I could not satisfy my dreams of unexplored cave in the well-combed hills of West Virginia. So I looked for other areas that might offer the promise.

Roger Brucker was a caver who had also written a cave book. He wrote with the same passion and feeling of infinity that I was seeking. I could identify with this unseen person, Roger Brucker. He was constantly looking for a missing link that would lead to the cave of unimaginable length that was lurking behind some unseen corner. In his book *The Caves Beyond,* Roger wrote of heart-wrenching efforts during a week-long expedition into the depths of Floyd Collins' Crystal Cave in Flint Ridge, right next door to the famous Mammoth Cave (said then to have 150 miles of passage). Nearly seventy men and women had dragged in supplies for an underground camp and had frustrated themselves trying to decipher the labyrinth of complex cave. That was 1954, the "beginning" of modern exploration in the passages of the Mammoth Cave region of Kentucky. Floyd Collins' Crystal Cave later proved to be the nucleus of the longest cave on earth, the Flint Ridge Cave System.

I combed the literature to put together what had happened since *The Caves Beyond* but found only fragments. In the fall of 1954 (the same year as the chronicled NSS expedition), a side passage not pushed on the expedition proved to be the key, the sought-for missing link. As recounted in a copy of the *NSS Bulletin,* a semiannual publication of the National Speleological Society, this passage led to immense vertical shafts and an underground river. It was this discovery that opened the way to the real Flint Ridge Cave System and the connections that followed.

In 1972, I opened my mailbox to a headline on the cover of the December *NSS News* announcing that the long-elusive link between Mammoth Cave and Flint Ridge had been found. The Flint-Mammoth System was truly vast. With the new link, the longest cave on earth had more than 144 miles of surveyed passage. No more information was offered, just the byline—a stop-press news flash dramatic in its effect. There would be more later, the editor promised. I brooded.

Sure enough, the next month the promised details appeared. As I began to read the dramatic account, I was shocked to find the name Roger Brucker as an author. Roger Brucker! It had been nearly twenty years since *The Caves Beyond*, and I thought he would surely be dead by now. Not only had he spent a lifetime exploring the Flint-Mammoth System from its very beginning to this culmination, but he had also realized what must have been his dream, a dream that was evident in his writings almost twenty years before. And he was still going. Wow, twenty years in one cave—that guy must be a relic by now.

I was jealous.

As I read with envy the account of the discovery of Hansons Lost River through a hopelessly tight, long-overlooked side passage miles into Flint Ridge, my jealousy slowly gave way to anger, an anger that boiled from deep within me. Hansons Lost River had led the explorers to Echo River in Mammoth Cave, the connection. Yes, a great discovery, but this event was being compared to the first ascent of Mount Everest, the tallest mountain in the world. To complete the metaphor, the link between Mammoth Cave and Flint Ridge was being hailed as "the Everest of Speleology," the pinnacle event of cave exploration. This was something, they said, that would never be exceeded. This was an obvious tieback to the year 1954, the birth of organized Flint Ridge exploration just one year after the first ascent of Mount Everest.

Bullshit!

This was sheer arrogance. The explorers I had been casting as heroes were instead self-indulgent egotists who thought that they had the greatest thing going anywhere. More than once in the long history of exploration, discoveries had ceased because of smugness that all had been done. Blinded by glory! The pedestal I had placed the explorers on turned to dust. I was determined to shatter their comfortable complacency. I would strike out and discover a great cave of my own, a cave to rival Mammoth Cave.

But how? Where?

As I sat in the darkened library listening to Joe Saunders spin his droll tales of sprawling prostrate on his belly, scooping out sand and gravel the consistency of chunky stew to enlarge crawlways too low to pass through, I started to see beyond the horrors that were flashing before me on the screen. To go on, you had to have a vision! I began to respond to Joe's dream. He, too, had visions of a great cave that would sprawl across the vastness of the Mammoth Cave Plateau. One of his maps showed the many ridges next door to

Mammoth Cave, an area that dwarfed the aerial coverage of Mammoth Cave and was devoid of any known cave. I could see what he was really describing, his arms flailing in enthusiasm as shadows across the bright slide screen. Caves, vast and seemingly endless, waited to be found. But only the diligent and patient would find them. I now realized that I *needed* to find them. I became a born-again caver. I now was looking to the caves of Kentucky as a holy grail in my quest to satisfy my thirst for adventure and discovery.

In the short time that Joe and I chatted after his presentation, we forged a friendship that would link me to the caves of the Mammoth Cave region of Kentucky. In the room full of cavers who sat listening intently to Joe that night, I doubt that anyone else saw beyond the slides that he had used to describe his passion for the caves, seeing only the belly crawls and the mud.

Joe never presumed that a new recruit was hooked. In the following months, my mailbox filled with trip reports from a dozen excursions into Crump Spring Cave. The narratives vividly described a trail of discoveries that had led to finding "the" large passage called Crump Avenue. Cavers, after crawling for hours, had dug in blowing crawlways that led to the discovery. These were the true heroes, not the arrogant cavers that thought they had a monopoly in Mammoth Cave. Joe's letters kept coming. Joe was not going to let his promising prospect get away.

In March 1974, I drove to southwest Virginia to meet Joe Saunders. This was the beginning leg of my first trip to the caves of south-central Kentucky. My route led me down the farm-studded Shenandoah Valley where the bright, warm sun forced the spring buds to burst. Joe lived on the east side of Blacksburg, a small college town nestled in a cave-rich valley in southwestern Virginia. The university there, Virginia Polytechnic Institute, had a rich caving tradition of its own.

Joe lived in a motel. How could anyone live in just one small room, especially with muddy cave gear? I was looking forward to seeing this arrangement for myself.

I arrived at the motel at noon and drove slowly along rows of parked cars, squinting to read the small brass numbers on the doors. My hastily scrawled directions wove around the crumpled scrap of paper like a snake coiling to fit in a confined space.

"I should learn to write so I can read it," I muttered as I struggled to decipher the scribbles.

Among the last few doors, I spotted a likely prospect for the correct room. The first number had fallen off, but the unweathered paint approximated the shape of the missing numeral.

I pounded on the door, hoping I was not disturbing a stranger. If Joe was in his room (if this was his room), he would be glad to see me. The door swung wide. Joe was twenty-seven years old, stocky, and over six feet tall. He beamed at me with his cheerful smile.

Besides working on his doctorate in agronomy, Joe was employed as night manager at this motel. Fringe benefits included his quarters, a room in the back of the complex overlooking the main highway. Stacks of papers, magazines, and rolled-up maps filled his room. Off to one side was a small kitchen area. Discarded empty cans, half-eaten bags of cookies, and an open jar of peanut butter evidenced a bachelor's diet: simple, quick, and cheap. And no cleanup! He spent his time and probably most of his money on caving. Joe also loved cave science. He prided himself as a self-taught, practical hydrologist.

"Look at this map," he said as he reached deep into a stack of paper and unrolled a dog-eared diagram of a nearby drainage basin. He pointed out the catchment areas, the swallow points where the water went underground, and the springs where the underground streams reemerged. Joe explained that he used fluorescein dye to trace the water.

"Tracing underground streams is like detective work." Joe glowed with excitement. "You never know what you are going to find! One time, I didn't know if the sinking stream would come out here," he pointed to a spring shown two thousand feet away on the map, "or way over here," now pointing five miles distant. "So, I decided to use ten pounds of dye—ten of those bottles over there." He pointed to a corner where containers of red powder were stacked. "Well, the dye came out in the nearby spring and I had to buy some woman's wash because I turned it all bright green! I didn't realize that was her water supply, but I found out."

Joe rolled out more maps, lovingly explaining each in detail. I felt honored and lucky to be accepted as a caving companion by him. This big, enthusiastic caver wanted to take *me* on a trip to Crump Spring Cave. Inside, I was beaming. Later on, I realized that he was just as pleased with himself to have so easily sucked in a new recruit!

Joe broke the bad news: We could not leave yet. He had some unfinished business to attend to, and we would not be leaving for several hours. The good news was thrust into my arms: a stack of United States Geological Survey (USGS) topographic maps, cave location data, and cave maps.

"Here, keep yourself busy by doing a bit of caving the next few hours. You might like these caves," said Joe. "We'll be leaving around six o'clock, so don't be late."

He had marked each cave entrance on the topographic maps with a dot surrounded by a small circle. These topographic maps were two-dimensional representations of the three-dimensional hills and valleys that make up the surface of the earth. Contour lines connected points of equal elevation, making features such as valleys, sinkholes, and ridges easily discernible. Joe annotated these maps to record the locations of long-searched-for and obscure cave locations.

Dumbfounded by his suggestion that I go solo caving, I said, "Sure, I'll wander around and take a look at them." I actually intended no such thing. On the other hand, I did not want to lie. Above all, I did not want to appear chicken. One of the first rules hammered into my head as a caver was never to cave alone. To do so was to invite disaster, for an incapacitating accident offered no margin of safety. Something as simple as a loss of light could spell doom if one were alone. It was just common sense to have a companion. Better yet was a party of four (one could stay with an injured caver while two left the cave to summon help). This was a rule I was not willing to break—well, not yet.

No beginner to caving, I had been involved in a large caving project for almost a year. Our D.C. Grotto project of exploring and surveying the Organ Cave System in West Virginia was big-time caving. We had charted more than thirty-five miles, and I had led many survey trips myself. The camaraderie of our group effort pleased me immensely. We relished the feeling of familiarity that came from systematically exploring the growing cave month after month. We enjoyed the surveying. Consequently, I hated the idea of wandering around in just any cave—it seemed a waste of time. No tourist trips for me!

Nevertheless, I had become restless. I had entered the Organ Cave Project in its twilight and wanted new cave, a project to call my own. The lure of virgin cave was too much to resist, and Organ Cave offered limited opportunity for new discovery. I soon found Lewis Cave, a mile from Organ Cave. A few companions and I charted seven thousand feet of water-filled crawlway leading to a waterfall. The cave was terrible, but we gleefully conspired to keep it secret from imagined cave pirates eager to scoop our newly found prize. We were proud but then later dismayed when nobody wanted to go in the cave, even when we leaked its secrets! Keeping the cave under wraps had prevented the necessary critical mass of enthusiastic cavers from forming. There was no pool of cavers from which to recruit. The short-lived project died, a valuable lesson to me.

To kill time, I drove around the Virginia countryside looking at the scenery, parked, and napped. When I returned to the motel room around five o'clock, Joe was back, stuffing a duffel with dirty cave gear while talking to someone sitting on the floor. I learned that we would be joined by this third caver.

Joe's red '69 Opel Kadette was piled high with gear that rearranged itself around us as we squeezed in. His buddy, Keith Ortiz, shoehorned himself in the back while I eased into the front next to Joe. My feet were jammed uncomfortably beneath the dashboard with two rolled-up sleeping bags. Keith was a student at VPI and had been a member of the local NSS Grotto for several years. While Joe engaged me in continual conversation about cave hydrology, Keith slipped into a fitful slumber. My eyes drooped and my neck ached. Joe's words became a meaningless drone. When Joe and Keith swapped places and Keith began his conversation from the driver's seat, I realized that sleep was not on the agenda for me on this twelve-hour drive to central Kentucky.

In the early morning, while fog hung low over the Kentucky hills, my exhaustion and heavy eyelids made the surrounding rural countryside appear surreal. Towns with names such as Cave City and Horse Cave showed how important the caves had been to the area's economy. As we drove into ridge country, we passed old farmhouses with smoke lazily drifting from their chimneys in the still morning air. Occasionally, I spotted a farmer starting his morning chores. It was a new day. We turned off the main road onto a mile-long narrow, rocky, rutted road that jarred all of us to weary consciousness. The car's muffler and pipes scraped and clanked as we jounced over bumps to Crump Spring. My head was pounding from lack of sleep. A loud ear-grinding noise emanated from the underside of the severely overloaded car.

Scr-aaa-pe!

"Crap! That was Oil-Pan Rock—always gets me," Joe swore to himself as the car continued to lurch and bounce into the valleys edging the Mammoth Cave Plateau. After many years of traveling this route, he knew all the hazards, rock by sharp rock.

We paused while I opened a gate in a barbed-wire fence and looked around. Cows grazed in pastures dotted with clumps of barren, leafless trees in steep-sided sinkholes. Patches of bleached, white limestone made this bottomland unfit for farming. This was real cave country. Anyone could see that miles and miles of cave had to honeycomb the earth beneath this tranquil countryside. There were no creeks draining these valleys, no water anywhere. Just minutes after arriving, I knew I belonged here in Kentucky. Yes, this was home. Couldn't I feel in my bones the energy of the vast caves of Kentucky? It was 1974—to me an eternity after the modern exploration of Mammoth Cave had commenced in the early fifties. But now, the real exploring could start. I was ready to go. Although Mammoth Cave was 150 miles long, this was just the beginning. There would be a lot more than 150 miles!

We turned the last corner at the bottom of the hill to come upon a small, neatly kept house with green shingles for siding. The yard was well manicured. Such a yard was seldom seen in rural Kentucky. Its beauty struck me, more evidence that this was "home." But we were not alone. Sleeping bags filled with cavers were sprawled everywhere on the front porch. They had driven from Illinois and Indiana to be a part of the adventure. The owners occupied the house only during the summer months and allowed cavers to use the porch the rest of the year.

Down the hill to the left of the house was a small, tree-rimmed sinkhole. On one side, a spring formed a noisy waterfall. Water fell into five-gallon buckets before flowing across the sinkhole bottom to disappear into a dark recess in a small cliff. This was the entrance to Crump Spring Cave. To me, this was the Garden of Eden. Could this incredibly beautiful place lead to the hellhole that Joe had described?

Part of me felt petrified. I had never been in any cave with such a bad reputation. My longest cave trip until now was "The Seventeen-Hour Fiasco" in the Organ Cave System near Lewisburg, West Virginia. On that trip, I had felt farther out than ever before. During the exit, I had accidentally dropped a rock on somebody's foot, breaking the caver's big toe. The trip out seemed to take an eternity. My exhaustion tortured me as I ascended the series of ropes leading to the surface. I was horrified a week later to see my victim in a full foot cast. It had been a trip to remember and a trip that many would never let me forget.

Crump Spring Cave would be different. This trip would be longer, and I would be on my hands and knees or, worse, belly for most of the trip. There would be hardly any walking cave here. For the first time, I would wear kneecrawlers, unruly looking hard rubber pads that would strap around my knees. Would I measure up against these hard men?

After napping for a few hours, we prepared our gear and crouched into the low entrance. A couple hundred feet inside, the stream flowed over the lip of a short drop. Joe uncoiled a twenty-foot length of cable ladder, tied it off to a rock column, and lowered it down the pitch.

We took turns climbing down the aluminum rungs, the spray from the waterfall soaking our clothing. As Keith belayed Joe down the climb, I peered into the only open passage leading from the bottom of the pit.

"Yech!" I cried. "We have to crawl into that?"

Joe snickered as he stepped off the ladder but said nothing.

Joe subsequently led Keith and me through an endless succession of low, wide crawlways beginning with the five-hundred-foot slog on our bellies through wet gravel—the drain of the entrance pit. We then moved along on our hands and knees through the sandy-floored C Crawl, wiggled our way on our sides in Thin Man's Misery, and dragged our bodies through the Long Crawl. There were thousands of feet of low, wide elliptical tubes: endless foot after endless foot. My arms ached. Every hour or so, we would stop to survey some tiny passage that did not go anywhere. They were low and excruciatingly tight crawls. Joe sat outside with the notebook while Keith and I struggled with the survey, shouting numbers back to him. Joe loved a complete map; he surveyed everything.

After twelve hours of this routine, we emerged into a tall series of vertical shafts: Five Domes. Here we surveyed eighty feet, giving us a little over two hundred feet for the day. It was late, we were tired, and it was now time to head out. I had expected the trip to be the closest thing to hell imaginable, but I found it to be far less difficult than I had feared. I could feel the allure of these caves. The passages went everywhere and seemed boundless, especially compared to the caves in West Virginia. I watched as Joe and Keith drooped during the long trip out; they were as tired as I was. I was holding my own against these seasoned Kentucky cave veterans.

We climbed out of the cave into a beautiful, sunny Kentucky morning after twenty-two hours underground. The wind that howled through the entrance snuffed out our lights. We had surveyed only a few hundred feet and found nothing new, but my first trip into the caves of central Kentucky was special. I had fallen in love with these caves. I was trapped. I would return.

Two months later, Joe and I prepared for a much longer trip into the cave. We would travel through the nefarious Whimper Route into Crump Avenue, the cave's "main" passage. We faced a one-way travel time of six hours. The total trip would last at least twenty-four hours. Keith Ortiz could not get away from classes at VPI (or so he said), so only Joe and I went.

This time, Joe's Opel was comfortable. The two of us sat in front, gear randomly tossed in the now vacant back seat. The passenger seat could now be reclined to afford at least a little sleep. Unfortunately, the only time Joe was silent was when I was driving. Otherwise, he babbled in the continuous conversation he said was necessary to keep him awake.

I had purchased a set of a dozen USGS topographic maps that covered the entire Mammoth Cave Plateau and had already begun to study the areas around Crump Spring Cave. Few caves were known beyond the boundaries of Mammoth Cave National Park, but those many large ridges suggested myriad opportunities for new discoveries. During the long drive, I quizzed Joe about all the other caves in the area, draining as much information from him as I could.

Then we started into the crawls of Crump Spring Cave, and it seemed that our trip was doomed from the start. Both our primary lights flickered and failed. Joe cut his hand trying to open a stubborn carbide lamp, and my cave pack zipper closure failed, resulting in the pack repeatedly spilling its contents. Delays plagued us at every turn. In four hours, we had gone a distance that should have taken an hour. We had barely begun and were already routed. Six hours after entering the cave, we emerged in the hot late-afternoon June sun, defeated. That cave could injure and frustrate.

A couple of hours later, after a hot meal in Cave City, we drove north on the interstate back towards Fisher Ridge. To the west in the dull haze beneath the sinking sun, I saw the outlines of a broad ridge paralleling the highway.

"Joe, what ridge is that over there?"

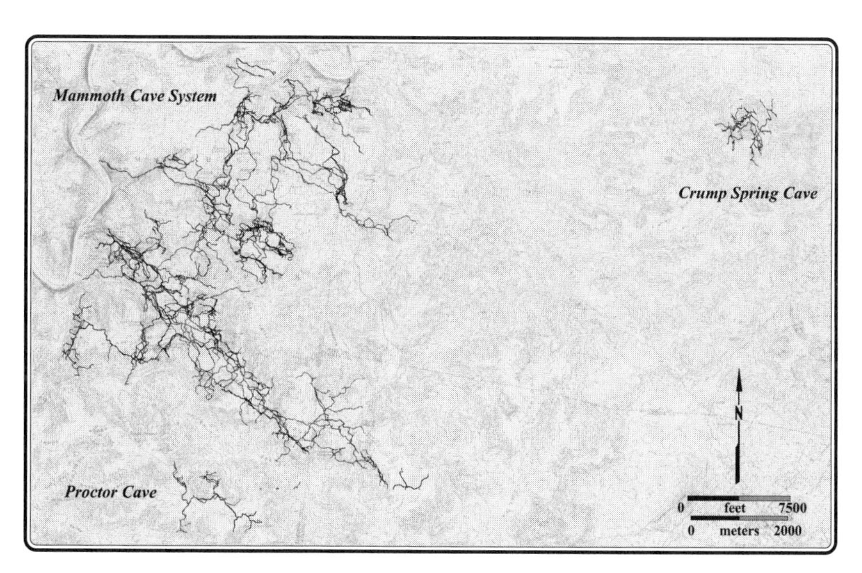

Caves around the Mammoth Cave area, ca. 1975

He squinted in the direction I was pointing. Thinking for a second, he answered, "I think that's Toohey Ridge."

"Any cave under it?"

"I don't know of any."

"Do you think there might be some decent cave there somewhere?"

He paused, then answered thoughtfully, "It has a lot of potential. Ought to be a big cave under there."

"Has anybody looked?"

He shrugged, "Not that I know of. Who would look? There isn't anybody else who could look."

"Why haven't you looked?" I said.

"No time."

I pondered his answer. Incredible. A major ridge right under our noses and untouched by cavers. "Joe, I'm going to find cave beneath Toohey Ridge."

"Sure, go for it!" he said with a smile.

Would it be a big cave? A cave system, maybe? At that moment, I decided I would lead the search for the undiscovered Toohey Ridge Cave System.

The Roppel Entrance

3

Beginnings

Jim Borden Discovers the Search for a Long Cave Is Not Easy

Wind shook and rain pelted my car as I vainly tried to sleep, my body stiff from being crammed into the front seat. Hours ago I had been routed from where I was sleeping on a picnic table beneath the shelter of a pavilion next to my car. I peered into the gloom of this fierce December storm. Silhouettes of looming tombstones surrounded me. Were the dead watching me? I shivered back down on the cramped seat as another blast of chilly wind shook the car. Who could sleep through this?

If anyone had noticed, it must have seemed an eerie scene. Every morning and every night, my yellow Datsun was parked in the Monroe Cemetery on the southwest end of Toohey Ridge. The curious would have thought that cavers were pretty strange people. (It's true.) Most of the rural residents had met this kid from Washington, D.C., who asked them if they knew of any caves in the area. They surely would think that anyone who would sleep in a cemetery lacked common sense.

In my exhaustion, I slowly drifted off to sleep, the driving rain creating a backdrop for vivid and violent dreams.

I awoke to a pitifully wet and gloomy dawn under a blanket of dense fog. It was New Year's Day, the beginning of the seventh day of my solo hiking and cave-probing expedition on the surface of Toohey Ridge. This was my second search for the cave system beneath the ridge. Earlier in the summer, I had spent a day exploring the lay of the land, learning what I could about the known caves and the folklore about them. During this past week, I had interviewed more than twenty residents who owned cave-promising tracts of land. I had walked miles of rocky, steep ridge sides, cramming my body into any hole that might lead to cave. The weather was terrible; the conditions appalling. Days of relentless rain had turned the hillsides into muck. It was always cold and gloomy, the fog hanging for days at a time, cutting into me like an icy knife. My diet of peanut butter and jelly sandwiches did little to relieve the cold. This was not fun, but it was necessary. Finding caves was tough work that one had to undertake like a detective.

I kept detailed notes on everything I found. I had accomplished a lot, and my topographic map was covered with little pencil dots with names and descriptive information crammed into the margins. But I was slowly realizing that discovering the Toohey Ridge Cave System would be a difficult task. No hole I found led to real cave or even showed much promise of doing so. I found mostly isolated pits leading apparently nowhere. Also, I never found the hoped-for blast of air associated with big cave.

Unlike the other major cave-bearing ridges (Flint Ridge, Mammoth Cave Ridge, Joppa Ridge), Toohey Ridge lacked a historically known cave to start cavers on the way toward major discoveries. In Joppa Ridge, it was Proctor Cave; in Flint Ridge, Floyd Collins' Crystal Cave. Toohey Ridge had nothing and so had largely been ignored by cavers. Early in the century, locals had searched diligently for a cave to commercialize but had found nothing. I had learned quickly that I was not the first to be thwarted.

It was the end of the week and time to head home.

My optimism dashed, I finally realized that I needed help. It was time to find a partner. Frustrated but not discouraged, I returned to Potomac, Maryland, and brooded about the future.

A month or so later, a copy of the *NSS Bulletin* arrived. Flipping through the pages, I came across an article about a Kentucky cave. Jim Currens, the author, had written an interesting scientific description of a small cave in Rockcastle County, south of Lexington in central Kentucky. It was not the article that intrigued me, however. Here was someone, a competent caver unknown to me, who lived near the Mammoth Cave area. And from the looks of his expertly drawn map accompanying the article, he was an accomplished cartographer. Being "unknown" could be important: cavers were usually well-known because they were involved in a major cave-mapping project. The caving network is tight. A well-known caver would be preoccupied and have too little time to become involved in something new. But maybe I could interest Jim Currens in Toohey Ridge.

I wrote him a carefully worded letter offering him a piece of the action, what I thought could be a real lifetime adventure. Without revealing too much information, but still presenting enough where he would not summarily reject my request, I offered him a chance to help discover and explore one of the longest caves in the world, the Toohey Ridge Cave System. My statements were bold, the vision farsighted. Could he see it, too?

I paused over the finished letter, contemplating its effect. I did not count on receiving a response. After all, this was a letter out of the blue from a complete stranger. I had omitted my age, thinking being seventeen would not work to my advantage. Better let that ride.

I sealed the letter and mailed it.

Six weeks later, in March 1975, I tapped on the door of a house trailer on the north side of Lexington. I had never been lost in confusing cave passages, but I was lost in this maze of a mobile home park. Obscurely marked addresses and the random layout of the roads baffled me. I was not even sure I had the right trailer number. The lights from the adjacent drive-in movie theater cast gloomy shadows as I stood patiently waiting on the metal steps. I wondered what kind of person this Jim Currens would be as the muffled screams from the violent movie bounced through the trailer park. After an eternity (fifteen seconds), the door was opened by a tall man with thick, plastic-rimmed glasses and a closely cropped mustache. This was Jim Currens. We shook hands as he smiled and invited me in. His wife, Deb, greeted me. Their home was comfortable and well kept.

I had driven from Maryland with a caving buddy, Bob Hummer. The three of us spent the next several hours poring over maps spread on the living room floor while Deb continued with her sewing. We talked geology, hydrology, and any other "-ology" that supported our conviction that there was an enormous cave beneath Toohey Ridge. Jim had also once looked at Toohey Ridge as a prospect for finding big cave, eventually abandoning the idea with the assumption that someone had surely already staked out this claim. I assured Jim that I had heard from good authority that we were probably the first to give Toohey Ridge a serious look. The discussion went deep into the night.

"Look at this anticlinal structure here," Jim said. "Big passages should form along this." He pointed to Monroe Sink, an especially large sinkhole nearly three-quarters of a mile in diameter, on a map showing the geology of the area. "The sandstone was thinner here because of the anticline and provided a quick route for the water to form lots of cave."

I nodded in agreement and offered my own argument for big cave. "All the water that formed Mammoth Cave from the sinkhole plain," I pointed to the west, "had to flow through Toohey Ridge!" I slammed my fist on the map for emphasis.

"Flint Ridge, too!" Jim added. "The main Salts Cave trunk points like an arrow directly towards the anticlinal structure at Monroe Sink." His finger swooped from the approximate end of Salts Cave to the northwest, past its entrance, and far to the southeast to where it would obviously have to run through Toohey Ridge.

Our determination grew and our confidence swelled at each fact we tossed at each other. There was a giant cave in Toohey Ridge; we were sure of it. And we would find it!

Jim Currens was a professional geologist working for the state. His job was to measure and inventory coal reserves in coal-rich eastern Kentucky. He had been caving for several years with the Blue Grass Grotto of the NSS based in Lexington. He had gained project experience in the seven-mile labyrinth of James Cave in the knobs south of Mammoth Cave National Park. He, like me, had bumped around for several years looking for the "right" project. This search was the common thread that was bringing us together.

The next morning over breakfast, we rehashed the details on some of the leads I had found during that wet Christmas week when I had camped in the cemetery on Toohey Ridge. Many did not continue, but a few seemed promising enough to need a second look. Bob Hummer and I planned to be on Toohey Ridge the whole of the following week. Jim had to work but would be able to make it down on Friday. I knew how important it was to get Jim on the hook, so we told him that Bob and I would continue our ridge walking to find other promising leads, saving the best ones for Jim's arrival.

Toohey Ridge is typical of the broad, sandstone-capped ridges of the Mammoth Cave Plateau. About four miles long, oriented north-south, it broadens from a half-mile to about a mile in width at its south end. Several large hollows cut into its flanks, making the ridge look like the outlines of a giant mitten. The sandstone crest of the ridge rises about two hundred feet above the surrounding valleys. Hydrologically, Toohey Ridge lies between Mammoth Cave National Park to the west and the sinkhole plain on the southeast. The sinkhole plain has been recognized as the principal water source contributing to the development of Mammoth Cave. All that water from the sinkhole plain flowed underground beneath Toohey Ridge on its way to Mammoth Cave and just had to have created a big cave system like Mammoth Cave.

The old-timers realized the potential of Toohey Ridge. Over the years, they had done what we were now doing, albeit using cruder equipment. However, their determination undoubtedly matched ours if the remains of makeshift ladders of cedar poles and tailings of dogged excavations were any clue. In the years preceding the arrival of us "modern cavers," only a handful of real caves had been found on Toohey Ridge, one of them exceeding a thousand feet in length. All ended in narrow slits or ancient collapses of sandstone or limestone breakdown. The system that must be there had not been found.

At the appointed meeting time at the Monroe Family Cemetery on Friday, Bob and I waited impatiently, looking at our watches every few minutes. Jim was late. I did not know Jim well and still was not confident he would come. The day was sunny and pleasant, and I was eager to get going. I gazed at the tombstones surrounding the pavilion where we waited. In the bright sunshine, they looked far less foreboding than they had that rainy December night. Maybe this was an omen of good luck.

After another fifteen minutes, my impatience led to discussion about leaving without him, but then I heard the high-pitched whine of Jim's dark blue Volkswagen Beetle as it came speeding down the road.

"Sorry I was late. Got a later start than I wanted." Jim smiled. "And you can't speed in these things," he said, gesturing towards his car.

"No problem. I was a little worried you wouldn't show."

Jim frowned. "I told you I'd be here."

I smiled, trying to diffuse his obvious annoyance. "Oh, I just worry."

"So, what's up?" Jim pointed at the map roll on the picnic table beneath the pavilion.

I unrolled the map and placed rocks on the corners to keep it flat. I had made notes in the margin of the caves Bob and I had checked. All of them had ended quickly. As I had promised, we had saved the more promising leads for Jim. I pointed to them on the map.

We started out, Bob and I leading the way with Jim following in his Beetle. We drove from lead to lead along the narrow, winding roads that zigzagged across Toohey Ridge. First we checked an open pit on the top of the ridge called Buzzard Pit. From above, it looked like a window into large passage twenty feet below. However, once descended on cable ladders, we found only cul-de-sacs off a circular room. We then drove to the north edge of Toohey Ridge to check a cave whose owner had reported that the entrance led to two levels of passageways. Our imagination had painted a glowing picture, but once there, we found little. Yes, there were two levels (by some stretch of imagination), but the cave went a total of only thirty feet before the ceiling closed down. The remainder of the afternoon went this way, lead after lead ending as abruptly as the others. Our results discouraged us.

By five o'clock, we had checked the last lead off my list. What a bust! In the remaining daylight, I showed Jim some of the other leads I had checked the previous year. These did not have a definitive end; exploration had stopped at blowing breakdown collapses or low digs. The caves might go, but significant effort would be required. As we reevaluated each obstacle, Jim nodded in agreement that I had come to the correct conclusion.

Jim was most fascinated by a cave nestled in a small hollow at the bottom of a large valley. By far, this was the most promising of all the caves we had visited that day. Jakes Cave— as it was locally called—blew a lot of air and looked as though it could lead into a broad section of Toohey Ridge. Unfortunately, a hundred feet inside, a pile of rocks extended from floor to ceiling with wind whistling through the remaining voids. We were blocked again, but possibilities existed for working through the massive collapse. If we were successful, we thought we would be into what we were now both calling the Toohey Ridge Cave System!

Jim Currens and I would return with tools to begin what would obviously be an extended project. We were in this for the long haul.

Six months later, the cave roof shuddered and groaned as I rammed the heavy, six-foot steel bar into the choked jumble of sandstone boulders that threatened to fall at any moment.

"This must have been what it was like at Q87," I said. Q87 was a frustrating sandstone choke that thwarted efforts for years in connecting the Flint Ridge Cave System with Mammoth Cave. Explorers in Flint Ridge had also rammed steel pipes into a collapse much like this one, hoping to open a way to climb up into Mammoth Cave. The cool air teased them about what lay beyond; repeated attempts were unsuccessful.

Straining, I wiggled and pried the bar in an effort to dislodge a few more sandstone boulders. What were our prospects for success?

"All I know is, this is pretty scary," Jim remarked from behind me. The ceiling groaned again. We heard the dull sound of rocks shifting and rolling far overhead. Intriguing.

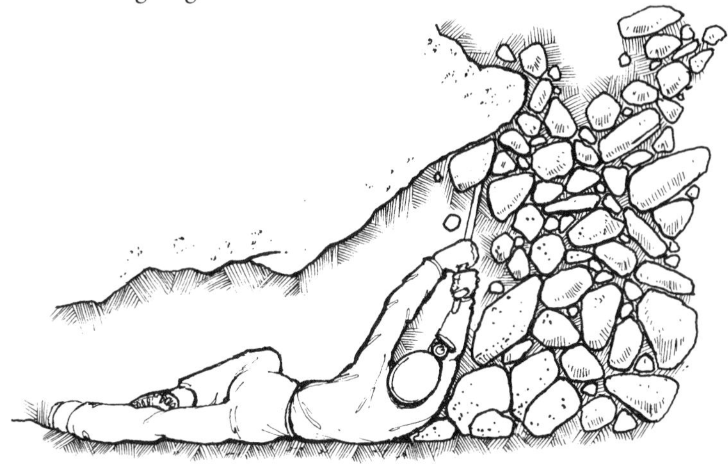

Poking a bar at a breakdown

I continued tugging at the heavy bar. Rocks began to fall from the hole I was digging upward. An avalanche! I recoiled, fearing the worst. Unfortunately, the only way to dig was straight up into the foreboding pile. Our engineering plan was to lie on our backs in the cold, damp sand—one at a time because the passage was so small—with our faces immediately below the jammed sandstone boulders. We would eat rocks if the pile shifted unexpectedly. In the nick of time, before the rocks did fall, we were to scoot out of the way. We did this work in shifts.

I pulled my head out just as dark sandstone boulders rained down into our sheltered crawlway. Exhausted now from hours of moving rock, we rested briefly beneath the cold, gray, flat limestone ceiling. A cool breeze refreshed us. This was hard work!

We were in Jakes Cave. We had returned trip after trip, our progress thwarted at this funnel of loose rock. Nevertheless, we were still brimming with optimism. The cave's gaping, arched entrance, six feet high and thirty feet wide, promised good things ahead.

Over the past several months, Jim and I had dug with great determination. We lugged in piles of Jim's tools: sledgehammers, stone chisels, picks, and longbars. By moving rock, we worked our way farther into the cave beneath ledges and well into the collapse zone. After maybe fifty feet, the ledges gave out, and we squirmed into a low room with a ceiling composed of a single large block that had peeled from the ceiling. At one end, the strong breeze blew from an upward-leading chute filled with grapefruit-sized sandstone rocks. We removed these rocks easily, but the supply seemed inexhaustible. At the end of each trip, we cleared a large cavity in the rock pile, apparently stabilizing it. The pile always resettled between trips, so each time we returned we faced the same situation over again. We took turns prying and hammering at the blockage. To show for our labor, we had only our carefully placed spoil piles, thousands of pounds of rock crammed into every niche we could find. Slowly, but surely, the room was filling up.

Two months after that first trip in March with Jim, after graduation from high school in May 1975, I moved to Lexington to attend classes at the University of Kentucky. I declared a major in geology, a fitting profession for a caver, I told myself. All my caving buddies were studying geology at various universities. I told my parents how important it was to become acquainted with the campus environment, and summer school would be a good head start. I did not tell my parents that taking summer classes was one way to move to Kentucky as soon as possible to continue the search for the Toohey Ridge Cave System unimpeded by a six-hundred-mile drive. My Datsun loaded to the roof with all my belongings, I headed off to school.

Jim Currens and I zealously poked around all the flanks of Toohey Ridge that summer and fall. We found passage fragments and cave openings, but we discovered very little the old-timers had not already found. Every hole we probed had signs of earlier efforts—cedar ladders, fragments of ropes, broken bottles. The old explorers seemed every bit as diligent as we were. One of our most notable efforts was in a dig in the south end of Renick Cave, a three-thousand-foot-long cave sandwiched between two large sinkholes. Around 1925, the old-timers had dug into a passage beneath a sandstone ledge in one of the sinkholes. Although not very extensive, the large passage was adorned with stalactites and stalagmites. But Renick Cave was too small and too remote to offer any promise as a commercial cave. Cave-onyx miners soon stripped the stone decorations for sale to tourists, and Renick Cave was forgotten.

We explored Renick Cave. Renick's large passage was a sample of what we knew could be found in Toohey Ridge: a thirty-foot-high and twenty-foot-wide borehole. The borehole ended, but we crawled into an obscure side passage and found a narrow rift blowing air. We climbed down the crack—optimistically named Hurricane Crack—on a cable ladder. Below, we followed a series of canyons nearly completely filled with fine, dry sand. It was good, going cave; yet everything we had explored had been tracked-up by some unknown explorer in the past. Finally, after several trips spent sorting through the confusing canyons, we lay on our bellies, scraping away sand in a low passage in an attempt to follow Renick Cave's wind. At least *something* had been left.

Breakthrough! We slithered through the dug-out passage into open space. Elated, we scampered through the continuing canyon; smooth, virgin sand stretched out before us. After seven hundred feet, the telltale sandstone boulders and dripping water told us Renick Cave might be finished. The walking canyon terminated in a floor-to-ceiling wall of broken sandstone rock, air blowing through the gaps. Damn! On the way out, we checked one small side passage and, to our delight, emerged into a tall canyon. Although the canyon was short and filled with sand at both ends, a wide crawlway led off to the west. After fifty feet, we stopped. The crawl continued ahead. We left it to survey later, savoring the thought of the hoped-for discovery until we could return.

The following weekend, while on a trip to Organ Cave, West Virginia, our anxiousness got the better of us. On Sunday morning, after a long caving trip the previous day, Jim and I climbed into the car just after dawn and made the ten-hour drive back to Kentucky. We cleverly planned a longer route of travel in order to avoid Lexington so that any second thoughts as we passed close to our home base would be removed. At 10:00 P.M., we entered Renick Cave, weary after the long drive and the lateness of the hour. We returned to the dig and began surveying, making our way up into the crawlway we had left behind the previous weekend. One hundred feet beyond the previous week's scuff marks, the sand fill rose within two inches of the ceiling. Shit!

Defeated and drooping, we headed to the surface and started the drive back to Lexington, narrowly averting disaster. I woke up to blaring tractor-trailer horns to find Jim glassy-eyed, driving thirty miles an hour without lights on an interstate highway. Sleep driving. Unfortunately, I did not do much better when it was my turn to drive. Twice, I was shocked awake by the sound of flying gravel as the car headed off the highway toward the ditch.

Renick Cave was finished.

Over a period of many months, Jim and I had explored fifty or so vertical shafts—moss-lined holes from 30 to 160 feet deep on the sides of the ridges. These vertical shafts are the downspouts to the master cave for water that flows off the sandstone caprock down to the water table. There had to be thousands of these shafts, but only a few would be enterable by humans. Narrow drains in the bottoms of most of them would stop cavers from continuing. Nearly all of them blew cold air. We climbed down some without rope; in others, we used cable ladders or a rappel line to reach the bottoms. In all of them, the water gurgled into too-narrow drain slots or bedding-plane openings or just disappeared into the gravel floor. No way to follow the water here.

At every turn, we were being defeated. At first, we thought our persistence, technology, and experience would beat the odds of finding the big cave. After all, were we not better than those old-timers with kerosene lanterns and homemade ladders? Perhaps not. We still had no doubt that the cave was there, but we started to become uncertain that we would find a way in. I would not admit that the odds against us were mounting. Given time, I was sure we would be successful. We continued to persist in our crusade.

One cave in particular tortured our patience beyond endurance. Wildcat Hollow Cave frustrated us with an interminable crawlway two feet high by four feet wide. It was a classic death trap. The entrance was at the very bottom of a long valley. During rainstorms, every drop of rainwater drained into it. Frequently, the gully leading to the mouth of the cave turned into a frothing torrent of foam and crashing water. In the cave, we squeezed through windows of car doors and over twisted pieces of abandoned farm implements, rusted tin cans, and discarded bedsprings that had washed in during the frequent floods. We could not survey this cave because the metal junk deflected our magnetic compass.

I was determined that a long crawlway would not defeat us. I returned with Joe Saunders a few weeks later, but we were stopped by a new pool of water before we had even reached the limit of exploration. Finally, I made a solo attempt at pushing the cave. The pool was now dry. Being alone, I moved quickly into unknown cave that trended on a sinuous path toward the heart of Toohey Ridge. Crawling over jagged rock made movement difficult, and after a few hours, I finally succumbed to exhaustion. I had crawled perhaps a thousand feet since leaving the surface. Ahead, the low tunnel rounded a corner and continued out of sight, the cold breeze continuing to blow. I sighed, laying my head on the cool floor.

As I lay there resting, the years of frustration at not finding the Toohey Ridge Cave System swept over me like a giant wave. Where was it? Damn that teasing breeze! I was so sure that Wildcat Hollow Cave would lead us to big cave, but it was almost as if the cave was playing with us—an infinite crawlway unrolling before us. I was at the end of my endurance, the end of my will. I wiggled my way back to the surface, emerging just a few hours before a violent thunderstorm dumped two inches of rain on Toohey Ridge, filling Wildcat Hollow Cave with water. When I returned later, I found its entrance sealed from mud by the rains, an ironic symbol of our success on Toohey Ridge. The valley bottom showed no indication that a cave had ever been there.

Would we ever find the big cave? Standing in the sinkhole on the flat floor that now covered the entrance to Wildcat Hollow Cave, I could not help but feel that this was a harbinger of how things would go in our future search for big cave in Toohey Ridge: lots of work and only mud to show for it. But the wind . . .

Wind, the lifeblood of a caver's enthusiasm, is one of the seductive indicators of going cave. Throughout the caves of the Mammoth Cave area, air moves through the passages like blood flowing through veins of a body. The lack or presence of wind is not a sure-bet indicator, but with other clues, wind can guide explorers to the discovery of new cave. Probably more than any other clue, wind has driven cavers to push into otherwise unthinkable passages and holes in search of big cave.

Airflow in the caves of the Mammoth Cave area is triggered by the chimney effect of warm air rising and cold air sinking and, to a lesser degree, by changes in regional air pressure. In principle, warm air moves from lower to upper openings on cold days and cool air moves from upper to lower openings on hot days. During times of extreme temperature or pressure changes, wind speeds through the cave can exceed thirty miles an hour. The number of entrances and the extent of the cave complicates the airflow.

Many of the caves we found on Toohey Ridge had this magical airflow, a further sign that there was a big cave somewhere.

Through it all, we kept returning to the sandstone rubble of Jakes Cave, where the wind howled through the breakdown obstruction. This still seemed our best chance of success.

Jim and I reclined wearily on the damp floor, waiting for the last of the rocks to roll from the chute. Better safe than sorry. Soon, the avalanche of rocks diminished to pebbles. Time to move again.

I returned to the rubble face where I poked and prodded with the steel bar. This time, the round sandstone rocks shifted easily with each jab. I quickly pulled back as a few of them tumbled from the chute.

More pokes, more rocks that rolled out like melons from a spilling cart. One large rock struck my shoulder.

"Ouch!"

I waited again for the pile to settle.

I eased back to the base of the chute and cautiously stuck the bar back up toward the face. I stretched but could not feel the base of the pile.

"Damn!"

"What's wrong?" Jim asked.

"So you *are* there?" I was annoyed he had not said anything when the rock crashed into my shoulder.

"Humph."

"I can't reach any more rocks." I scooted farther up the slope to assess the problem. I swung my light upward.

Suddenly, open space!

"We may have something!" I cried.

I knocked away the last remaining boulders that blocked the route up into the void.

"Careful, watch for loose rocks." Jim was always more wary than me.

I was eager. The excitement of potential discovery overshadowed my caution. I eased into the hole. Above me was blackness. Small rocks fell from somewhere above as I squeezed up, feet scraping and pushing at the sloping floor for purchase. Once through, I looked around, searching for the expected big passage. As my eyes adjusted to the openness, panic shot through me as my carbide lamp's beam revealed the shadowy objects above. The ceiling was thousands of pounds of black sandstone rocks held up by imagination alone. Small rocks bounced off my helmet. Visions of Floyd Collins trapped and dying in nearby Sand Cave in 1925 raced through my head. Trapped or crushed—which is worse? I quickly slithered back to safety where Jim waited.

"What did you find?"

"Looks pretty good, actually. It's a big canyon or something, about three feet wide and twenty feet high. I can't see very far. There's shattered rock everywhere."

Watching Jim, I continued, "But I'm not going. Give it a shot, if you want."

He wiggled passed me and jammed his torso up into the chute. Silence. Moments later, he slid back out of the hole.

Jim was no fool. "No way. Let's get the hell out of here."

Open cave, and no one wanted to check it. After tens of hours of digging, we had finished our last digging trip. It was just too dangerous, even for me. We gathered up the tools and headed toward the surface.

As we left the cave for the last time, we heard behind us a low rumble as the pile resettled once again.

Would we ever get a break?

All our efforts on Toohey Ridge led nowhere. In the shadow of the longest cave, we could not find a going cave. We had only worthless mud and scary stories of sandstone boulders. The disjointed half mile of new passage we had found beneath Toohey Ridge was not a cave system. Who was kidding whom? Where was the Toohey Ridge Cave System?

The vastness of Mammoth Cave lay across the valley from Toohey Ridge, and at Mammoth Cave was the Cave Research Foundation. I had grown up envious, yet respectful, of the CRF. The organization had what I wanted (and now searched for)—the longest cave on earth. Yes, its members were arrogant, but they still had it. Once I turned eighteen, I participated in a number of expeditions with the group, quenching at least some of my thirst for big cave. Even when I was not caving with them, I visited to find out what was happening. I also wanted to brag about our exploits in Toohey Ridge. I sought approval; I wanted their endorsement.

One summer evening while visiting the CRF Flint Ridge Field Station in Mammoth Cave National Park, I met Steve Wells. At the time, I was still seventeen, ineligible to cave with the CRF. He was one of the participants on the 1972 exploration team that connected the Flint Ridge Cave System with Mammoth Cave. His friendly smile put me at ease. Steve was a graduate student working on his master's thesis on the hydrogeology of the Central Kentucky Karst. Both qualifications impressed me—an expert in cave connecting and an expert in cave science! Steve's wisdom would be valuable, and his acceptance of me, everything!

I chatted with Steve for awhile but quickly got to the point, not mincing my words. "Steve, what do you think of the potential for big cave in Toohey Ridge?" I was confident he would share my confidence.

Steve looked at this eager teenager thoughtfully. "I don't think any cave of significant length will be found outside of Mammoth Cave National Park," he said. "The plateau is way too fragmented—the cave will be all chopped up. Besides, Mammoth Cave is probably unique—just an anomaly."

He smiled again and turned away to the others. End of conversation. More arrogance! He was one who should be open-minded, but he was like the others. Was it ignorance?

I was seething. These damn CRF cavers had it all. They monopolized the longest cave. Now they were claiming no other cave could ever come close to being a Mammoth Cave! Didn't they see that the *rest* of the Mammoth Cave Plateau was there for the taking? Blindness! Of course they were wrong!

After that evening, I never saw Steve Wells again.

In the fall of 1975, I concluded that I needed a life change. The months of disappointment and frustration had taken their toll. I was dismayed that my energy and Jim Currens's determination had not been enough to guarantee our success. My confidence that we would ultimately prevail was wavering. Overall, caving was looking like a dead end. I decided education would now take a new priority, and I would begin again to explore my other passions—rock climbing and mountaineering.

In December 1975, I packed everything in my yellow Datsun and drove to Colorado to enroll in the Colorado School of Mines. I yearned to challenge myself intellectually at what was one of the top mineralogical engineering schools in the country. In my spare time, I would seek out and climb the granite and sandstone cliffs in the mountains west of Boulder.

School was everything I thought it would be. I was working hard and loving it. But I also missed the caves. Jim's last comment when I left kept gnawing at me: "Don't worry about it. You gave us a good start. I can take care of things from now on." His attempt to console had had the opposite effect. He had taken so smoothly for himself what I had left behind. Was he happy to see me go? I could not leave my dreams so easily.

In late January 1976, I caught pneumonia, which laid me up long enough that I had to drop out of school for the semester. There would be no way I could catch up. After returning to my parents' home in Maryland and working at odd jobs for a short while, I called James F. Quinlan, geologist at Mammoth Cave National Park, to ask if he could give me a job. Quinlan's caving program would not start for several months, but he conducted year-round hydrological research. Surely he could use some help, and I needed to be near the caves again.

Indeed, Quinlan did need a field assistant. So in early March, I spent my days monitoring the water level in wells in the sinkhole plain south of Mammoth Cave National Park or changing dye traps in the dozens of springs along the Green River. The work was fun, but unfortunately, I could not find a place to live. For the first few weeks, Don Coons provided floor space in his living room—too crowded, considering another of Quinlan's field assistants lived on his couch.

During off hours, I continued to walk the hillsides of Toohey Ridge, poke into holes I found, and talk to landowners. One day I found an unoccupied shack that looked ideal for a cavers' fieldhouse. Because it was isolated, unruly cavers could not disturb neighbors. More importantly, it was dilapidated, so no one wanted to live in it. The back room ceiling was falling in, and a chimney had toppled. The local people knew me quite well by this time, so I had no trouble getting permission to use the shack.

Jim Currens was impressed with my resourcefulness, and we immediately made improvements. We bought a wood stove for heat, threading a long length of stovepipe up through the broken-down chimney. We built bunk beds for the main room, and in the small central room we set up a small kitchen to heat cans of food. We installed a hand pump on the shallow well in the back and a pit toilet in the front yard. We laughed as we dragged the crummy outhouse with a moon and star carved in the front door and placed it over the hole. We spent many evenings on the front porch drinking beer or just talking. It was home. By the end of March, I moved from Don Coons's house to the new Toohey Ridge fieldhouse.

The house was a cozy firetrap. A running joke was that in winter, you had to "water the wall." Often, cavers had awakened, eyes burning, from the smoke. Across the room, they would see through the smoke that the wall behind the wood stove was aglow, ready to burst into flames. One of them would jump out of bed and race over to waiting buckets of water and empty them against the smoking wall. To avoid this situation, we assigned a "watch" where someone was in charge of throwing water on the wall to keep it cool. His job was to water the wall. We were lucky we were not incinerated.

A fieldhouse was fine, but we needed a cave to go with it. My determination returned.

In the spring of 1976, Jim Currens and I redefined our objectives. Toohey Ridge was too stubborn to reveal its secrets. We needed new territory in which to work. More ridges surrounded Toohey Ridge; some were just a few hundred feet across the valleys from Toohey Ridge. Surely an entrance to the system we sought did not have to be *on* Toohey Ridge, since such a system probably encompassed quite a large area. We marched out, fanning into the valleys, continuing the now broadened search, finding new holes to check and caves to push.

My annoyance with the Cave Research Foundation's air of superiority continued, but my respect for them and their success remained. I wanted to hedge my bets so I could work in the longest cave one way or another. With my pride swallowed, I tried my best to conform with the CRF's program. I wanted to belong. I waved stories of my experience as a project caver as a patriot waves his flag. "Here I am, honed and ready to take command at Mammoth Cave." I was a self-proclaimed protégé.

My own arrogant attitude pissed off those old veterans. A few patronized me, and some even enjoyed my company, but most seemed basically intolerant. I could visualize the vectors of a collision course between the CRF and me. A poster on the wall of the CRF Field Station said it all: "Those of you who think you know it all are annoying to those of us who do."

Trip after trip, party and expedition leaders told me I was screwing up. If I did not straighten up my act, they said they would throw me out.

This was just prejudice, I thought. They resented someone with my energy, understanding, and desire. I rocked the boat, and they did not like boat rockers.

It was a bittersweet pill. Although I admired the CRF's organization and liked most of the people, I could not stand their lock on leadership to which I was so capable of contributing. Where was their sense of appreciation? I slowly withdrew to the security of my own project a mile to the east.

On 3 April 1976, I led a group of cavers to look at holes along the northeast corner of Eudora Ridge, while Jim Currens took a group to search for cave entrances in some of the still unchecked valleys surrounding Toohey Ridge.

Eudora Ridge is another broad expanse of sandstone just to the northeast of, and almost adjoining, Toohey Ridge. Eudora Ridge was in our new search area, and we found caves there to explore. Jim's group walked the valleys and climbed up each hollow on the flank of Eudora Ridge looking for holes that water flowed into during rains.

A stranger emerged from the undergrowth, startling Jim and his companions. "Hey! What are you guys up to?"

Had they trespassed? Jim usually made great efforts not to. They must not have noticed they had crossed from one farm to the next. He answered in his most professional tone, "I'm Jim Currens. We are cave explorers looking for caves around here." He added, "I apologize. We didn't know we'd crossed onto your property."

The stranger looked over the ragtag group he had stumbled upon. Satisfied with Jim's story, the stranger continued, "No problem. I just wanted to make sure you guys weren't hunting or something. We've had a lot of problems with hunters." The stranger stretched his arm out to shake hands with Jim. "I'm Bill Downey."

Jim clasped his hand. "Pleased to meet you."

"I own the farm up the hill." Downey pointed behind him with a nod of his head. "I know where a few holes are you might like to look at. Why don't you guys come with me and meet my partner, Jerry Roppel? All of us are camped up on the ridge."

Without waiting for an answer, Downey turned up the hill and led the way through the woods. After walking up an old road, they emerged into an open field. Across it was a small mobile home and a bonfire. A half dozen people were cooking hot dogs on sticks.

Bill Downey was from Louisville. His open and outgoing style disarmed people. He and Jerry Roppel, from Bowling Green, thirty miles to the south, owned this farm, a weekend retreat where they could hunt or just escape the pressures of work and city.

After an hour of socializing, the entire group squeezed into the back of a pickup truck. They sat among coolers of beer and empty oil cans. The battered Chevy truck pitched along an overgrown and rutted road that meandered along the bottom of the valley. The passengers in the back clung to the sides to keep from being bounced out. Roppel soon pulled into a pasture past an abandoned log homestead where cattle grazed by a small pond. Sinkholes dotted the bottom of the narrow valley. Roppel stopped at the edge of the woods below a shallow draw in the flank of Eudora Ridge.

"You boys come take a look at this. Found it fencing the other day."

They scrambled after Roppel up the draw, through low-hanging branches that made walking difficult. They crawled under a new strand of barbed wire and felt a blast of cold air. A narrow gully coming down from the top of the hill disappeared into a three-foot hole in the side of the draw. The gale blew out from its unknown depths.

Jim had not seen anything like this in two years. Real cave! This was not just one of the dozens of moss-lined holes we were accustomed to seeing, but a cave that still took lots of water and blew lots of air. He crouched, straining his eyes as he peered into the dark. He tossed a stone through the low opening. Silence, then a faint echoing clatter from below. A pit! They might be able to squeeze through the eight-inch constriction beyond the initial opening, but they had no ropes or lights. They would have to get some gear and return.

That evening, Jim told me about the new discovery, Roppel Cave. I ached with envy. All this work, and I had missed the big moment! They had even decided upon the name!

The next morning, we gathered Jim's tools together and all the vertical equipment we could find. Our plan was to send a small party ahead to rig the pit inside the entrance. The rest would check some other leads on the surface. Herb Scott, a jovial, round-faced caver from Hopkinsville, Kentucky, was eager to lead the party checking the pit. This was fine with me. If Herb succeeded, he would be sucked in to our caving for life. If the hole did not go, I would not be there to suffer the disappointment of yet another defeat.

Jim Currens, Steve Stewart, and I fanned out to search for Isenberg Cave. This was another on our list. Local folklore painted a picture of grand cave, but nobody could tell us where it was. Two hours of bushwhacking resulted in nothing but shredded clothing. Anyway, my mind was back with the others. Suddenly, Herb Scott's party emerged from the brambles.

"Herb, didn't the cave go?" I asked apprehensively.

"Too small to squeeze into," he said. "Tried hammering. Rock's too hard to break. Damn near busted our hands trying, but it just won't give. Any ideas?" They had rigged a rope to descend the drop found yesterday, but nobody could fit under the first low opening into whatever lay beyond.

Jim smiled, and with a twinkling eye responded, "I have some `instant cave' back at the fieldhouse."

We returned quickly to Roppel Cave. Jim removed a bag of white powder and a small plastic bottle of rose-colored liquid from his pack. These were the two components of his "instant cave." Separate, they were harmless; together, they were an explosive. After biting the top off the plastic bottle, he squeezed the liquid into the plastic bag of white powder and kneaded the mixture together like cookie dough. Herb Scott, meanwhile, pulled out the rope they had rigged earlier, coiled it, and stashed it behind a large tree. I had heard tales of expert blasting opening the way in nearby Jesse James Cave, which was where Jim had learned this skill. Jim's precautions impressed us all.

Jim carefully packed the explosive mixture between the problem ledge and the cliff face, using mud and rocks to tamp the charge. From his pocket he pulled a blasting cap with two protruding wires. Energized, it would detonate the primary explosive. Meanwhile, we payed out a roll of silver-colored wire to reach a battery Jim had placed behind a large tree.

"Borden, bite off the insulation from the far end of your wires. Then twist them together," Jim instructed.

As I obeyed, he explained that this would guard against latent static electricity prematurely detonating the explosive he was about to wire.

"Done?"

"Yeah," I said.

"Are you positive?"

"Yes!" I shouted. Suspicious, untrusting son of a bitch, I thought.

Jim completed the wiring of the charge by twisting the wire to the blasting cap leads. He was careful not to let the two exposed leads touch each other. Done, he climbed out to join us. We all ducked behind trees for cover.

Jim authoritatively surveyed the scene. Once satisfied that all was well, he shouted, "Fire in the hole!"

He touched the exposed ends of the blasting wire to the battery terminals. A burst of white smoke shot from the opening. We hunkered down.

Crack!

Rocks ripped through the leaves around us. Seconds later, silence returned, smoke blowing away in the wind.

Our ears rang. We eagerly ran over the sharp, freshly shattered rocks. The ledge was gone, turned to rubble and dust. We cleaned away the loose rock and admired our fine entrance to Roppel Cave.

We re-rigged the pit, lowering the end of the rope into the drop. Herb Scott clipped onto the rope, and after testing it with a stout tug, he rappelled into the cave.

Rigging a rope off a tree

"Take a good look around and let us know what you see." We would wait until we were sure the cave went somewhere.

From below, Herb shouted, "Off rope!"

Herb was in a complex cluster of six or seven vertical shafts about thirty feet high and ten feet in diameter. Leaves and branches littered the floor. The way to continue was not obvious. If he stood very still, he could hear the faint thunder of a distant waterfall. Dark holes high on the wall hinted at passages but were unreachable. Herb knew about the necessity of reaching that water if Roppel Cave was going to amount to anything. Flowing water is a better indicator of going cave than dry leads. Some holes that looked promising turned out to be blind alcoves or cracks too narrow for cavers. Air blew from all the cracks, but there was no way through. These shafts were alive; they carried lots of water to somewhere. Something must go.

"Send Rod down," yelled Herb. "So far, I haven't found anything. I do hear a waterfall and I need some help here."

Rod Metcalfe, from Lexington, Kentucky, was a geology student. A new recruit, he had provided strong help on a number of trips.

Rod and Herb rechecked everything, looking under ledges and behind boulders for the elusive way on. Under one small ledge just fifteen feet from the base of the rope, they saw a crawlway, overlooked at first because it was so tiny. Herb peered into it and felt a breeze blowing in his face. He could see ten feet into a slightly larger canyon. Cave.

Herb jammed his body into the small hole, grunting as he inched forward. The walls were lined with lumpy calcite deposits with sharp edges—popcorn—that grabbed and tore his clothes. He stopped as his chest got hung up on a constriction. The passage was too small. He wiggled back out, stripped off his shirt, and forced his exposed torso through the chest compressor, leaving bits of ripped skin behind.

"Ouch! That's a real booger!" he gasped. "Rod, come on through. I see a narrow canyon ahead . . . might need your help."

Rod was smaller, but he still had trouble forcing himself over the sharp mound of popcorn. They belly-crawled along the tight, body-sized tube over a deep, narrow canyon. Thirty feet farther on, they peered down into a lower-level room. Ahead of them at the same level, they heard the sounds of water. Fifty feet farther, they were over a larger, deeper hole. They heard the shower of the waterfall directly below them, too far to see. They returned to get more help and a rope.

After we listened to the twosome's animated report coming up from the bottom of the entrance drop, we knew we were onto something. I grabbed the two rolled-up sections of thirty-foot-long cable ladder to take with me as I rappelled down the rope. Cable ladders were the mainstay of our vertical caving in the many short vertical shafts on Toohey Ridge. Constructed of aircraft cable with quarter-inch-diameter aluminum rungs six inches wide every twelve inches, they are fast and convenient and often make climbing gear unnecessary. Unfortunately, they are not as strong as rope and should be used with a safety line, and they can be awkward, if not dangerous, in some climbing conditions.

We snap-linked the two ladder sections together and unrolled them into the first hole. A jutting horn of rock provided a place to rig the ladders. I could not see the bottom and there were no rocks to toss, so I had no idea how deep the pit was. I tied a safety line around my middle. Rod jammed his hips into the canyon and looped the other end of the safety line around his back. With my feet in the ladder rungs, I squeezed through the canyon and into open space where I could begin climbing down. As I descended the ladder, Rod slowly fed out the rope. If I slipped, he would catch me.

I counted thirty rungs, then stepped onto the cobble-covered floor.

"Off belay!" I shouted.

"Belay . . . off!"

I untied and stood among the tangled heap of the excess thirty-five feet of cable ladder. The room was a bell-shaped vertical shaft floored with sandstone cobbles, broken rock, and more storm debris. Water dripped into a pool at the far end and flowed out beneath a ledge beside me. I had climbed down from its highest point, twenty-five feet above my head.

Descending a cable ladder

"Anything there?" came an impatient shout from above. Two hopeful faces were crammed into the space, staring down.

I crawled through the stream under the ledge. "Checking."

I squeezed on my belly on the gravel floor of the stream into another much larger vertical shaft with a ceiling forty feet above my head. Above, I could see the shadow of a canyon snaking across the ceiling, apparently the one from which Herb and Rod had seen the waterfall. Just in front of me was a jagged gash in the floor. Another pit! I scooped up rocks and tossed them through the hole. *Whoomph!* The crash echoed up seconds later. The echoes confirmed that the pit below was large. I threw in rock after rock, pleased at every crash. Any kid who has smashed pop bottles knows the feeling.

Several minutes later, muffled voices announced the arrival of Jim, Herb, and Rod at the lip of the pit.

They had heard my racket and shouts and knew I had found something. We sat gleefully on the edge of the pit, feet dangling into the abyss. Every loose rock we could find whistled down the pit.

"Wow! This monster is huge!" I said.

"I can only see the one wall here." Jim pointed to one side. "Everything else is out of sight. It must be at least seventy-five feet across!" His voice echoed in the open space.

Jim's broad grin reflected my feelings. Would this lead into Toohey Ridge? This *had* to be the way; it felt so right. But we were out of rope and time. We had burned the day getting this far. We began making plans for the next weekend.

First things first. In order to get out, we had to face the ladder climb, the narrow canyon, and the Chest Compressor. And we found that rerolling the ladders in the narrow confines of the canyon above the drop was impossible—another disadvantage of ladders. With one person pulling the ladder up and the other three stationed along the canyon, we threaded the ladders out to the shafts at the entrance where I rolled them up.

We climbed up the rope to the surface, where the wind snuffed out the flames of our carbide lamps. Three hours after throwing the last rock down the new pit, our happy, tired party walked back to the waiting vehicles under a beautiful full moon. On 4 April 1976, we had found a cave.

Negotiating a canyon in Roppel Cave

4

A Way In

Jim Borden and Jim Currens Find a Cave

The week following the discovery of Roppel Cave crawled by. Wondering what was down that pit became an obsession for me.

On Friday evening, I paced, watching for headlights to appear on the rutted road leading to our fieldhouse. My excitement spilled over into impatience and fear that no one would show up. After what seemed an eternity, Jim Currens arrived. He had recruited a large crew of cavers for the weekend, all interested in the promise of Roppel Cave. The excitement was contagious. Far into the night, we drew up our plans. We would survey what we had found the previous week, rig the new drop, and continue exploring beyond.

Jim and Herb Scott would spearhead our advance. Jim brought explosives to enlarge the Chest Compressor as well as the tight ladder drop beyond it; then they would rig the drops. My job was to survey the route from the entrance with the remainder of the party. A sensible plan.

However, it turned out more like "The Three Stooges Go Caving." After Jim blasted open the hole at the top of the ladder drop, I ordered the survey crew through the former Chest Compressor before Jim and Herb could repack their tools and gear and head toward the big drop.

"You idiot! We need to get past you," railed Jim.

"If you hadn't been so damn slow, we wouldn't be ahead!" I shouted. "What did you expect us to do—wait around for you guys to get ready?"

After the exchange of insults, my survey crew squeezed down into the canyon, and the rigging team crawled over us, their knees digging into our backs. They rigged and descended the ladders, and we resumed the survey.

When we reached the rigging team at the new pit, Jim was hammering a drill bit into the rock to place anchors. Two artificial anchors would be necessary, since there were no natural tie-off points. He had already driven one Rawl stud—a steel bolt slightly larger than the receiving hole—into the first hole; then he placed a hanger over it for rigging, using a tiny crescent wrench to snug down a nut to hold the hanger on. We threaded our rope through a carabiner hooked to the hanger.

Jim moved swiftly to place the second bolt as we uncoiled the one-hundred-foot length of nylon rope and lowered it into the pit. These were Jim's bolts; he would go first.

He rappelled through the narrow rift just below the anchored tie-off. "Yahooo!" He celebrated his gliding descent into a very large shaft. On the bottom, seventy feet below, he unclipped himself from the rope.

"Off rope!" echoed from below.

I rigged to descend while Jim checked for leads. Herb would follow me with the survey gear.

Shortly, Jim bellowed, "Borden! Come on down!"

"On my way!" I yelled. "On rappel!"

As I slid down the rope, I saw that this was an immense, flat-floored room, by far the largest we had yet found. A waterfall thundered into a large pool. Water ran out a small passage on the opposite side of the room. Jim's light flashed from within a large side canyon fifty feet across the room.

Jim met me as my feet touched the bottom. I said, "What's over—?"

"There," Jim pointed at a black void, seeming to know what I was going to ask, "a bunch of tall shafts. I didn't follow them out, but they look good."

A typical vertical shaft in Kentucky, formed by vertical flowing water seeking master base level

He then pointed to the base of the rope.

"That big pool flows into a canyon. I went just a couple hundred feet down it. Blows a lot of air. We'll survey it."

The excitement in his voice told it all. We had found real, going cave!

Herb lowered the survey tape as a crude plumb line to measure the depth of the pit. Water from the spray made the tape shimmer in the light like a silvery ribbon rising the full height of the drop.

"Seventy-one point five!"

The loose zero-end hung freely, just touching the gravel base of the pit. "Seven-one point five feet—got it!" I recorded the number in the survey book and sketched a profile of the drop.

Herb and Rod Metcalfe joined us. Now our four carbide lamps fully revealed the impressive volume of the shaft. The circular-shaped walls funneled up to an apex seventy feet above the flat floor. The rope came through the apex and fell free all the way to the floor.

Jim made a proclamation: "We should name this pit Coalition Chasm!" Jim and I had planned the organization and structure of an enduring project. Jim had suggested the name Central Kentucky Karst Coalition, and it stuck. That's the beauty of an organization of two: there is not much debate. Unlike the more regimented CRF, we were building a loosely affiliated group of individuals to work cooperatively towards a common goal—a true coalition. We wanted to express recognition of that fact. Coalition Chasm it was.

Roger Brucker's well-proven cliché came to mind: "Follow the water to find big cave!" We ran our survey line into the drain canyon. Soon, we were walking between tall, clean, light-gray glistening walls. Wind blew in our faces. Water gurgled cheerfully below us.

"This is great!" Feeble words to express a heart full of elation. The passage headed due south. I knew from the topographic map and what two years of effort had etched in my brain that we were following the flank of Eudora Ridge toward a saddle that joined it with Toohey Ridge. If this kept up, we would soon be walking beneath Toohey Ridge!

"I believe we found something this time," Jim offered.

No such luck.

As we continued our survey, the canyon not only deepened, but its sides squeezed together. We wedged our way along the top, over the unreachable stream that we could still see far below. At survey station B29, we could squeeze no farther as the walls met. The dancing water beneath us merrily continued.

Disappointment.

I retreated to where I could jam myself down into an intermediate level of the canyon, hoping that I could continue forward from there.

"I'll check ahead," I told the others. Gypsum coated the walls of the narrow canyon, small flakes that fell off the wall at the slightest touch. As I determinedly moved forward, rocks clattered down through the deep canyon, hitting the now quiet stream with a final *kerplunk!* I fought my way forward to reach a wide spot in the continuing canyon. The gentle breeze cooled my

Water from vertical shafts draining through crawlways and canyons or smaller shaft drains

perspiring face. Ahead, the canyon walls briefly widened to six feet, an insecure purchase for me to continue safely alone. I returned to the others who were waiting at our last survey station, B29.

The next weekend, Herb Scott and Bob Cook joined me at our field station to discuss strategy.

Bob spoke: "Herb, I don't think we should survey until we see if this cave goes." I was frustrated by his suggestion. Only weekend spelunkers explore without surveying. We were cavers with a system.

"Yeah, I think so, too," said Herb. "Usually, places like this just get too damn tight. I remember a canyon like this in Brushy . . ." He again told his account of his exploration of Brushy Knob Pit, a horror hole south of Mammoth Cave. I cringed at his discouraging tale. "It was exactly like this. It got taller and narrower as we went downstream. Finally, we could go no farther." His conclusion: "This will end the same way."

I did my best to silence him with mental projection, then pretended to ignore him.

I offered a leadership cheer. "We can make it go if we just push hard enough!" They scowled.

It took all of us an hour of hot, strenuous effort from the entrance to squeeze, climb, and rappel to the base of Coalition Chasm, a horizontal distance—our new map told us—of only one hundred feet. The entry passages to Roppel Cave were formidable and dangerous. The cave wore out the cavers.

We quickly moved through the now familiar cave, survey numbers passing in a blur: B20 . . . B22 . . . B25 . . . B29. An arrow marked the spot where we squeezed into the lower level of the canyon. Moving ahead in unsurveyed cave, flakes of rock fell from the friable walls.

We stood at last week's farthest penetration: the wide spot in the canyon. I paused, studying the problem before me. After going over the movements in my mind, I slowly leaned out over the fifteen-foot-deep canyon, my left hand on a projection above me, my right arm reaching toward a small handhold on the far side. My body formed a bridge, fully extended across the six-foot gap. I stretched my right foot to a small ledge along the wall. I felt the rock hold break and fall from beneath my foot. Panic! Instinctively, I thrust my left foot over to the opposite side and, in a swift motion, heaved my body across. Heart-pounding success.

Now it was Herb's turn. I held my breath as he tried to duplicate my moves. However, I was taller than Herb; he could not reach the ledge with his right foot.

"Shit! I can't reach it!" He was wide-eyed, his adrenaline surging.

After a few more attempts, he retreated. When he tried again, he used a layback move to gain enough friction to press his feet against the wall. His face reddened from the exertion of holding his body close to the wall with his arms as his legs pushed out. His sudden lurch was more like a controlled fall. He was over on my side.

"You sure had a funny look on your face when you made that last move!" I quipped.

Sweat poured down Herb's face. "Jesus. A scary bitch!" He panted.

Bob Cook, still on the far side, had taken in our acrobatics of the last few minutes. He looked pale.

"I'll just wait here. You guys look ahead."

Arguing with Bob would be a mistake. Even I knew not to push people to climb when they do not feel comfortable; it's a ticket to disaster. The Step-Across Pit was a formidable challenge.

Herb and I advanced down the unexplored canyon. Gypsum covered every rock surface of the passage. Its roof lowered to four feet from the floor. A hundred feet farther along, we peered down a small crack into a comparatively large room.

"This killer doesn't let up!" I said. We forced our bodies through the widest part of the passage, a slot in the ceiling. Our feet dangled and thrashed in the space below as we moved inch by inch.

The Floorless Crawl

We grunted and scraped, sweltering from exertion in the floorless crawl, wedging our hips and shoulders to avoid slipping through the crack. Mercifully, after fifteen minutes and fifty feet of traversing, we were able to climb down into the small room. We collapsed on the dry sand floor, breathing heavily. It was the first place we had seen where we could rest in comfort.

Looking up, we studied the narrow snaking canyon in the ceiling we had just traversed. There had to be a better way. But the walls were smooth; the only way was the route we had come. I shook my head in wonderment: the Floorless Crawl was going to be quite an obstacle for future explorers if Roppel Cave went.

The canyon, now narrower than before, led out under the wall we were leaning against. "I don't know if I can squeeze through any more of this shit," Herb said. He was breathing hard.

Rested and still hopeful, I lowered myself into the continuing passage. Soon we could move no farther. The walls squeezed in again just like at Station B29. Maybe we could move down again.

"Herb! See that below us?" I gazed down at what appeared to be a large passage. My spirits rose.

He cocked his head, trying for a better view. "How the hell do you expect to get down there? The canyon's only six inches wide here!"

We backtracked, probing wider spots that might yield a route down. I found a subtle widening in the canyon that I could force my body through. I took off my pack to shed excess baggage. It was tight, but I managed to wiggle through.

"I'm down at the stream. Looks like I can follow it."

I waited for the scraping sounds of a companion following me.

Nothing.

"Are you coming?"

Herb answered, "I saw you go through there. I'll get wedged for sure. No way I'm even going to try."

"I'll just see where the canyon goes," I told him.

The passage ahead was the smallest yet. There was no sign of the large passage I had seen minutes before. Where was it? I bent my body like a contortionist to make progress along the stream, now rushing in the narrow channel one foot below me. In several places I had to lie in the stream to squeeze around the tortuous bends. I moved for a half hour, thinking that around the next corner would be a room, hoped-for relief from this tiny passage. But I had been gone longer than either Herb or I had expected. Herb might worry. I turned around, discouraged by my failure to find going cave.

Bonk! My head hit a protruding ledge as I tried to slide around a particularly annoying bend. Darkness. The impact had extinguished the flame on my carbide lamp. It takes two hands to light a carbide lamp, usually an easy task. Now, however, it was a struggle to reach it, my body jammed in this crack.

Pop! The passage flooded with light as I drew my hand across the flint striker wheel, igniting the acetylene gas. I looked up to see what my head had struck. There, beyond the projecting rock, was blackness. A room! Adrenaline rushing, I compressed my body, stuffing myself into the opening. I felt my skin tear as I pulled and scraped my way through into the space above. Then I collapsed, struggling to catch my breath.

As I rested, my eyes adjusted to the gloom. I was in a low room about ten feet around and four feet high. There were three passages leaving the room, not including the way I came. A low, wide, sand-floored tube looked promising, but I did not feel like crawling just now. Another three-foot-round opening blew a lot of air. I walked out the third and largest passage, five feet high and three feet wide. After a hundred feet I saw a narrow canyon snaking across the floor. This was caving. I mentally fit the passageways together in my head, like a computer manipulating a wire diagram of a complicated object. If I had persisted in pushing the stream, I would have made it here without all that struggle, I thought—a valuable lesson.

The light was dimming, shadows looming larger. My light was fading, my carbide nearly spent. Pleased, but now concerned about my failing lamp, I returned to the room, squeezed back down the slot, and struggled out along the stream. I should not have left my pack. The CRF party leaders mercilessly chewed out cavers who left packs behind. But I did not need a safety lecture now; I needed my pack. Finally, I saw the glow from Herb's light above me.

"Herb, where did I go down?" The featureless canyon offered no clues. It all looked too tight.

"You went down right where you're at."

I looked around, trying to force my body up at the more likely spots. No way. Worry seeped through me. Where was the climb? I was sure that it was right here. Getting up through a squeeze is always a different matter than coming down. Was this the spot?

Eyes alert for clues—scrape marks, loosened pebbles—I walked along the bottom of the canyon toward the Floorless Crawl. A small vertical shaft dripped water on me. With my dying light, I evaluated the possibilities: I might be able to chimney up here to reach known cave.

I placed one foot on one wall and the other behind me. I jammed my body between the water-soaked walls, using my hands for balance. Alternately, I moved one foot, then the other to the top, sliding my back up the opposite wall with each step. I squirmed through the squeeze at its top with feet dangling.

"Having fun?" Herb's tone dripped sarcasm. He had wedged himself between the walls of the canyon to make himself as comfortable as possible during his long wait for me. So, he was pissed off? I would show him.

"Well, we have cave." My lamp had died.

He pointed at my pack that I had left behind, "Didn't you think you'd need this?"

No answer was appropriate. I changed my carbide in silence, knowing that if I had taken my pack, I would have been worn out carrying it and would have never reached the larger cave passages.

We retraced our steps, bone-weary. The obstacles we had overcome on the way in seemed doubly difficult now. We passed the Floorless Crawl and came to the Step-Across Pit where we had begun our exploration. I was not able to reverse my acrobatic moves to go back but found some other small holds near the ceiling, and with feet flailing below me, I scrambled to the other side.

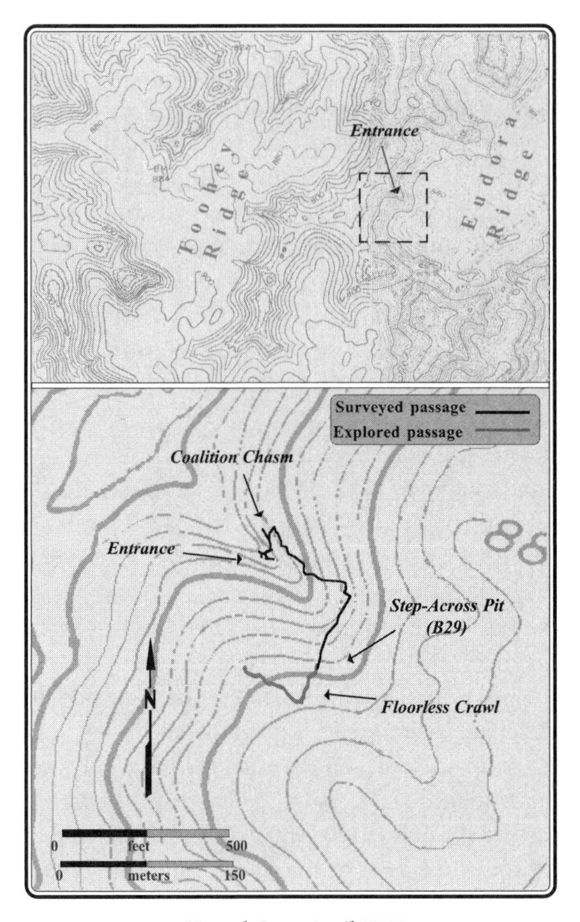

Roppel Cave, April 1976

I thought Herb was going to fall, but somehow he tenaciously clawed his way across.

"Shit!" was his final, single, red-faced judgment as he reached the safety of the far side.

At Station B29, we pulled ourselves out of the canyon floor, expecting to see Bob Cook waiting for us, but he was gone. He had given up on us coming back and was probably topside sleeping by now.

The entrance passageways—two rope drops, the ladder climb, and the Chest Compressor—finished us.

Neither Herb Scott nor Bob Cook returned to Roppel Cave.

On rappel

5

Small Rewards

The Roppel Cavers Grind It Out Through Tough Cave

The cave did not give up secrets easily. Yes, it did go, but we had grunted and struggled for every foot. Our zeal overcame each obstacle as we hoped to break into the big cave. Didn't hard work count for anything? It had to, we continued to tell ourselves. We ignored the atrocities we inflicted upon our bodies. But the pain and suffering did convince many cavers never to return after their first trip.

By fall 1977, the large-scale map of the cave had grown, but it was still just a small blob when plotted on the small-scale topographic maps. I saw that minute squiggle on the topographic map as part of an immense cave, two miles of it tucked into the southeast corner of Eudora Ridge. Most people saw a cave that really didn't go anywhere.

We continually looked for fresh opportunities to replenish our caver ranks. Those from far away, relatively caveless places such as Michigan were easy cannon fodder for trips into Roppel Cave, but we were not surprised when these new cavers rarely returned. But then Ron Gariepy, already a veteran of several trips, and I recruited John Barnes, a Michigan caver who now lived in Lexington. With several phone calls and a face-to-face meeting, I had fed

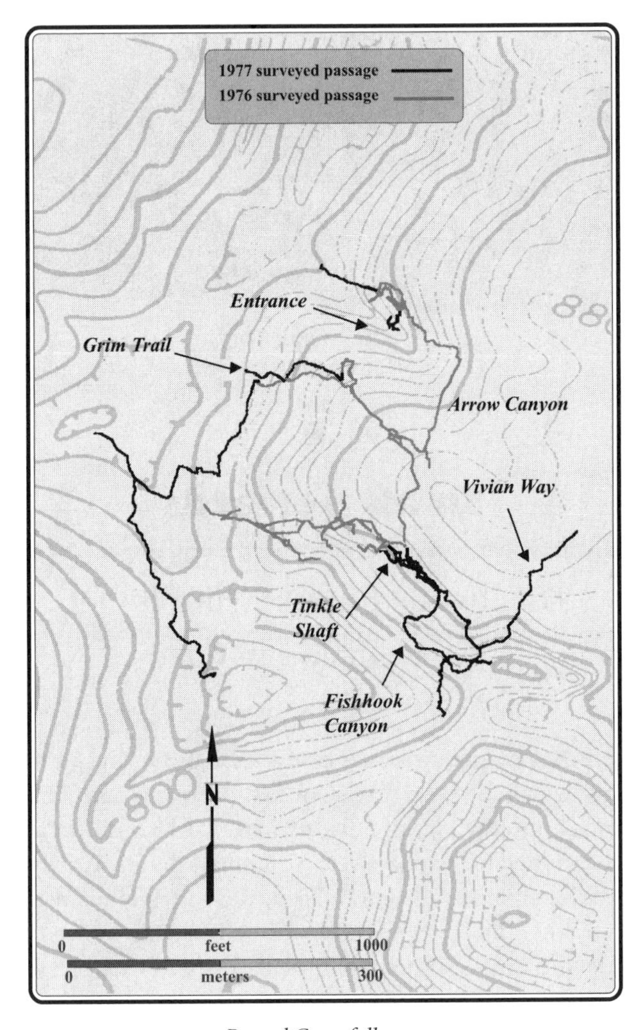

Roppel Cave, fall 1977

his cave hunger with descriptions of elusive wind in tall, complex canyons that headed into the blankness of Eudora Ridge. They beckoned the intrepid explorer to push around the next corner for the big discovery. It was not a lie. I extended morsels of fact and hope. His imagination formed them into a feast of images he wanted to taste for himself. A classic suck-in.

Fifteen minutes after John Barnes entered the cave, he became one of us. He squeezed and grunted, cursing the closeness of the walls of the "world-famous" Roppel Cave. It was 2 December, and the cold, howling wind through the entrance had convinced him: miles of cave must exist to cause such a gale!

John, Ron, and I pushed for two hours along the now familiar route through Arrow Canyon, across the Step-Across Pit and the Floorless Crawl, and on to the low, wide room I had found on my solo push, the B57 Junction. Ron and I sat quietly on the cool sand floor of the room, steam rising from our bodies. Soon, a sweating John dragged himself up through the hole in the floor leading from the confined canyon below. He collapsed into an exhausted heap.

"A killer!" he gasped between short breaths. "And it's so hot!"

Although the caves in this part of Kentucky have a temperature of fifty-four degrees Fahrenheit, high humidity and overexertion make the over-dressed caver hot.

I could see his denim shirt and wool underwear poking out from beneath his coveralls. He had dressed for the cooler Michigan caves he was used to.

"Peel off a layer or two of clothes before we go on," I insisted. I pointed to the low, wide elliptical tube. "It gets worse."

John took the advice, struggling with his clothing as Ron and I climbed into the C Crawl, the second passage explored off the B57 Junction. The first lead we had followed nineteen months earlier appeared to be the main passage. We had surveyed west in walking canyon. The fact that it headed west toward Toohey Ridge had caused us to brim with optimism about its prospects, but it soon degenerated to a seemingly endless, low and muddy belly-squirm. We named it Grim Trail.

Later, Bill Eidson and Ron Gariepy explored the C Crawl, lured by its dryness, pleasant oval symmetry, and gentle breeze. The C Crawl headed due south along the flank of the ridge. After crawling six hundred feet through passage often less than a foot high, the two cavers emerged at the base of an enormous vertical shaft two hundred feet long and at least one hundred feet high. At one end, a tall breakdown mountain led teasingly up into blackness; at the other, a narrow ribbon of water fell from the distant, unseen ceiling. This immense void was named Tinkle Shaft.

The shaft's blackness hid a junction of upper-level passages that were subsequently explored over several trips. To the east, toward the heart of Eudora Ridge, we dug through sand and rock to follow the breeze through a crumbling maze of canyons. At three different points the passage closed down, the way ahead buried under broken rock. The way on was never clear, but each time, we managed to find a path and scraped our way through. The passages were hot and dry and the sand sucked away our moisture as we struggled to dig onward. This was the K Complex.

Today we were going to follow the path blazed on the previous trip, during which we had spent four hours lying on our bellies pulling rock, piece by piece, out of a low stretch of crawlway, a cool breeze wafting over us. Eventually, we had struggled through to follow another few hundred feet of the canyon maze. There were openings at every turn, but we found only one route ahead. Grudgingly, the cave was yielding its secrets to our persistent efforts.

In the C Crawl, Ron and I pushed our packs ahead of us and pulled ourselves along on our bellies. The best approach for us was to roll our packs like a log as far as we could, then crawl up to them and gave them another good push. This rhythmic movement made the long crawl pass more quickly. Stopping to cool off occasionally, we could hear John scraping along behind us. No chance of him getting lost here. Inevitably, the gaps between us increased, and we moved through the crawl as three solitary individuals instead of as a party.

After four hours of traveling, we climbed Tinkle Shaft to weave through the maze of canyons and broken rock of the K Complex, finally reaching the end of the survey. The final survey station was marked with a small cairn of rocks with "K30" smoked on a flat piece of rock beside it. We lay in the dry sand looking at the recently opened dig extending into the darkness. That cool breeze suggested discoveries ahead.

This was why we went caving—the anticipation of discovery, the satisfaction of finding something we knew had to be there. I basked in my certainty that we were exploring in the longest cave in the world, that this was just the beginning of adventures that would last a lifetime. To be sure, a little over two miles is hardly the longest cave, but I just knew it was all here.

Ron dragged the survey tape behind him into the dig, building piles of rocks to use as survey stations. Surveying through the constrictions of this dug passage was difficult. Soon we were moving faster through the intricacies of the cave beyond the dig. At almost every station, we had to scout ahead to find the main route. The air took many more routes than we could. The survey went up, down, and through piles of rocks, a confusing series of twists and turns. I sketched quickly to keep the survey book up-to-date with drawings and descriptions of the many leads we checked. Occasionally, I instructed the rest of the party to slow the survey until I finished sketching. I was falling far behind the crew.

After several hours, we had traversed to the highest level of canyons we had found so far. Ron disappeared over a pile of rocks with the tape to set the next station.

An echoey word floated back to us: "W-o-www!"

Ron's manner of speech is deadpan. He couples it with dry humor, a combination that makes caving with him most enjoyable, if sometimes puzzling. His voice characteristically conveys no emotion.

"What's that, Ron?" I shouted.

An even more distant shout than before: "It's really big!"

"Big?" This could mean anything. After all, this was Roppel Cave.

"Yeah, *big!*"

I dropped my survey book and scrambled up the slope to see what he had found. John Barnes was sitting at the last station, changing carbide. Below, I could see the faint yellow pool of Ron's carbide lamp swallowed up by the gloom. I slid down the breakdown slope and stopped next to Ron. A fine elliptical tube opened before us. The passage was twenty feet high and thirty feet wide, shaped like a subway tunnel, with large pieces of rock on the floor. There were no footprints.

"Wow!"

We climbed back up the hill to the last survey station where John waited.

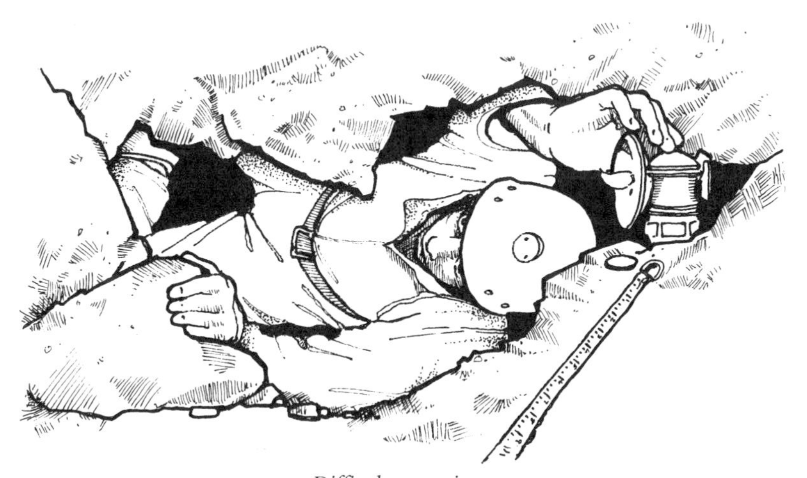

Difficult surveying

Savoring the moment, we ate lunch, a feast of cold beans and wieners with a chocolate bar for dessert. Few cavers ever found cave like this. We speculated about where this passage might lead. More big passageways? An underground river?

We surveyed into the large passage. The survey line turned from the southeast to the northeast, aligned dead up the center of the ridge. We surged on in an adrenaline rush.

Two hours later, we stood, horrified, at K72, nine hundred feet beyond where we had eaten lunch. At our feet, the ceiling plunged into the sand floor. This was the absolute end of the passage! Upon examining the terminated passage, we recognized it as an ancient lift-tube that had carried water from a lower passage when these levels had been active with flowing water. Now the lift-tube was plugged solid with sand. Whatever passages lay beyond the lift-tube were buried beyond our reach.

How could this wonderful passage end? Our energies drained away.

It was a bittersweet defeat: great cave, but no go. We gathered our gear and headed out, checking for side leads. Maybe we missed something.

We squeezed into every hole we could find, but they all either filled with sand or became too small.

Eighteen hours after entering Roppel, our weary party dragged out with spirits badly beaten.

The 2 December trip on which we discovered the large, nine-hundred-foot dead-end passage, which we named Vivian Way, was a turning point for the project. Cavers' enthusiasm vanished. Many of the "regulars" in Roppel Cave stayed away and offered only excuses. Over the next several months, the results of a series of unproductive trips painted a bleak outlook for the future of Roppel Cave.

Two parties revisited Vivian Way but failed to turn up anything except a few feet of survey and a miserable, long crawlway that dead-ended in a puzzling, round room, the epitome of our project. Frustration was bad; defeat was worse!

Another memorable trip was to leads at the far south end of the C Crawl, beyond the intersection with Tinkle Shaft. On an earlier trip, Bill Eidson, Don Coons, and I had surveyed nearly a thousand feet through foot-deep water and thigh-deep, oozing mud troughs. The C Crawl was no longer a dusty, dry tube once it passed below Tinkle Shaft. We found a new underground river and another complex of intriguing canyons and tubes. It held great promise, but it was so far away. I mounted a return with Jim Currens and

Chuck Thomas, and our crew spent all day getting there. Jim railed at the difficulty of the trip in and was ready to head out before we barely got started with the survey. With cajoling, we did manage to set thirty survey stations in eight hundred feet of low belly-crawl before Jim mutinied. We were not even able to look at the new underground river and its leads. Perhaps this new area, Fishhook Canyon, was too far from the entrance for productive work. Six hours of crawling through mud and water to get out there was hard work. Without a willing party, I did not return to the watery passages of the C Crawl.

By spring 1978, we had surveyed a little less than three miles. Three lousy miles! Two years of effort and dozens of trips had yielded probably the most difficult series of passages ever put together in a single cave. A trip to the distant reaches of Roppel automatically became a twenty-hour ordeal. We were kidding ourselves. How could this be the beginning of the longest cave in the world? We were wearing down, and our will was draining away.

By summer, Jim Currens decided he had seen the writing on the wall. Ever practical and realistic, he wrote trip report notes detailing his coming abandonment of the project unless he could figure out a different approach. I lambasted him as a pessimist. My glowing optimism was diminishing, but it still shone.

Jim focused on the areas around the entrance. He told himself that if Roppel Cave was going to amount to something, two things had to happen: First, going cave had to be found relatively close to the entrance. Jim did not like long trips, and he frequently lectured that any breakthrough made along the cave's frontier would prove too inefficient for exploration of miles of cave. Thus, a "promising" lead that took six hours of crawling deep inside the cave would not qualify as a good lead to him. Second, and most important, the going cave must be in Toohey Ridge, six hundred feet west of the Roppel Entrance. Up to now, we had no luck finding any leads with these constraints, so six hundred feet may as well have been six hundred miles.

Since the beginning of work in Roppel Cave, we had been puzzled by the discrepancy between the volume of air flowing through the entrance and the lesser volume of air that passed through Arrow Canyon. We were missing something. Despite diligent searching, we had found no clue to the missing airflow. Jim's strategy was to check out methodically every nook and cranny in the Coalition Chasm complex for the elusive airflow. Where there was airflow, there had to be cave. We had already found several canyons that fanned out to the west beneath the valley towards Toohey Ridge. All had terminated, yet their presence suggested that other canyons might be there and success might still be possible. Find the missing air, Jim insisted, and we find our way into Toohey Ridge. He was driven by this single goal.

Until this time, only one noteworthy lead had remained unchecked. John Hiett, from Lexington, had written a report about a trip on 8 May 1976 and described this lead as a popcorn-encrusted crawlway traversable for three hundred feet through some small shafts to a blowing breakdown collapse. Jim had searched for it a number of times but had failed to find it. I had seen it once but had forgotten it. Its location was poorly described and the area was highly complex, but this elusive passage in Hiett's trip report tantalized us:

> Exploring onward, two smaller shafts are reached. The third and presently final shaft drains to the OPPOSITE direction to the other two. The final point where the water drains is under some chert breakdown. It appears to be easily removable. This is a good lead; it blows air.

It blows air! That lead might be the key to Toohey Ridge! Jim was sure of it. Now, if he could only find it.

Bill Farr, a caver from Texas, had written Jim a few months earlier to arrange a trip into Roppel. He was going to be in the area and wanted to take advantage of the opportunity, so they made plans. Uncharacteristically, Jim promptly forgot them, and then Bill neglected to call to confirm, as agreed upon. One Saturday morning the phone rang. It was Bill, wondering where

One of the Mammoth Cave region's karst valleys, the bottoms of which usually lie 100 to 300 feet below the ridgetops

everybody was. Oops! In a flurry, Jim threw his gear into his car and raced down to the cave.

With nothing else in particular to do, Jim decided to poke around again in the high canyons and shafts of the Coalition Chasm complex to look for the elusive lead. The two cavers chimneyed up and down the ninety-foot-tall canyon on this impromptu trip, investigating every pocket and lead they could find. The height and breadth of the canyon were dramatic—passages that led off could be hidden anywhere. They spotted several black holes that they could not reach. The two studied them carefully, wondering if these holes were the source of the wind. Over several hours, Jim and Bill systematically checked every hole they could find and reach. Nothing.

Finally, Jim found the entrance to a small, popcorn-lined crawlway hidden in shadow in the wall of a large canyon. They had overlooked it on their first pass. The crawl was neatly tucked into a shallow recess in the wall. Bill and Jim crawled in the first few feet to a larger area where they could stand up. Footprints. This might be the lost lead.

After more crawling, they stood up in a small dome and inspected the pile of rocks that practically blocked the continuing canyon. This had to be the end of John Hiett's exploration; the description matched perfectly. A small trickle stream lazily flowed beneath the pile of rocks, and a steady breeze flowed through an opening at the bottom of the pile. Jim peered beneath the rocks. It would take elbow grease to move enough rocks to get by. The two made the quick trip back to Coalition Chasm to pick up the digging tools that were routine for each trip taken to this part of the cave.

Twenty minutes later, they were back and beginning their attack on the pile of rocks that guarded the destination of the westward-trending stream. They soon cleared away enough debris for a caver to squeeze through between the top of the pile and the roof of the passage.

"Bill, you're smaller than me. Take a look," Jim said.

Bill looked dubiously at Jim but good-naturedly said, "Sure, I don't get up this way often."

Bill squeezed over the pile of rocks and down to the floor of a narrow canyon. Jim could hear grunts and the scraping of shifting rocks.

"I think it's too tight!"

"What's it like?" Jim asked.

Bill glanced around, then said, "Wait a minute. I see something here. Maybe it isn't too tight. I'll give it a shot."

Bill groaned. The low ceiling pressed his body flat on the floor in a pool of cold water. He pushed with his feet, his belly dragging through the wet gravel. The passage was too low to turn his head to look in front of him, so with a measure of faith he continued his push through the squeeze. A protruding rock-horn fossil dug into his back and scraped past his hips as Bill advanced.

After six feet of this squirmway, he climbed to his feet in the narrow crack. It was big enough to stand—barely. A narrow lead continued.

"Jim! Come on through. It looks like it might go."

Jim could barely make out Bill's voice, muffled by the cavities among the rocks and the wet floor.

"Okay, I'm on my way," he shouted back.

Indeed, Bill was smaller than Jim. Once over the rock pile and lying prone in the stream, Jim was wedged tight, both arms outstretched in front of him. He yelled through the crack ahead.

"I'm stuck!" He rested, panting in the constricted space.

"Not much I can do from here to help you," Bill said.

Jim breathed deeply to regain composure. He forced himself to relax and slowly backed out of the squeeze.

"I'm back out!" shouted Jim.

"Want me to go ahead and check this lead out?" The open passage ahead tempted Bill.

Jim wasn't going to miss this. "No. I'll make it."

Now Jim tried to figure out how to get through the obstacle. The geometry of the passage—a corner combined with a few strategically located rocks—meant that he would have to lie in the stream with one shoulder jammed under a low ledge. It was narrow, so he would have to put one arm forward and the other back in order to angle the shoulders and decrease the cross-section width of his body. Choosing which arm to put forward was all-important; once started, he couldn't change his mind. If the choice was wrong, he would have to reverse and try again—unless he got stuck. After calculating his series of moves, Jim began his second attempt.

Jim could feel and hear his clothing shred as he shoved himself past the sharp popcorn protruding from one of the walls. He squeezed through the stream at the floor of the canyon. Sharp rocks jabbed into his chest, his muscles straining at the contortion. He made minute progress by digging the toes of his boots into the wet gravel. Jim squeezed past the spot where he had gotten stuck before, then, after a few more tenuous pushes, he was through.

"Damn!" he said. "I barely made it!" Sweat dripped from his brow.

"Looked like you were having trouble," Bill said, smiling. He was lying on his belly on a gravel bank above the stream and had watched the struggle.

Jim was panting. "I am absolutely the largest person who can get through that squeeze."

Resting, Jim watched the steam rise from their bodies and drift back towards the entrance. Although small, this lead looked promising: it had airflow, it headed west towards Toohey Ridge, and it was headed downstream. Other than a bit more room to move, what more could a caver want?

After cooling off, they pushed forward into unexplored cave with Bill in the lead. The passage alternated between slow walking and slower crawling. The walls twisted tortuously and bulged with popcorn that broke off in showers as they passed. The sluggish stream gradually incised deeper in the floor of the canyon. After about three hundred feet, Jim and Bill decided to turn back, saving this passage for the survey crew. Ahead, the passage continued, the stream still cutting deeper into the floor, air blowing from somewhere unknown. The compass showed that they still were heading generally westward. Wow! A real live, going lead.

Was this the way to Toohey Ridge?

Bill Walter proving to be a strong caver on a long, tiring trip

6

Reunification

Bill Walter Returns to Lead the Way to the Big Cave

I listened to Jim Currens describe his push of the breakdown in the new lead off Coalition Chasm. Simple and to the point—"Borden . . . it *goes.*"

This was great news—a going lead so close to the entrance and heading downstream toward Toohey Ridge! Jim was sure that this lead was the key to the big break we were waiting for.

I went home to Maryland for the summer school break but planned to return to Kentucky over the Memorial Day weekend to cave in Roppel. My initial plans had been to return to the wet, distant passages of Fishhook Canyon; those leads had to go somewhere! Miraculously, I had succeeded in putting together a willing (and able) crew. One of my recruits was Bill Walter, a caver unknown to me. Jim had introduced us briefly. Bill was small, wiry, and in his forties. Jim had assured me that his reputation as a caver in Tennessee was second to none. Still, I was dubious of new people. They could be a risk—especially if they were senior citizens.

After ten seconds of considering Jim's account of his push of the lead near Coalition Chasm, his enthusiasm rubbed off on me. I decided that it was a superior alternative to Fishhook Canyon. I had been dreading that tough, wet, six-hour trip beyond the C Crawl: a hot crawl followed by cold water. I decided that I would have no trouble justifying a change in plans to the rest of my crew.

But what about this Bill Walter?

Bill's first trip to Roppel Cave had been with Jim a few months earlier; they had gone to the end of Grim Trail. Bill had read the December 1977 *NSS News*, which had included a feature article about our work in Toohey Ridge. The article had been yet another fishing lure dangled by Jim and me to hook any prospective caver, for we had been losing cavers in droves and were desperate for new blood. Our passionate story emphasized our dedication and steadfastness in our efforts in what was surely destined to be a great cave. Our article worked on one person: Bill Walter read the story with interest and empathized with the determination and frustration we had been experiencing. He wrote Jim a letter offering his help, backed up with an impressive compendium of his caving history. He could sense that we might be on to something at Toohey Ridge and that we were looking for any help. As soon as Jim had read Bill's offer, he dialed his phone number.

At the time, Jim had not yet given up on the main part of Roppel reached via Arrow Canyon. Beyond the junction with the C Crawl, Arrow Canyon continues a western trend. Eventually, the stream is lost into a lower level, and Arrow Canyon degenerates into a low and muddy crawlway: Grim Trail. We had surveyed through the goop for two trips, eventually halting at a small room. The way on was through a tight canyon, beyond which we could see a larger room. I had grunted and squeezed for an hour, digging on my side with the one free arm that was not uselessly pinned behind me. Through this constriction, the passage extended as before—a low crawlway. No breakthrough, but the cave still continued. Despite my urging, no one else in the party was willing to follow. So I wiggled back through the narrow slot, leaving the lead for another day.

Grim Trail headed directly toward Toohey Ridge at a low enough elevation to pass easily underneath the overlying valley. Jim was eager to return, but he needed someone small as well as strong—a proven caver who would not abort the trip. Bill Walter was the candidate of choice.

In January 1978, a party of four, including Bill and Jim, returned to the end of the B Survey at B105, the tight constriction that I had squeezed through. The trip to the lead was slow and deliberate as the other two cavers, who were unfamiliar with the cave, labored through the rigors of the Entrance Series and the obstacles in Arrow Canyon. The snail's pace, combined with mud the consistency of printer's ink, made Bill shiver. His smaller size allowed him to move quickly through caves, but if he did not keep moving, he got chilled.

While Jim and the others worked on enlarging the slot, Bill squirmed through to see if the lead went anywhere. It was obvious that it would take hours for the rest of the crew to make it through—if they ever made it through, that is. Beyond the room on the other side of the slot, the crawl indeed continued, but it was an excruciatingly tight belly crawl over gravel. Bill slid in, grunting and groaning as he inched forward. After about a hundred feet, the ceiling rose, and he moved quickly through the four-foot-high, ten-foot-wide passage.

Meanwhile, Jim and the others managed to enlarge the canyon just enough to get their smallest caver through. They had heard Bill squeezing through the low belly crawl; it had been many minutes before his scrapes had trailed off in the distance. It was obvious that even if they all made it through, the way beyond would be no easier. The three cavers were cold and depressed, routed and ready to return to the surface. Having surveyed two stations, they sat back and waited for Bill Walter.

Forty-five minutes later, Jim and the others heard Bill's grunts as he squeezed his way back through the belly crawl over gravel. Shortly, a warm-looking Bill was sitting with the rest of the badly chilled group, elated about his push. His shredded shirt suggested how difficult the passage ahead was.

"Looks good. Once I got through the squeeze, it opened up pretty well and the mud disappeared. I went about five hundred feet before turning around."

"Which way was it going where you quit?" Jim was focused on its prospect for getting into Toohey Ridge.

"Not sure; I didn't have a compass and it wound back and forth a lot. But it looked good where I stopped—nice elliptical tube thirty inches high and ten feet wide. Good airflow. We ought to survey it."

"No . . . too cold. Let's go out."

Bill looked around at the party members and sighed. Jim appeared cold, but the other two looked totally vanquished, trashed by the difficulties of the cave and mud of Grim Trail. There was no sense arguing.

Before the party packed up and left, Bill drew a sketch of the passage in the survey book.

The trip out was even slower than the trip in. The pace was interminably frustrating. It took over four hours for the tired group to struggle to the base of Coalition Chasm, each of the obstacles of Arrow Canyon seemingly taking forever to cross. Bill was shivering and Jim was disgusted at the pitiful pace of the other cavers. Once up the seventy-foot climb of Coalition Chasm, Bill and Jim could stand no more and bolted for the surface, confident that the short distance remaining would prove to be little problem for the slow cavers.

As the two sat topside in an unusually balmy January evening waiting for the sounds of cavers reaching the bottom of the forty-foot entrance pit, the western sky was aglow with brilliant flashes of lightning. Wind started to roar through the leafless trees.

"Shit," said Jim, "I think we ought to get back to the fieldhouse. Looks like we're going to get hammered by this storm."

Bill looked at the light show. During the bright flashes, he could see the frothing dark clouds rolling towards them. "You think they'll be okay down there?"

Jim started gathering his pack. He was not going to wait. "Sure, it's just a few feet. How could they get lost?"

Bill was not familiar with the mile-long hike over the hills and valleys back to the fieldhouse, especially in the dark. He shrugged. "Okay, if you're sure."

Two hours later—after cleaning up and eating dinner at the fieldhouse—the storm continued unabated. Still no cavers.

"Jim, don't you think we should go back and look for them?" Bill had brought up this question before but had met with resistance each time. Now, it was apparent something was very wrong. No one could take this long to get out of the cave.

Jim looked at the foreboding weather and frowned, finally agreeing that they had better do something.

An hour later, at 2:00 A.M., they arrived at the Roppel Entrance. A small stream formed a waterfall in the normally dry entrance pit.

Jim crouched into the entrance to the top of the pit. "Hey!" he yelled at the top of his lungs.

He and Bill strained to listen and were greeted by a distant, "Help!"

Jim looked at Bill. "You know, I don't feel too good—I must have the flu or something. Would you mind going in?"

Bill had already decided that he would be the one to go in and get them. He put his vertical gear back on and rappelled into the pit through the chilling waterfall.

Bill found the two cavers lying in the crawlway just thirty feet from the base of the entrance drop. Both of the cavers had decided they were lost and would wait for someone to come. Neither had the initiative to push on to find the rope.

Bill looked at the nearly hypothermic pair and smiled, "Come on! Let's get you out of here!"

Bill Walter's first trip to Roppel Cave was certainly memorable. Bill had demonstrated his ability as a caver and team player. As far as Roppel Cave was concerned, the cave was certainly tough, but he was sure there was something to be found. He had left behind a good, going lead that might reach big cave. He promised himself that he would return.

Bill Walter's return to the caves of the Mammoth Cave region was a legacy from earlier days. Raised in Louisville, Kentucky, he had begun his caving adventures while a teenager. The newly found caves of Flint Ridge near Mammoth Cave were close at hand, and the year was 1953, one year prior to the famous National Speleological Society's C-3 Expedition in Floyd Collins' Crystal Cave. The caves of Flint Ridge were ripe for discovery. The lure of integrating all the caves of Flint Ridge into one vast system was too much for Bill to resist.

He checked around to see who was caving and soon was following Bill Austin around the lower levels of Crystal Cave. Crystal Cave was vast, and Bill was immediately hooked. With Austin, Jack Reccius, Charlie Fort, and others, Bill Walter became involved in the clandestine explorations of the lower levels of Salts Cave and Unknown Cave.

Both Salts Cave and Unknown Cave are located within the boundary of Mammoth Cave National Park on Flint Ridge. At the time, the National Park Service did not permit cavers to explore caves within the park. As a result, all the prohibited trips into Salts Cave and Unknown Cave within the park were secret operations. Cavers would sneak in and exit the caves under cover of darkness, using furtive drop-offs and pickups. All this deceit was to avoid the vigilant eye of the unfriendly feds. The cavers loved it and so, I have heard, did the park rangers who occasionally caught them.

The lower-level passages of Salts Cave ranged far out beneath Flint Ridge. Salts Cave was Bill Walter's true love. On trips of up to twenty-four hours and sometimes alone, he unraveled the mysteries of Salts Cave. On one memorable trip, he and Charlie Fort spent seventy or more hours in the lower levels, snoozing in sleeping bags they had dragged in for bivouac. This was the commando-type of dedication to exploration that later made Bill famous.

His relatively short career as a Flint Ridge caver came to an abrupt end when he and Richard Wheeler were surprised and apprehended by Joe Kulesza, Park Service ranger, as they emerged from Salts Cave. A stiff fine and prospects of jail discouraged Bill from returning. He still wonders if he wasn't set up for this arrest.

Bill later left the Kentucky caves behind and settled in McMinnville, Tennessee, where there are caves everywhere in the surrounding countryside. His prowess and zeal as a caver continued to the point of legend. He got deeply involved exploring nearby Cumberland Caverns, where he was one of the spearheads of exploration that eventually unraveled twenty-eight miles in Cumberland, one of the longest caves in Tennessee.

However, Bill did not limit himself to just one cave. He was a caver's caver who spent nearly every spare moment tramping the high, wooded limestone flanks of the Cumberland Plateau. To him, there were caves everywhere, as if he possessed X-ray vision. I believe that single-handedly, Bill has found and explored more miles of cave than anyone ever will. Although a loner, he always welcomed anyone who wanted to go caving. He was endearing to all, pleasant, humble, and soft spoken.

On 27 May 1978, Bill Walter, Bonnie Butler, Chuck Thomas, and I were setting survey points in Jim Currens's new passage off Coalition Chasm in Roppel Cave. I had romantically expected glistening, smooth-walled canyons with the babbling music of running water beneath my dangling feet, wind in my face. What we found was clothes-shredding popcorn, tacky mud, and excruciatingly difficult surveying. Even the wind had pooped out. We were sandbagged. Was this why Jim was a no-show for this trip?

We seldom stretched the survey tape farther than ten feet in the tight, winding canyonway. Reading the compass bearings was an almost unbearable struggle. As the stations passed by, the canyon grew taller, the popcorn became sharper and more dense, and the walls pressed closer together. Just moving forward was agony. After eight hours of exhausting effort, we thankfully set the last station, S64, and I smoked the number on the wall using the three-inch flame of my carbide lamp.

Ahead, things looked even grimmer. The canyon was twenty feet high but barely eighteen inches wide. Just beyond survey station S64 was a shallow pit. Bill wiggled to the top of the pit and, by spreading his feet wide enough to press on the walls, managed to chimney to the bottom. He looked around. Only a minuscule drain, just barely large enough to squeeze into, led off.

He climbed back to the top, then grunted his way ahead along the high level for 150 feet before giving up. That short distance took him nearly twenty minutes to traverse. It was painfully narrow. Below, where he had quit, more

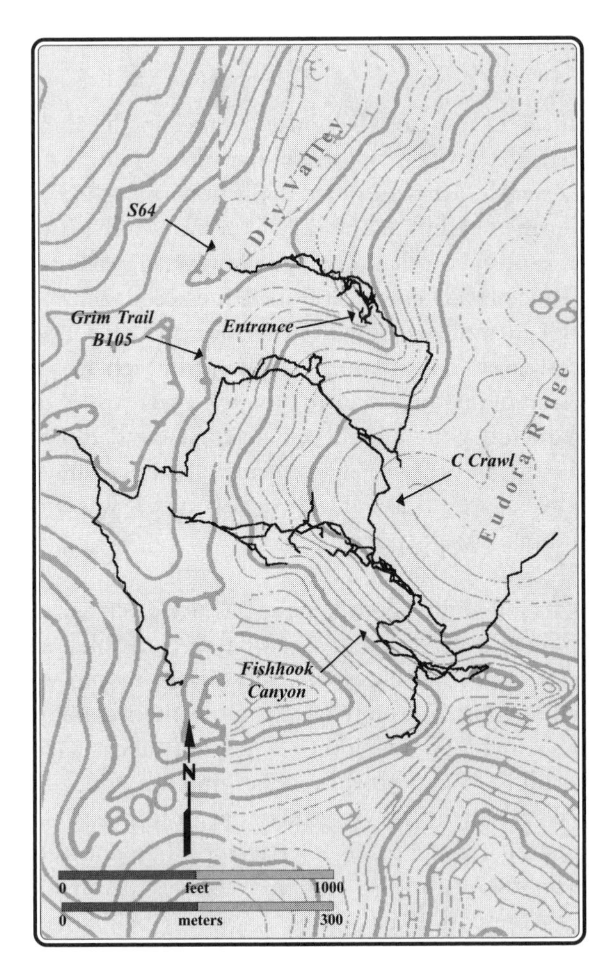

Roppel Cave, summer 1978: penetrations across Dry Valley

small pits led down. Physically and mentally wrecked, we left the cave. Bonnie Butler and Chuck Thomas never returned. Would Bill Walter come back?

Back home, we plotted the survey to discover that the S Survey at S64 had penetrated half the distance across Dry Valley, only a few hundred feet from Toohey Ridge. On paper, the lead looked more than promising; it screamed for a return trip. However, none of us who had been on the survey trip was eager to go back because the memory of the passage burned fresh in our minds. Jim Currens's eyes blazed when he saw the map showing this passage. He already envisioned himself reaching the big ridge beyond the end of the new survey.

Six weeks later, Jim took one person with him to the end of the S Survey. They were armed with formidable tools: a hammer, rope, and high hopes. The pit reported at the end of the survey intrigued him—down is always good, especially when beneath valleys.

Once they arrived at S64, they found out why we had terminated the survey. And try as he might, Jim could not squeeze past the constriction at the top of the pit to get in position to descend. There would be no going down this trip, and Jim did not know of the leads beyond the pit on the high level. Discouraged, Jim left the cave with nothing accomplished. He summed it up in his CKKC trip report: "Net result of trip: one thirty-eight-foot rope delivered where it can be picked up and delivered to where it might be needed!"

By midsummer 1978, I was dreadfully discouraged. A string of aborted trips and dashed hopes had taken its toll. The previous year, I had reinvolved myself in the Cave Research Foundation and had begun a systematic resurvey of Ralphs River Trail off Pohl Avenue in Flint Ridge. This had led to the discovery of an underground river that nearly connected Flint Ridge with Great Onyx Cave. In fact, it passed beneath surveyed passages in Great Onyx Cave, but despite several trips, we found no connecting passage. The discovery recharged me. Caving with the CRF also gave me a respite from the rigors of Roppel Cave. For the latter half of the summer and into September, I attended every CRF expedition and all but ignored the caves to the east near Toohey Ridge. No, I hadn't quit, but the thought certainly crossed my mind. The CRF had plenty of things to do, while we had slim pickings. But there was too much of myself in Roppel Cave to quit; I was forever linked with the cave.

Labor Day weekend was coming up. I had just returned from Maryland to Kentucky to attend fall classes at the University of Kentucky. I was a senior and was preoccupied with the future—what would I do after graduation?

Jim Currens asked if I were ready to go caving in Roppel again. I had just arrived at school and was thinking about attending the CRF Labor Day expedition. I didn't feel up for punishing caving in Roppel.

Jim wanted to return to the S64 Pit. Surely, I thought, his mad desire to get to Toohey Ridge had blinded him to the wretchedness and punishment of the S Survey. Moreover, I was shocked by his statement that the S Survey was the first in his countdown of leads to check before he personally abandoned the project. Such pessimistic talk! I could not bring myself to face the finality of any discussion of "abandonment." I told him I wasn't interested. Jim said he would find somebody. I wished him luck.

That somebody was Bill Walter. Bill continued to be intrigued by the end of Grim Trail; after all, he had pushed it farther than anyone, and the cave continued to grow. Despite the mud, Grim Trail was promising—it could be the way to Toohey Ridge. In the confusion of planning the trip, Bill thought he was agreeing to return to Grim Trail, not the S64 Pit.

Bill knew something was terribly wrong when, after arriving, Jim started talking about pits. There were no pits off of Grim Trail! His fears were confirmed when Jim unrolled a yellowed, dog-eared map of the cave and pointed to the squiggly line that led from near the entrance toward Toohey Ridge.

"Bill, did you see this?" Jim asked.

Bill approached the map on the hood of Jim's car and carefully studied the thin line-plot of passageway at which Jim was pointing.

Jim continued, "I think if we can push the bottom of the pit, we might just get into Toohey Ridge."

"Well," Bill began, "do you know how tight it is?"

"Yeah, but it's close to the entrance and the best thing we've got going right now."

Bill disagreed but good-naturedly went along, now resigned to the order of the day. "Okay, but don't forget to bring a hammer."

"It's already in my pack."

Bill concluded, "Well, it's heading in the right direction."

The pair readied their cave gear. Bill and Jim presented dramatic contrasts in caving equipment. Jim, always fastidious, carefully cleaned and maintained his gear after each trip. His carbide lamp's gleaming yellow brass shone, and he was dressed like a sportsman from the pages of *GQ* magazine. Bill wore clothing recycled from years of use on his farm. Holes were patched, but shreds of clothing hung off him like a scarecrow. Some of his equipment was older than I was! For Bill, equipment was a personal extension of his own soul, tenderly treasured, lovingly repaired, and never discarded. For Jim, equipment was a precision tool, to be honed, maintained, and organized. In either case, their equipment always functioned, each in its own way.

Two hours later, they were grunting through the S Survey. Sweat rolled off their faces as they forced their way past the popcorn that stuck onto and pulled at their clothes like Velcro. Jim's zipper on the front of his coveralls had long since come apart, leaving a wide gap that exposed his middle. Moving was hot work.

At S64, they stopped to cool off. It had taken them only forty-five minutes to get to this point. As they sat resting, the sounds of their heavy breathing and dull, rapid heartbeats were amplified by the close walls.

Bill reflected, "Well, we're here, so we might as well get started."

Over the years, Bill had found miles of cave, and he had a firm rule that he often said aided in his success: No lead can be said to end unless investigated and declared hopeless no less than two separate times. Often, cave that "ended" did so only because of lack of will, exhaustion, or overlooked possibilities. It took a fresh eye, another look, before one could write off a lead once and for all. He had found cave many times where others had given up; often, he would find cave where he himself had previously turned back. At S64, he put his own rule in action: he would check the pit once again.

Jim rigged the rope he had left after the previous trip, tying it to a natural bridge that spanned the narrow passage just before the pit. Bill wrapped the rope once around each outstretched arm and across his back, then slid down the rope into the pit. This arm rappel is a hasty but hazardous way to descend since one uses only rope friction to arrest the descent. Jim struggled over the top of the pit, having discovered that the best way for his larger frame was to climb above the squeeze before getting on the rope to rappel the usual way.

Jim urged Bill to lead on. Bill was smaller; it made sense for him to go first. After just a few feet, it was apparent that the drain was extremely tight. Jim retreated back to the S64 Pit to repair his malfunctioning zipper. Bill continued alone.

Bill pushed ahead and immediately confronted a sharp turn to the left. This turn could be negotiated only if he squeezed along on his side while using his arms to keep his body from slipping into a small crack. After a few more feet, he stood up in a small room to regain his breath.

In ten minutes of struggling, he had progressed just six feet. Ahead, the passage was narrower than before, this time with sharp ledges jutting into the passageway. Bill pounded at ledges with the hammer. Rock chips splintered and flew in all directions, and he forced his way inches forward. The cool breeze teased him as he shattered yet another ledge.

Bill had previously encountered similar situations in Kentucky. Canyons, such as the S Survey, often become narrower and deeper in the downstream direction. If the passages could be pushed far enough, they often intersected small pits. Pits in turn are the route to the underlying lower levels. The problem was, routes to the lower levels are almost always tiny. Sometimes they are too small for anyone.

Bill struggled around a corner to see the canyon narrowing to an impassably small size. He was disappointed, but in the past he had managed to continue on in less likely situations. The stream had cut down and was now flowing beneath a layer of broken rock that had wedged between a ledge and the wall to form a false floor. Bill thought there might be passable cave below.

He worked on the floor with his feet, stomping at the rocks wedged in the canyon. A cavity soon began to open, and he redoubled his efforts. A few minutes later, a hole led down to the hoped-for lower level.

Squeezing down through this hole, Bill found an eight-foot-diameter room, two feet high. The only lead was a tight keyhole passage less than a foot high and two feet wide, with a narrow canyon in its floor. Shining his light down the improbable lead, Bill couldn't judge if it got too small. There was only one way to find out.

This was Bill Walter at his caving best. Alone and far out, he basked in the solitude as he pushed into the unknown, no one to slow him down. He jammed his body into the small keyhole. Sharp protrusions ground into his side as he inched his way forward. Foot after foot, he fought his way in. After a final desperate squeeze that pressed his shoulder blades painfully together, he fell into a small room where he could stand up on a flat, sandy floor.

"At least I can turn around now," he uttered to himself as he rested prone on the floor.

More often than not, such small passages are pushed to their ultimate conclusion just so the caver can find a place to turn around. It is a caver's worst nightmare to push a small lead and find that the only possibility is to back out the entire way. More than one major passage has been discovered by cavers going on in search for that larger spot.

After a brief rest, Bill crammed his body into a vertical slot three feet high and a foot wide. Ahead, he could hear the sound of falling water. After eight feet in this slot, he was rewarded at last; he could see the inky darkness of open space. It might be nothing, but he felt sure it went on.

But it was time to return. He pushed back through the slot, the keyhole, and the other tight spots, swinging his hammer at offensive nubbins and sharp corners. Some of the tight spots were more difficult in reverse, since he now was moving uphill. After twenty minutes, he reached Jim, who was waiting in the chilly passage.

"Well, how did it go?" Jim asked.

"It opens up a few hundred feet ahead into a large room." Deadpan.

"How tight is it?" The important question.

"It's real tight. You'll need the hammer to get through. Even then, you might not make it."

Bill reentered the narrow canyon, giving the hammer to Jim. After a few minutes, Bill could hear the faint pecking sounds of hammer blows against rock as Jim attempted to enlarge the passage.

"At least it will be easier for me to get through now," Bill chuckled.

Twenty-five minutes later, a hot, winded Jim fell out of the last squeeze. Bill was sitting on a ledge, fresh and well rested.

"That keyhole just about did me in!" Jim said between long breaths.

Bill wondered if Jim would be able to get himself back out.

They climbed down into the room Bill had seen. It was a large, elongated vertical shaft with dripping water and a small pool; a large boulder covered most of the floor area. The water gurgled out a small crack in the floor. Up on the walls, several narrow canyons led off. The dark, damp walls swallowed the pale yellow light of their carbide lamps. After their hours of squeezing, the room felt as big as all outdoors, although it was little more than forty feet long and high. From S64, they had descended over seventy-five feet into this dome. An open crawlway led off to the west. Maybe they had broken into the lower levels?

They refilled their carbide lamps from the pool, then stuck their heads into the crawlway. There was a stiff breeze. After losing the wind in the high canyons around S64, they had found it again. The crawlway was slightly lower than hands-and-knees height. With Jim leading, they bellied in. There was a small stream in a narrow canyon below the hard, damp floor. Off to one side, they looked into a small room. That could wait. They crawled on, following the cool breeze.

After a couple hundred feet in this passage, they climbed up through a slot into a slightly larger crawlway passage. The wind was stronger now, and the passage drier.

"Wow," Jim exclaimed, "maybe we're on to something here!"

Bill continued crawling. Better not press your luck by thinking you had found something before you actually had. Bill had been fooled before, but he smiled at the promising prospect.

The canyon in the floor snaked off to the right out of sight. The mud had yielded to dry sand. Gypsum flowers sparkled on the walls and ceiling. Solitary cave crickets leaped out of the way of these fast-moving human giants.

After a few hundred feet, the crawlway intersected a walking-sized canyon passage.

Jim shouted, "We must be beneath Toohey Ridge!" He pulled a compass out from inside his pack. "Due north, straight up the flank of the ridge!"

"Let's see where this thing goes!" Bill replied.

They sidestepped along the tall, narrow passage, following a strong wind toward new discoveries in Toohey Ridge. This time it did feel right. This was unlike anything they had seen in Roppel before. Instead of shaft drains and tight crawlways, they were in ridge cave, passageways that the Roppel cavers had been searching for since the beginning.

They walked briskly, almost running. Sweat dripped from their faces. They stopped at a broad intersection. A fine, twenty-foot-wide, four-foot-high elliptical tube led left and right—a major junction. They stopped long enough to feel which way the air was blowing, then loped onward in a crouching stoopway, wind in their faces.

The sand-floored elliptical tube was almost walking height. They scooted with their backs hunched over, hands locked behind their backs for balance. They were now heading west, directly into the heart of Toohey Ridge.

Long silent, Bill now howled, "This looks more like Unknown Cave than Roppel Cave!"

After twenty years of absence, Bill had recaptured in a mind-numbing rush the feeling of infinity that he had felt when exploring the seemingly endless corridors of Flint Ridge caves in Mammoth Cave National Park with Bill Austin. Since leaving the caves of Kentucky, he had mostly lost touch with that special feeling. It was good to get it back.

The two sped onward, farther into Roppel Cave than anyone had gone before. In two years of exploration, in over dozens of trips twenty or more hours long, no one had advanced more than a mile from the Roppel Entrance.

Now, in just this one trip—and they were still going—Jim and Bill had far surpassed that mark.

As the minutes ticked by, the pair passed through hundreds of feet of cave. They advanced as dim pools of light running along the virgin cave passage. Suddenly they encountered blackness.

They stumbled out of the crouching elliptical tube, the distant wall barely visible. They stood waiting for their eyes to adjust. The image of a large railroad-tunnel-like cave passage materialized. They had realized their caving dream.

The Toohey Ridge Cave System lay at their feet.

A large elliptical passage in Roppel Cave

7

Roppel-Mania

Roppel Cave Explodes North

Bill Walter and Jim Currens stood quietly, stunned, staring in disbelief. The image was surreal. Yes, they were in Roppel Cave, but they were standing in borehole. Two years of gut-wrenching effort in small passages made it difficult for them to accept what they now saw. In two directions, the pristinely smooth sand-floored borehole disappeared into darkness. They had entered through a side passage, one that had seemed large when they had found it but now seemed small in comparison. A cold breeze rapidly chilled their sweat-covered bodies. Big cave! Instinctively, they turned to the south toward the broadening expanse of the main body of Toohey Ridge and perhaps tens of miles of cave.

The soft, glittering sand preserved their new footprints as they quietly plodded down the enlarging passageway. Within a few hundred feet, a large, ten-foot-diameter tunnel led off to the left.

Jim, agape, stared at the beckoning passage, "Boy, this is real cave!"

Bill disappeared down the passage, returning a few minutes later.

"The main way fills after a few hundred feet, but a great looking canyon lead takes off at the end. It's blowing a lot of air. What a lead!"

On and on. They soon reached an area where the large passage they were following ran steeply downhill and became even bigger. Mud covered the floors, and mudbanks rose up to meet the dark walls. They were now obviously in a section of cave that flooded periodically. Rotting sticks and leaves were subtle reminders that this was no place to be during threatening weather.

Cautiously, they worked their way ahead in the main passage, the mudbanks affording slick footing at best. At one spot, they forded a stream that crossed the passage between two high banks. A waterfall splashing on rocks in the distance told them of its source—a dome and perhaps a way to higher-level cave.

Soon, they stood in a tight crouch, mud oozing up around their boots, straining to see down a low passage. The magnificent trunk passage had withered down to this—a three-foot-high, five-foot-wide mud crawl. The strong wind blowing into their faces fluttered the flames on their carbide lamps, which made seeing what lay beyond that much more difficult, but with that wind, they knew this crawl was an important lead.

Bill, resolute on pushing leads to their end, began to step into the muddy passage with his back and legs severely contorted to avoid having to kneel in the mud.

"No!" Jim cried. "We don't want to get wet, not with all this big, dry cave we have to look at! This can wait."

Bill looked at Jim thoughtfully; he had a point. "Okay, let's go see where the other direction leads."

The twosome retraced their steps back along the slippery mudbanks and back into sand-floored borehole. They covered a couple thousand feet in just fifteen minutes, then walked on past the elliptical tube from which they had first entered the borehole and moved on to more new cave.

Here, the passage was more regular. They walked north down a large oval-shaped tunnel with dry sand on the floor. Along the wall at floor level, spectacular displays of pure white gypsum cotton provided a striking contrast to the clean, gray walls they had been seeing for the last several hours. At several points, they could see a wide, deep canyon underneath the wall ledges. They ignored the inviting passage and continued their steadfast march deeper into Roppel Cave.

They reached a wide junction of three passages, including the one through which they had entered. The two choices they now had before them were equally inviting.

Jim pointed to his right. “We'll leave this for Borden,” he said with a hint of sarcasm. “He's gonna croak when he hears about this. I can't believe he's missing it.”

“Why didn't he come?” asked Bill.

“He's over at Flint Ridge. Serves him right! He doesn't even deserve for us to save this for him.”

The passage they were saving was a five-foot-high and twenty-five-foot-wide elliptical tube. Bill walked into it a few paces. Yes, it did go—it was a wonderful lead.

They continued straight ahead. A shallow pit cut through half the passage. They paused to peer for any openings below, then continued walking around it on a ledge to the right. Soon, a dark void loomed ahead of them. They emerged into the largest room yet on this already amazing trip. Above them the ceiling reached nearly fifty feet high. This was another junction—a tall, wide canyon led left and right. Car-sized boulders littered the floor, fallen from some unseen point far above.

Jim and Bill turned to the right, climbing over large breakdown blocks that spanned the wide canyon passage. Soon, they were walking along the smooth floor in a passage twenty feet high and eight feet wide. The unmistakable sound of falling water ahead spurred them on.

A pit!

They stood on a precipice where the floor dropped away into the space of an immense vertical shaft. They heard a large waterfall off to the left, but at their vantage point it was hidden from view. They could not climb down the twenty-foot wall but could see the distant outline of a large pile of rocks that made up the far wall, suggesting that this was where the canyon might intersect a hillside along the edge of the ridge.

Jim was beginning to feel the effects of the difficult push through the narrow passages beyond S64 and the many hours of adrenaline-surged exploration. He sat down at the last, most impressive junction, drained. Bill, who never seemed to tire, ran out the opposite, still unchecked, direction. He crawled under a pile of rocks to find large, going cave with good wind.

Two hundred feet farther, a narrow rock ledge jutted from the wall leading around the edge of a pit—a beautiful vista. The round, ten-foot-diameter pit dropped twenty feet to a clean-washed floor. Below, Bill could see a walking shaft drain leading off. Across the pit, out of reach, a round hole led off. From the hole issued the sound of a large waterfall.

Bill continued down the main passage another few hundred feet. The cave continued as before—big and walking-size—but he was out of time. Getting out would be a lot of work.

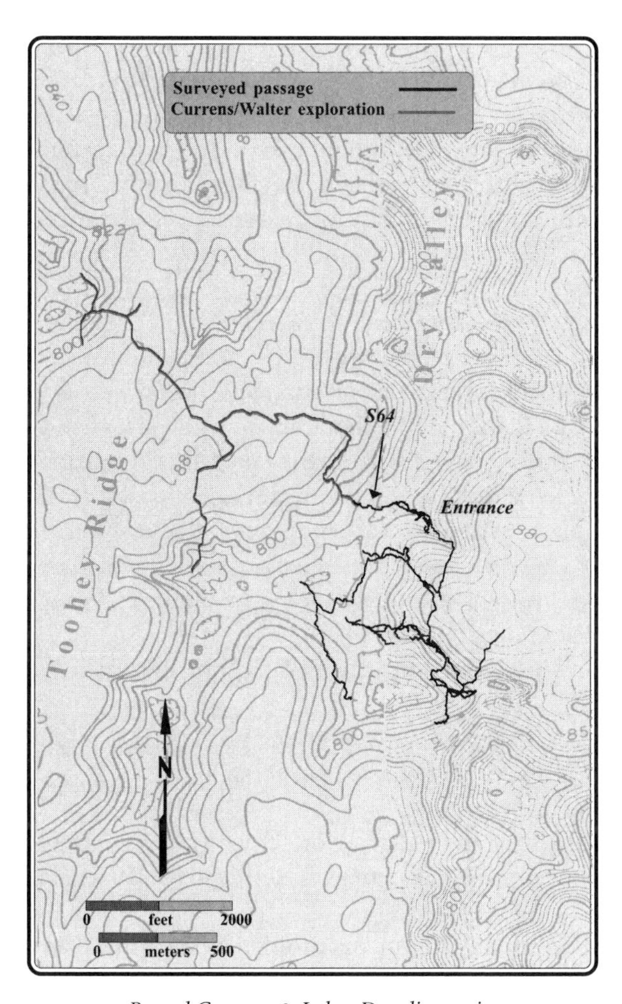

Roppel Cave, 1978: Labor Day discoveries

The pair walked back down the large, sand-floored tunnel and began crouching through the low passage leading toward the surface. The cave they had run through on the way in from S64 now led on interminably. Could they have really traveled this far? They passed by bend after bend, the numerous landmarks reduced to a blur in their memory, their order and significance lost. They moved silently, preserving all the energy they could for the rigors they knew lay ahead. Jim's coveralls with the broken zipper had long ago been reduced to tatters, catching on every projection. Ahead, they still had to squeeze through the drain series below the S64 Pit and ascend the twenty-foot climb to the main level of the S Survey canyon.

After what seemed an eternity, they stood looking up at the rope that hung down from the S Survey canyon at S64, the last worrisome obstacle between them and the entrance. Jim had been thinking about this problem spot for the entire trip. Would he be able to get up? They did not have the vertical gear to climb directly up the rope. Instead, they could use the rope only as a safety line. They would have to climb the wall of the bell-shaped room.

Bill was first. He was confident, since he had done the climb before, and once up he could belay Jim. Bill worked his way up the far wall to an unseen ledge where he could sink his hands deep inside a horizontal crack that led across the side of the pit to the rope. Methodically, he shuffled his feet across a series of small projections, moving his hands from hold to hold deep inside the crack. An occasional loose rock made the traverse unnerving. Bill was leaning back dangerously, the overhanging walls pressing his body outward. If he should fall, he would land on his back on the rock-covered floor fifteen feet below.

After four or five steps, he grabbed the loop at the top of the rope and pulled himself up out of the pit.

"That wasn't too bad. I don't think you'll have any problem with it," Bill called.

"Yeah. Right. You looked pretty hung-out up there. I'll take the belay if you don't mind," Jim said. "Better safe than sorry."

"No problem."

Bill set up the belay and wedged himself tightly in the floor of the canyon. He would have no trouble catching Jim, although the fall would certainly be a jolt.

Jim made the climb without any problems. The belay provided the security needed for him to climb with confidence. A belay is never a bad idea.

I was attending the Labor Day CRF expedition and knew that Jim and Bill were going on this trip to Roppel. Although I had hopes for their success, I chose to cave in Flint Ridge that weekend. I needed a break from the disappointments of Roppel Cave. However, I was curious about what they had found. John Barnes, who was also caving with the CRF that weekend, and I made the twenty-five-minute drive from the Austin House on Flint Ridge over to the Toohey Ridge fieldhouse, the base for the CKKC operation, and waited for them.

Idle conversation whiled away the hours. We sat on the front porch in rickety pastel-green chairs that creaked as we rocked back and forth. We joked about catching some exotic disease from the unidentifiable black fungus on the cushions we sat upon. Furniture nobody wanted was abandoned here. Beggars can't be choosers.

As the lengthening shadows gave way to darkness, we spotted two figures walking through the gap in the woods at the far end of the field, moving slowly toward us. We could see tattered shreds of clothing dangling from one of them. From this distance, the person could pass as a scarecrow. Each had a helmet dangling from one arm. Cavers.

Jim Currens and Bill Walter walking toward the fieldhouse

Soon, the two battered explorers stumbled to the front porch of the fieldhouse, sweating profusely from the mile walk from the cave. It was a warm September evening; caving clothes made for a hot walk. The smile on their faces was a tip-off.

"Well, what did you guys find?" I asked.

The two glanced at each other with a sly grin. "Not too much," Bill said.

He was lying.

Jim asked Bill, "Do you think we should tell him?"

More smiles.

Silence.

Their levity was maddening. "Christ! Just tell me what you guys found!" I wasn't interested in fooling around.

Jim began, "Well, we just spent eight hours walking beneath Toohey Ridge . . ."

They spun their electrifying tale as John Barnes and I stood there wide-eyed, almost in disbelief.

Surely this was the big advance we had been seeking for so long. Indeed, if this breakthrough into Toohey Ridge only partially approached our grand expectations, there would be miles of new cave to survey. We weren't in Roppel Cave anymore; we were in the Toohey Ridge Cave System! We had at last broken the shackles that confined us to the drain system that had bound Roppel Cave for the first two years of its survey.

But, where were we? This lack of knowledge was killing me. Usually, a party comes back with a book full of survey data. But Jim and Bill had not surveyed; they never intended to. The need to find something—anything—had superseded our normal policy of survey-as-you-go.

They had explored through much difficult passageway before finally reaching the borehole. The effort required to survey what they had traversed would be time consuming at best, painfully difficult at worst.

I made Jim retell the tale of his and Bill's trip no less than a dozen times during the next week. I was hungry for detail, wanting to experience each section of passage as they had. As the days passed, Jim began to trim his estimate of the distance from S64 to the new trunk—now called Arlie Way, after a hospitable and helpful local landowner—down to around a thousand feet. Being super-conservative was a natural balance to the usual tendency of cavers to overestimate distances within caves.

I was subdued. After all the effort I had made, I had missed the trip of a lifetime. Still, I shared in the ecstasy of the discovery, the feeling of victory after the pounding frustration we all had experienced.

Dutifully, I volunteered to lead the first trip to survey beyond S64 toward the new discoveries. Jim was unable to go, but I had no trouble talking Ron Gariepy into the trip. This would be the most difficult stretch of survey on the route to Arlie Way. It was my self-imposed penance for not being on the discovery trip.

To ensure accuracy of our maps, survey teams in Roppel Cave took two sets of readings—foresights and backsights. With teams of three or more, this was easy. The compass reader could have a target ahead and behind—a lighted survey station—and only had to turn around to get the necessary sets of bearing and slope angle readings. A two-person survey required a more imaginative approach to accomplish this. The instruments were either passed back and forth, or the team surveyed with only one set of readings and completed the backsights on the return trip. On this day, Ron and I would have to do both.

We struggled with the survey through the tight drain series below S64. The difficulties of surveying in narrow passage cannot be overstated. Reading the compass with any degree of accuracy often involves unimaginable contortions. Getting one's head over the survey point is critical to being properly aligned along the survey in order to get accurate compass readings. The taped distances between the survey shots were so short in the sinuous passage that a stretch of the arm was all that was needed to pass the survey instruments back and forth to get both sets of readings.

It took three hours of energy-draining effort to reach the first large room discovered by Jim and Bill. Ron and I called it the Boundary Dome, on the boundary between the tiny canyons and belly-crawls of Roppel Cave and the grand walkways of the Toohey Ridge Cave System. This would later become a key landmark on the trade route in Roppel Cave, a route to be later traversed by nearly every survey party.

Ten hours into the trip, I marked with soot "S118" on a large slab of rock at the junction with the low, crouching tube that led into Arlie Way. Ron Gariepy and I had surveyed nearly eleven hundred feet but were still a long way from our objective. We were out of time. Certainly, it would be a lot farther to Arlie Way than Jim Currens's estimated thousand feet.

Ron and I had to see for ourselves the new passages beneath Toohey Ridge. We crouched through the long, low tube between us and Arlie Way, its beauty rejuvenating us.

Fifteen minutes later, I stood in Arlie Way, savoring the elation that Jim and Bill must had felt just a week before. Footprints led off in two directions, the only blemish on what was otherwise a smooth sea. I glowed.

"Christ, Ron, look at this!"

He stood beside me taking in the immensity of it all.

I looked one direction, then the other. Why hadn't I found this big cave?

I ran off to the left with Ron following close behind. We did not stop until we reached the muddy place where Bill had been stopped by Jim a week before. Bill was right: this windy portal ought to lead somewhere.

But not today.

We retraced the narrow track that Bill and Jim had followed. We stood in the large junction room with the big boulders, peered over the edge of the pit from the narrow ledge Bill had found, and explored a little farther in the big passage. I also climbed down the floor canyon below Arlie Way and walked five hundred feet to a junction of five passages before returning.

The following weekend, Jim Currens, John Barnes, and I continued the survey from S118 through Hobbit Trail ("'Cause you'd have to be a three-foot-tall Hobbit to walk through it!") into Arlie Way. On one spectacular survey trip, we charted all the cave that Jim and Bill Walter had explored. We finally closed the S Survey at station S163 at the south end of Arlie Way, marking the last station with red flagging tape at the edge of the wet passage that had challenged each of us. Our book was filled with numbers that totaled a mile, and we were shot. On this one trip, we had increased the surveyed length of Roppel Cave by over a third. The rigors of the other two-thirds were rapidly fading into a distant memory.

Survey parties began to pour in as we enjoyed an unprecedented surge of cavers. To be sure, many were rainy-day friends, there for only a few trips to skim off the cream of new discoveries. But a few would become valued leaders in the project, cavers to whom we could pass the baton of leadership. These individuals were the heart and soul of an enduring project. We had to let natural leaders lead, and we could spot natural leaders by their followers.

In just a matter of weeks, the cave had ranged across the northern lobe of Toohey Ridge through a series of spectacular elliptical tubes and large canyons. Wonderful cave never ceased to amaze us. One such passage began as a suspected cutaround but shortly blossomed into a beautiful passage festooned with gypsum and dripstone formations that defied the imagination. As the passage opened up and it was clear that this was not just a cutaround, the lead caver dropped the survey tape and took off, the cave too spectacular to believe. A thousand feet later, the caver was retrieved and the party retreated to continue the survey in the new find, finishing with another four thousand feet to add to the map. This passage, named Yahoo Avenue, shot off on its own to the west, leading far away from the main part of Roppel Cave.

By the end of September, Roppel Cave had grown from a length of 2.7 miles to over 5 miles.

After the intoxication of the discoveries had subsided, the slow and methodical plodders climbed to the forefront of new exploration. While everyone else was catching their breath after September's flurry of activity, Ron Gariepy and his longtime caving companion, Bill Eidson, started their own special style of poking. For them, nothing was sacred. They were nonconformists and did not hold to the notion that information is reliable and to be believed without question. If they decided they wanted to check out a passage for leads, they would do so, fully and completely, even if it had been reported that there were no leads. Being non-mainstream project cavers by choice, they did not have the benefit of the accrued knowledge that had been gathered. For them, caving with each other was an ideal situation. Alone and in the cave, there were no distractions. Theirs was caving for the joy of it—the pure adventure experience. They were not model project cavers, but their participation and efforts were as valued as any. They were fun to cave with, and I made a point to encourage them as much as possible. Friends and contributors to the project were not to be squandered.

Bill Eidson, like Ron Gariepy, was quiet and unassuming. Tall and neat, he contrasted strikingly to Ron in caving style. Ron, the mathematician, reveled in the topological relationships of the cave. He could lose himself—as he could in mathematics—in the vastness of the cave. Bill, the engineer, was analytical, yearning to decipher and understand everything about the cave. Together they made the complete caving party. I called the two of them the CKKC's "Lone Runners."

Their approach was not for everybody, however.

One of the first areas selected by the plodders Bill and Ron was an area known as the Rift, located at the end of Yahoo Avenue. Yahoo Avenue led four thousand feet to terminate under a valley far away from the rest of Roppel Cave. There was wind, and enormous canyons led downward. Here was an area crying to be explored. Surely, great discoveries must await.

Bill and Ron loved a challenge, especially when the prospects of discovery seemed bright. Armed with ropes, pitons, bolts, and chocks, along with their normal complement of climbing gear, they made their way through Yahoo Avenue, dragging a terribly misinformed and apprehensive John Barnes behind them.

Hours earlier, John had watched them suspiciously while Bill took a tangled mass of climbing gear from the trunk of his car. The shiny aluminum hardware glistened under the bright sun, catching John's attention.

"Hey, what's all that stuff?" he asked.

Bill looked over at John. "Oh, this is just the climbing gear. The rest of my vertical gear is in my other pack."

Bill motioned to the plump red pack lying against the tree in front of the fieldhouse.

"What do we need all that stuff for? All I brought is my rappel gear and prussiks."

Ron said nothing but continued putting cans of food and equipment in his pack, careful not to let on that he was eavesdropping. He had experienced this scenario before and knew the next scenes almost by heart. Bill would be the straight man.

"You never know what we might find; we want to be ready for anything." Bill pulled out something that most accurately could be described as a railroad spike from his pile of gear. "John, this is a piton." He handed it to John for his inspection. "If there are no natural anchors, we can hammer this into a crack in the wall and clip into it."

John examined the piton, inspecting the eye-hole in one end. If density and mass meant strength, it was probably safe.

"This is a chock." Bill held up a hexagonal nut-like piece of metal threaded with a loop of nylon rope. "This can be wedged in a tapered crack. Between all this stuff," he motioned toward the pile, "we can improvise any kind of rope anchor."

John handed the piton back to Bill. "Okay, I guess. But I've never worked with this kind of stuff."

Ron and Bill rolled their eyes. They knew it might be a long trip.

Hours later, deep in the cave at the Rift, Bill was hammering the four-inch-long piton into a crack. He was leaning wildly out over the abyss, long swings of the hammer testing his balance as Ron carefully belayed him.

Bang... bang... bang... The deafening hammer blows reverberated from the depths as the piton slowly sank into the wall.

Ron and Bill had first tried to rig to a large, secure-looking rock on the floor beside the pit. As Ron was uncoiling the one-hundred-foot rope, Bill had pushed on the rock with his foot to test its strength. Abruptly, it teetered over the edge and fell into the pit. The rock ricocheted off the walls with deafening crashes on its long trip to the bottom, finally ending in a loud *whump!*

"Oops!" muttered Bill. "I guess that wasn't such a good spot to rig."

Now, John watched warily as Bill ran his hands along the wall, looking for a suitable crack. The next choice was the piton.

The telltale *ping!* indicated that the piton was solidly placed, and with a couple of extra swings for good measure, Bill ceased hammering. One inch of the piton was visible—it was probably a safe placement. Bill finished uncoiling the rope, clipped it into the piton with a carabiner, and tied the far end around a second large boulder in the passage a few feet away, a backup anchor just in case.

Thirty minutes later, John, Bill, and Ron leaned out over a ledge and peered down another pitch. Their small, four-foot-square ledge was continuously pummeled by a shower of water issuing from somewhere high above the rope tie-off forty feet above their heads.

"Well, Ron, what do you think?" Bill asked. "Do we have enough rope?"

Hanging onto the rope, Ron leaned far out over the pit, straining to pick out any features in the gloom below. The din of falling water dashed any hopes of determining depths by the usual method. If they tossed rocks and counted until they heard the crash, they could usually estimate the depth, but with the deafening roar of water and no loose rocks of any size, there was no way to do that.

"I don't know," answered Ron. "It's pretty far."

They tried a different technique: they tied a heavy object, a hammer, to the end of the rope and slowly lowered it, trying to sense when the weight reached the floor. But they couldn't tell; the rope rubbed against too many spots.

John sat despondently, listening to the exchange of seasoned cavers.

Bill said finally, "I think we probably have enough rope. If we wrap it around this boulder, I think we can stay out of most of the water." He made a loop in the rope and draped it over the boulder—a half-hitch.

John's apprehension now gave way to near terror. "You're not going to rappel off that shit are you?" He pointed to the boulder with the half-hitch. It looked like it was just stuck in the mud.

"Sure, why not?" Bill replied. "The rope is still tied to the piton. This is just a rebelay to move the rope. Anyway, look at this." He flipped the free end of the rope in demonstration. "There is no way this can come off."

"Will the rock hold?"

"Sure!" Bill gave it a dramatic kick for added emphasis.

John's sober expression did not change. Watching Bill for the last couple hours had taken its toll, and he was skeptical. He had little experience in elaborate forms of rigging and absolutely no experience in rock-climbing techniques; Bill and Ron were well versed in both.

"I don't like the looks of this," said John.

Ron and Bill stared at John. A proclamation was coming.

"I'm not going down there," John announced. "I think this is unsafe . . . and so are you guys."

No hesitation. "Okay, we'll see you later." Bill clipped into the rope and rappelled into the unknown pit.

Ron and Bill played the scene flawlessly, Ron picking up right where Bill had left off. "We'll be back in a few hours. You can wait here."

This was not the answer that John had expected.

Smoothly, Ron continued, "We understand how you feel. We don't expect you to go down if you don't want to. But don't expect us not to."

Aghast, John watched as Ron backed off the ledge and rappelled out of sight.

The two seasoned cavers glided down to the bottom of the waterfall. Six feet of rope lay at the bottom—barely sufficient! They walked along five hundred feet of indeterminately high canyon fifteen feet wide, stopping where a wall of broken rock led up into the darkness. Enough. Although Ron and Bill could be perceived as unfeeling tyrants, they were not going to leave John up there too long.

They pulled out the survey gear and measured the new passage back to the base of the rope. Four hours later, they tied into the survey station at the top of the rope.

John Barnes was chilled and angry and said little as he watched them pack up the remaining gear. He had sat for six hours on his perch, midway on the wall of an enormous canyon, unwilling to go up or down by himself. His spirit was broken, and he would not return to Roppel Cave. Later, he would write in the grotto newsletter about "unsafe" cavers in Roppel Cave.

Ron Gariepy and Bill Eidson returned to the Rift many times. They descended numerous pits, made climbs into upper-level passageways, and fashioned daunting pendulums across wide gaps. The twosome untangled a baffling series of tall canyons, pits, and boreholes, filling in the northwest flank of Toohey Ridge with cave. However, the "big" discovery eluded them. They turned their attention to other parts of the cave, using their usual approach of poking anything and everything—discover and assimilate.

In August 1979, the pair sat eating smashed candy bars at a large junction in Arlie Way where the lead that Jim Currens had saved for me led off to the north. This was called Symmetric Junction, since all directions looked the same. Three hours earlier, they had entered the Roppel Entrance with nothing particular in mind to do; they would just see how things played out. Of late, most of the big cave known in Roppel had been surveyed. Roppel Cave was a little over eight miles long now. The best lead known was Black River, a wet stream passage that flowed north from a complex of canyons. Bill and I had discovered Black River six months earlier off Pleiades Junction in Lower-Level Arlie Way—the room with five passages I had found below Arlie Way on my first trip. Later, Tommy Shifflett and I had pushed a tall canyon with the Black River running in its bottom to a point where the main stream disappeared under a ledge. I had looked under the ledge at the lead. Mudbanks rose to the ceiling with muddy, deep water flowing between them. Yech, I thought, probably a sump.

At Symmetric Junction, Bill asked Ron, "Do you know where that goes?" He was pointing at the broad, elliptical tube with a splendidly flat ceiling and floor.

"No, not really. I think Borden said that it got low or something."

"Why don't we take a look?" Bill asked.

Ron was easy-going. "Sure, let's go." They crouched their way north. The passage was perfectly consistent, five feet high and thirty feet wide, with a smooth, flat floor. They passed labeled Y stations from a survey made the previous year. On that trip, Pete Crecelius, Bill Walter, and I had put in a thousand feet of Y Survey, ending at a breakdown pile. I had crawled a few hundred feet farther to a pool of water. It was late, I was tired, and I said it looked like it might fill. Since the passage had so much crouching in it, we named it the North Crouchway.

Bill and Ron climbed over the pile of rocks at the end of the survey. Overhead, a tall canyon cut across the passage. They craned their necks, evaluating it. No way to climb up today. They silently continued in the low, wide passage, soon reaching a place where they had to crawl on their hands and knees. Wide pools and tacky mud made the going slimy. At the first pool, the scuff marks stopped, and they were in virgin cave.

"Ron, did you notice the footprints stopped back there?"

"Yeah, I did. Why'd they stop?"

"Well, it's wet, but this passage definitely goes. Feel that air?"

Ron sat on a dry bank of damp sand. Steam rose from his body and drifted down the passage in the direction they were headed.

The pair got up and moved ahead past the water in the sandy passage, not stopping until a large pile of breakdown blocked their way. For the better part

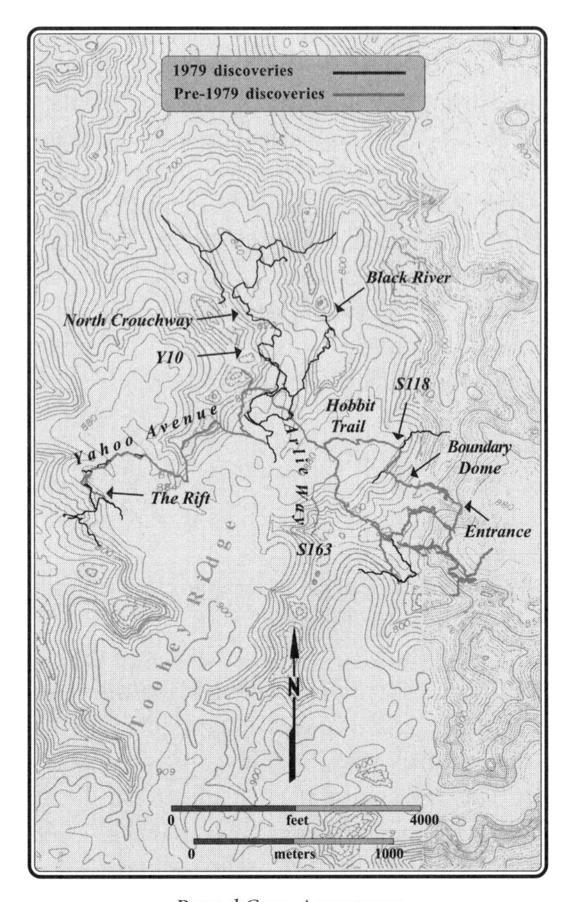

Roppel Cave, August 1979

of a mile, they had followed the low, wide elliptical tubes, mostly crouching, occasionally crawling. Many side leads beckoned, but they kept to the main route.

Their minds were numbed as they retraced their steps through the long passageways. They had made a breakthrough, but the cave they had found was too complex to comprehend.

At one rest stop, Bill propped up a small, flat rock and smoked a message on it. Anyone who followed their tracks would be mystified by the unexplained number on a rock: Bill Cuddington's NSS number, a caver's joke that celebrated the almost legendary travels of an old-time Tennessee caver.

Bill Eidson and Ron Gariepy grinned at each other and left the cave.

Survey notes and cave maps that tell explorers where the cave goes

8

Transformation

The CKKC Goes Big-Time

I graduated from the University of Kentucky in the spring of 1979, in spite of my focus on the exploding growth of Roppel Cave. Caving, I had discovered, was incompatible with school: grades, social life, friends. Nevertheless, I managed to earn a B.S. degree with a major in mathematics—a real feat considering I had registered for twenty-four credit hours that last semester. During my junior year, I realized that an undergraduate degree in geology was either a ticket to a job far from the caves or, worse, to unemployment. I saw caver after caver with a geology degree ending up in west Texas with big oil. I probably survived as a student due to caving companion Ron Gariepy who, as a tenured professor of mathematics at the university, served as my faculty adviser for the last year of my education. As a caver, he understood.

Prior to graduation, I realized that there was little opportunity for employment in the immediate area. Like my caving, I wanted my career to be meteoric, so to achieve this, I sought employment in large companies. I landed a job as a computer systems programmer in Silver Spring, Maryland, not far from where my parents lived, with a large company that promised many opportunities. So, in May 1979, with mixed emotions, I loaded my car to its roof with my accumulated possessions and headed back to Maryland. The moment was bittersweet. I was leaving something behind I had grown to love, but I was also returning to the earliest roots of my caving career. Despite this major change, I felt sure my involvement with Toohey Ridge would continue unabated. Moreover, there was a fresh new world of cavers clustered around D.C. to recruit for the intoxicating excitement of big-time caving in Toohey Ridge. Rationalizing, I viewed my move to distant Maryland as an opportunity rather than a disadvantage.

I quickly sought out my old caving haunts in West Virginia and old friends at the monthly D.C. Grotto meetings, which had been moved to a larger room in a nearby police station. Some people from four years earlier still attended, as they had for probably twenty years, their interests and abilities still enduring after all that time. As expected, there was also the influx of new blood: young, strong cavers who blazed the new trails of discovery in the nearby West Virginia caves. I saw that there were some strong candidates for Kentucky caving, so, with barely suppressed excitement, I began my recruiting crusade.

Fully reimmersing myself into West Virginia caving was another matter entirely. New discoveries were being made at an unprecedented pace, one not seen for better than ten years. Cave exploration in West Virginia was experiencing a rebirth; new extensions to long-known and presumably explored caves were the order of the day. The new caves were spectacular: deep and full of fast-moving water and surprises, they were formed in the steeply bedded limestone rocks of the Appalachian fold belt. The caves in the Mammoth Cave area, although significantly more extensive, were more predictable and less spectacular in relief and variety, formed in the shallow-dipping, flat-bedded limestones of the American Midwest.

I quickly became involved in this cutting-edge caving undertaken by a small circle of competent cavers, enjoying a spirit of accomplishment I had longed for early in my career as a caver. Those early years had been spent in a comparative drought of cave discoveries; now, in 1979, I had reached what I had long sought for in West Virginia. However, the magic of this revitalized climate was not enough. After Kentucky, everything seemed no better than second best.

I could return to the caves of Kentucky no more than once a month due to the length of the drive and the limited time off available from my new job, so it had been only natural to cave alternate weekends in West Virginia. It was then that I became aware that I had been seduced by the caves in Kentucky, even given their relative dullness. The feeling of boundless opportunities for discovery is like no other. I longed for the sense of limitless passages stretching to infinity, a feeling that I had experienced only in Roppel Cave. Surely, the caves are bounded—nothing is infinite on this earth—but Roppel Cave's boundaries seemed to offer a special vastness. The most appropriate analogy to this is the seemingly endless opportunities in the 1800s when the Central Asian steppes were opening for exploration. No, Asia was not infinite, but for all intents and purposes, it seemed so. Vast, with challenging terrain and diverse unknown peoples, there were obstacles and uncertainty aplenty—a romantic uncertainty—just like the exploration of Roppel Cave. Now the caves of West Virginia were to me more akin to the exploration of an island. In fairness, there was much to find, often spectacular, but the limits were almost always known. Caves in Virginia and West Virginia were usually confined by compact geographical features and diverse geology, such as local valleys and structural basins. These caves were bounded; any special sense of infinity was totally absent.

There was, of course, no reason not to cave in West Virginia. Moreover, I certainly would not belittle the caving pleasures of others. It was apparent that those who caved in West Virginia experienced something emotionally that cemented their own bonds to the underground in a way not unlike mine. But there was a difference. I kept my sense of a lack of fulfillment to myself, and I enjoyed the caves for their own sake. However, my resolve strengthened to continue my work in the caves of Toohey Ridge. I knew then that my bond with those Kentucky caves was unbreakable.

I spent considerable energy and time selling the caves of Toohey Ridge to dedicated West Virginia cavers. I prepared and peddled an excellent slide presentation to the several caving organizations, a card borrowed from Joe Saunders's deck. At each encounter, I found myself continually baffled as to why it was not obvious to all the cavers in the audience that an opportunity to explore in the Mammoth Cave area was one to be grabbed without hesitation; it had certainly been obvious to me back in 1973.

Despite general disinterest, a few of these cavers did take the bait. They had heard so much that they had to see what all the fuss was about. So, on a regular basis, I took a small contingent of cavers on the twelve-hour drive to Toohey Ridge, usually during three-day holiday weekends. It was a long way, but to me it was worth it—and, to my surprise, some of my new-found companions thought so, too. I had successfully recruited some of the strongest cavers from West Virginia. Although they were a headstrong crowd, I felt that their contributions could be significant. Their work and results would overcome any preconceptions about the Mammoth Cave region they might harbor.

Not everyone welcomed the new crowd from Washington, D.C., to Toohey Ridge. This sudden influx of strong, talented cavers forever changed the makeup of the CKKC. What had started as a small, close-knit group began to grow and became as diverse and unpredictable as the caves that we were exploring. No longer would Roppel Cave be the small, private project that Jim Currens desired and enjoyed.

"There's too many people!" Jim would often rail. "We're losing our control!"

"Control? What the hell are you talking about?" I would respond.

"You can't control that many people. Someone will screw everything up."

"But, more cavers ensure the continuity of the project!" I had learned that when I had secretly explored Lewis Cave in West Virginia, years before.

"The project was just fine the way it was. Why bring in all these people?" Jim asked. "They just want to scoop booty. They don't care about the project."

"They'll do good work and we can use the help!" I argued.

It didn't matter. The die had been cast and there was no turning back now. The only way to keep the cave was to give it away, so to speak. I had observed too many closed projects that faded into oblivion once the core contributors lost interest. Part of the challenge of a large project was to foster a continuing influx of new, talented individuals who would become vested in the effort. A small percentage of these would stick around, bonded to the project by the pride of the effort. To accomplish this, the current leadership would have to continuously shed responsibility to the new arrivals, a legacy that would continue indefinitely as the long-established leaders eventually moved on to other challenges. The CRF had long touted this strategy as a model for success. It had certainly worked there.

Jim's frequent complaints about how the project was being managed provided ample incentive for buffoonery. One evening, Bill Walter, Hal Bridges, and I traveled together from Bill's home in Tennessee up to the cave. Hal was a recent convert and had taken over the computer processing of cave data from Jim. He slept for most of the three-hour drive, waking to the sound of the engine roar as I sped up for the traditional "car-jump" on the Toohey Ridge Road. With the right acceleration, a car could make a satisfying leap off the pavement, remaining airborne for ten feet. The effect was best when an unsuspecting passenger was thrown against the ceiling of the car. Hal had detected our attempt and held on.

We laughed as we raced along the gravel road past Renick Cave, fishtailing around the familiar turns. It was very late. Suddenly, an idea struck.

We carefully drove down the rocky lane that led to the fieldhouse. As we came over the last hill, we cut off the headlights to preserve the element of surprise and gently rolled to a stop. It was a moonless night.

We eased out of the car and crept toward the dark fieldhouse, stopping fifty feet from the front porch. Only one car was parked next to the house: Jim's car. He would be asleep inside. Perfect!

I recoiled my arm and hurled a rock toward the house with the most powerful throw I could muster. I aimed high into the air to gain as much vertical velocity as possible. Seconds passed as we waited in eager anticipation.

Bam! The loud crack shattered the dark silence. We cringed as the rock loudly rolled down the corrugated metal roof of the fieldhouse, biting our fists to suppress our laughter.

We waited.

After about a minute, I hurled the second volley.

Bam!

Movements from the house.

"Who's there?" came from within.

It was time for phase two.

Hal sprinted off to the right across the road to the north side of the house. He would continue the barrage from the opposite flank.

Seconds later, we heard the impact of a hard collision.

"Shit!" Hal tried not to be too loud. In the gloom, we could see him tumbling head over heels. He had tripped over something big and hard. We winced at the spectacle.

Hal was trying to contain himself. He shouted a muffled, "Damn it!" while jumping around holding his right knee. "I think I hurt myself!"

He was disabled. Our plan was falling apart.

"What the hell is this?" Hal exclaimed, as quietly as possible. He was investigating the cause of his pain. "I hit a goddamn table! What is this doing here?"

We barely noticed as the screen door of the fieldhouse swung open.

Jim bellowed in his most authoritative tone, "I don't know who in the hell you are, but if you don't get out of here, there's going to be trouble!"

I let out an evil cackle.

"I'm not kidding!" Jim threatened, waving his arm up and down.

We strained to see what he was holding.

"I've got a stick of dynamite that I am ready to throw if you don't get the hell out of here."

"Uh-oh," I said to Bill, who was still standing next to me.

Bam! Hal had not heard Jim's dramatic threat and had resumed the attack of missiles.

"Okay, you guys are asking for it!" Jim was sounding more menacing.

"Oh, he's bluffing," said Bill.

"Maybe," I answered.

Uncertainty prevailed. But now that the jig was up, Bill and I broke out laughing.

Jim, realizing he had been the butt of a prank, shook the stick of firewood in his hand and fumed, "Assholes!"

With the large number of cavers from the Washington, D.C., area making the long journey to central Kentucky, it was not surprising that one or two would advance to the next plateau of involvement and be willing to receive the baton and work in a role that was far more than caving. Managing a cave project is a complex business: information must be communicated, trip reports logged, survey books tallied and copied, maps drawn, and a volunteer organization continually inspired. In fact, on a cave project of any size, more time is spent doing paperwork than actually going caving.

In the early days of the CKKC, when Jim Currens and I were pretty much the entire organization, it was simple—we split the tasks. I maintained survey books, trip reports, and did the logging; Jim drew many of the maps, did the computer processing, oversaw surface activities, and managed publicity. This was a good arrangement—fair and equitable. When I returned to Washington, my CKKC duties went with me. As the membership increased and efforts in the cave did also, the magnitude of the duties doubled and redoubled. It was a big job.

Roberta Swicegood and Cady Soukup easily took my bait and began caving at Toohey Ridge. Both women were young and had lots of energy; they also reveled in the bonding that a large cave project can provide. Their personal connection to the cave was sealed by a near tragic accident. On the pair's first trip into Roppel, Roberta stepped on a chert ledge that broke beneath her weight and sent her crashing ten feet down to the stream. Her forearm throbbed—broken. Their party of four was deep in the cave in a potentially dangerous situation. With a choice between a major rescue call-out or proceeding on their own, the group immobilized her arm and headed back toward the surface. The trip was slow and tedious as they helped Roberta through the more difficult sections. Through their teamwork and guts, they reached the surface after dawn the following morning. Roberta's arm healed and she returned.

The CKKC was perfect for Roberta and Cady: it was a large and exciting project full of dedicated people working together toward a common goal. They immersed themselves in the cave and fell in love with it and the people. They sought more involvement and offered their help in the growing management needs of the project.

I quickly turned over to them any opportunity and task they wanted. They were competent and boundless in enthusiasm. We set up regular work sessions and in short order had most of the administrative tasks smoothly under control. Productivity was unprecedented. Their performance and accomplishment drove me to a higher level also, a level I would never have achieved had I continued to work alone. We spent countless hours together talking about the cave, designing strategy, planning exploration, drawing cave maps, and seeking ways to attract additional, qualified cave explorers. We were close friends.

This was all wonderful, as far as I was concerned. The fact was, however, that there had been a subtle but major shift in the Toohey Ridge power base. The core of the CKKC's dynamic strength was no longer in Kentucky; it had moved five hundred miles away to the D.C. suburbs. This fact further eroded my partnership with Jim Currens. Not only was the project growing to a level that dissatisfied him, he now also complained that we had arrogated his responsibilities and diluted his power, all with the specific intent of isolating him from the project.

He thought we were pushing him out.

"Absurd!" we told him. "Are you accusing us of conspiracy?"

From our perspective away from the scene, we were only doing the work that needed to be done. Were we usurping him? In reality, maybe, but there was certainly no conspiracy. In a sense, his conservatism and insistence on control prerogatives were seen as just another obstacle to overcome among the many that loomed.

The cornerstone of this debate became the survey notes. Since the beginning, I had insisted on maintaining them. I was interested in drawing maps, and the notes were the keys to the project. I did not trust Jim with them. I probably didn't trust anybody. I painstakingly logged and filed each survey book, cautious to a fault about loss. Losing the notes would be a disaster. Jim had been drawing maps of Roppel Cave and argued that good maps could be drawn only from the original sketches; copies would not do.

"Why?" I asked.

"Copies are never as good as the originals," Jim said.

I furrowed my brow, considering this argument.

A ploy! I thought. This was just a red herring to wrestle the notes from me! In project caving, everyone knows notes and survey data are power. I was sure that Jim wanted that power.

But maybe I was just paranoid.

For years, I had, most of the time, been supplying Jim with complete sets of copies of the field notes, if for no other reason than to avoid any accusation of holding back. But he had been doing positive, productive work, so I would have given him copies anyway. Nevertheless, I was intractable. I would not split up the collection of original notes. That could lead to their loss, something, unfortunately, that was common in the history of project caving. There was no way I was going to give in to what had become Jim's insistent demands. With our opposed and hardening positions, the issue became the catalyst of a much broader distrust. The turbulence of this conflict spread quickly to embrace our other points of contention—the newsletter (on one occasion, both Jim and the D.C. contingent simultaneously published two different versions of the *CKKC Newsletter*), cartographic responsibilities, and magazine article preparations.

Roberta and Cady told me they saw little merit in Jim's position, but they could see the fires of conflict threatening to destroy the fabric of the CKKC. They mediated, vigorously offering various compromises. At times, Jim and I would strike an agreement, only to find weeks later that fine issues of interpretation would cause it to unravel. This was as bad as arms reduction treaty negotiations. Jim and I soon became adversaries. Our escalating disagreements and vociferous conflicts marked the point of no return to our six-year partnership. To Jim, the unprofessional opportunists, led by me, were seizing what he had so painstakingly built over the years. They were impulsive, insensitive, and irrational. To me, the only issue was who would come out on top.

And I already knew the answer to that question.

Richard Zopf climbing out of the river

9

Treasures Beneath Doyle Valley

CRF Cavers Try to Expand the Big Cave into Joppa Ridge

I—Roger Brucker—explored Mammoth Cave intensively while Jim Borden's Roppel Cave adventures unfolded. We in the CRF were making big discoveries and surveying miles of passage. The Mammoth Cave System sprawled across Mammoth Cave Ridge and Flint Ridge, linked by passages that snaked below Houchins Valley. Discoveries were everywhere. To the south of Mammoth Cave lay Doyle Valley, and beyond that, the vast expanse of Joppa Ridge. I was sure that the next frontiers of Mammoth Cave lay across that valley. The New Discovery section of Mammoth Cave was explored to within a few feet of Joppa Ridge, but like other leads that had promised to take us beneath Doyle Valley, it did not go.

Over the years, others had also looked to Joppa Ridge for new cave. One such discovery was Proctor Cave, the hoped-for key to the Joppa Ridge Cave System.

Proctor Cave is located precisely due south of the underground Snowball Dining Room in Mammoth Cave, across the mile-wide Doyle Valley. The valley floor is 240 feet below the tops of Mammoth Cave Ridge and Joppa Ridge, still 150 feet above Green River level, leaving 150 feet of limestone for connecting passages to Mammoth Cave.

In the last days of slavery in Kentucky, Jonathan Doyle was a slave of a nearby landowner. During the Civil War, he escaped and joined the Union Army. In 1863, he went AWOL from the military and returned home, hiding and living in the woods on Joppa Ridge.

One day he felt cold air pouring from a crevice where he camped. He moved rocks and dug his way into a cave. The landowner was so pleased to know of the new cave that he arranged to free the slave.

Doyle discovered a "large river, as yet inaccessible to visitors," according to William Stump Forwood's 1870 *Historical and Descriptive Narrative of the Mammoth Cave of Kentucky.*

In 1887, the Mammoth Cave Railroad was completed between Glasgow Junction on the L&N Railroad and Mammoth Cave, and its right-of-way passed within a few hundred feet of Proctor Cave. For forty-two years, visitors paid to see Proctor Cave. When the Mammoth Cave Railroad died in 1929, Proctor Cave died with it. Thirty-eight years later, CRF teams began to explore and survey in Proctor Cave.

In 1970, a CRF party led by Gordon Smith found "Mystic River," a puny trickle of water six inches wide. To reach it required crawling on knees and belly for nearly two thousand feet. No wonder cavers passed it up for easier pickings. Proctor Cave went back to sleep.

In 1973, John Wilcox became interested in Proctor Cave and its potential. He selected a strong party to open up its new areas of exploration, picking Richard Zopf, Steve Wells, Bill Hawes, and my son, Tom Brucker, to find something there.

When I heard Tom was going, I said, "Be sure to take along enough rope. Go down the pit where that river goes. No excuses."

Tom spared me the "oh-dad" tone: "Richard's picking out the rope now. Don't worry."

Richard Zopf interrupted: "The only rope left is this one-hundred-foot Goldline. Maybe we'd better just leave it—"

"That'll do just *fine!*" I said. They would spin and bounce on this elastic rope, but they would be able to handle it. I had descended pits on Goldline rope dozens of times.

Deep inside Proctor Cave, Richard's party split into two teams for the systematic checking of leads near the so-called Mystic River. Richard and Steve Wells followed the ceiling canyon that diverged from the floor canyon of the R Survey at R70. The canyon led them to a small dome. Richard climbed up the dome to investigate what looked like a shadow. He squeezed up through a narrow crack that led into a virgin elliptical passage fifteen feet wide by eight feet high.

Richard said he felt a sense of disbelief. The discovery was immense, and there were no footprints in the sand floor.

Richard called the others to see the new cave. They set off exploring, carefully confining their footsteps to a narrow track to keep the new passages as pristine as possible and to watch for any signs of aboriginal visits.

Richard counted his paces as they swept along trunk passages twenty to thirty feet wide by ten to twenty feet high, past intersections and pits. They found cave rat debris and exotic gypsum formations.

Altogether they counted eight thousand paces—between three and four miles of prime, spectacular cave passages in which they had pressed the first footprints. Some cavers call this "scooping booty," racing through big virgin cave without surveying; CRF leaders frown on this kind of irresponsible behavior by others but have hidden their own guilt.

Richard later insisted that they had not acted irresponsibly. (He lamely pointed out that his indoctrination to CRF caving had been with Tom Brucker in Hansons Lost River, looking for the Flint Ridge–Mammoth Cave connection, a trip where there had been no surveying.)

Expedition leader John Wilcox was ecstatic at their news. He grinned broadly but chided Richard a little. CRF teams usually bring back survey data rather than just tales of wonder, he reminded Richard.

When I heard the news, I asked Tom if I could make the trip into the new passages in Proctor Cave. "No," he said. "The crawlway is too tight. You'd never fit." I smarted at that.

Over the next few years, CRF's thinnest cavers poured into Proctor Cave. They surveyed about five miles of new passages, little more than the complex of trunk passages first discovered by Richard Zopf and Tom Brucker.

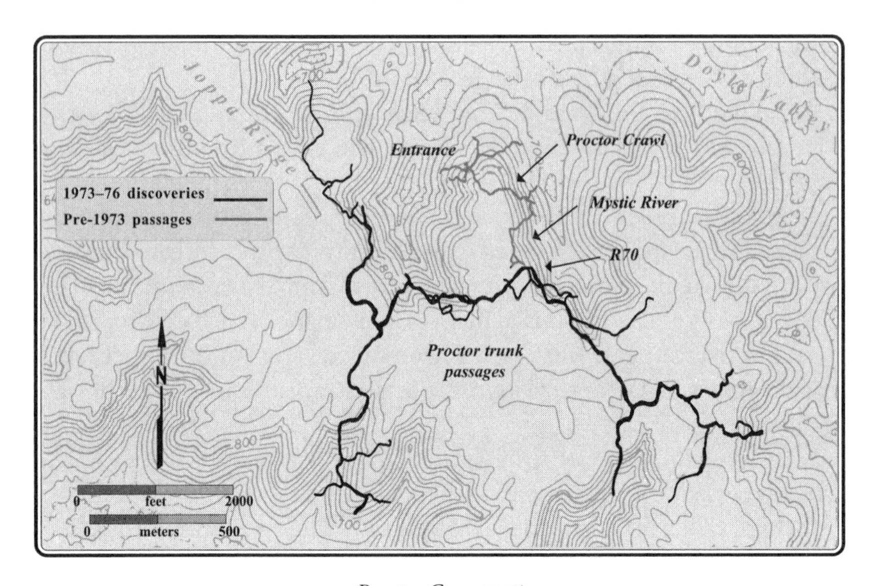

Proctor Cave, 1976

Leads were followed systematically, and most were surveyed to their ends. Exploration slacked off when only a handful of unremarkable leads remained unsurveyed. The Joppa Ridge puzzle remained unsolved.

Doyle Valley was like an insurmountable mountain peak. One could draw a line along the boundary between Mammoth Cave Ridge and Doyle Valley; no cave seemed to be able to cross it. I often studied the map looking for a route that would probe beneath the valley toward Joppa Ridge. Each time, nothing new presented itself. And since I was president of the CRF, between 1974 and 1976, I had many other things to worry about. Joppa Ridge would have to wait—or so I thought.

"Roger, skip the political shit and let's go caving." Diana Daunt—sharp-minded, determined, tiny, skinny—was extending more than a casual invitation to me. She knew my focus was on the frustrating politics of CRF relationships with the National Park Service but knew my heart was into cave connections. Her disarming invitation would reveal a cunning streak if I listened further.

She continued, "Back in August, John Wilcox led a trip into new passages under the south end of Mammoth Cave. We put in a lot of survey in narrow canyons that didn't end and found some domes and leads. There's a K Survey that takes off from a big dome at Station A57, forty-five or fifty stations of it . . ."

Instead of blurting out "Go on," I smiled, knowing she would reveal progressively wonderful details—and her real intention.

"On the Labor Day expedition, a party extended the K Survey from K47 to K89 where it ends in a dig. The survey goes south and then turns west." Diana lowered her voice: "It's the best and closest lead we've got to connect Mammoth Cave to Proctor Cave."

Bingo! She was recruiting me to scoop John Wilcox. Her desire to connect the caves was at least as strong as mine. With her determination and my zeal, the passage might be pushed. But why hadn't she recruited Wilcox? It didn't occur to me at that moment that Wilcox would be tough to con. He probably knew the possibilities out there better than anyone and would be skeptical of surplus enthusiasm. Besides, Wilcox seemed preoccupied with his own explorations. Yet, just maybe he had overlooked this one!

I was assigned to lead this trip to continue the K Survey. Previous trip reports described the unfolding discovery. On 1 September 1973, Ernst Kastning had led three others to the crawl that departed from Logan Avenue. They were barely able to keep their chests above the water, but the passage opened to a comfortable scrambling height, ideal for surveying. They carried their K Survey extension through several hundred feet of alternating crawlways and small stand-up rooms to Station K86, where sand on the floor sloped upward to within inches of the ceiling. One of the party dug ahead for ten feet; larger passage hinted just beyond.

The words on the page screamed for attention: "About five minutes of digging is all that is needed to go farther . . . it is the continuation of a very distinct passage heading south toward Joppa Ridge."

Jim Keith, Diana, and I were good diggers. We tucked a short wrecking bar in our pack, the necessary tool for loosening rocks or packed gravel, tested the compass and clinometer, and set off. Our objective: dig at K86 and carry the survey through to Proctor Cave. We closed the solid gate of the Frozen Niagara Entrance to Mammoth Cave behind us—all caves within the national park were gated and locked for security and safety reasons, restricting access to those with keys—at nine o'clock, a good, early start for a two-meal trip.

The sand and small gravel in the K Survey was loose and easy to shift to the side at places where the passage tightened. The passage then successively opened upward into a series of six dome-like rooms, each oriented diagonally to the direction of the passage we were surveying. The cross section of each room was that of the inside of a miniature gothic cathedral, with tan-gray walls six to eight feet high and about fifteen feet long.

Diana led the digging. She seemed able to wiggle through impossible places, while Jim Keith and I excavated a deeper and wider trench in the floor following her. In one two-hundred-foot reach of passage, we dug a sixty-foot-long V-shaped trench, ample for even the largest caver. (I had learned that if I expected to lure others, nothing conveys the credible promise of reward like a generous-sized dugway.)

We named the passage Snail Trail after the snail shells we had found in the sand—an unusual discovery—and because we had traversed the passage at the pace of a snail.

Near the end of our trench, we unsheathed the short bar where the fill rose to within an inch of the ceiling and a cold draft of air poured over our faces. The sandy dig led us downward at a twenty-degree angle, following the dipping ceiling.

Diana and I took off our hard hats and removed the suspension straps from inside. This was formidable digging and required proper tools. We worked the helmets as buckets, biting out brimful scoops of sand and passing them back for emptying. It was an efficient bucket brigade. For me, this was caving at its finest, chatting with amiable friends about man's adaptation, development, and use of tools while extending our earthworm tunnel downward for six feet, then to the left for seven feet.

We took turns at the face for several hours, then Jim complained of stomach upset and retreated to the room behind us. Diana and I kept digging. Suddenly I was aware that Diana was not passing back helmets of sand; she was gone. "Hey, don't go off without us!" I yelled.

"Hurry up. There's big cave here!" she demanded.

Was this how the seductive Lorelei had lured the sailors onto the rocks? I squirmed through the last seven feet of tunnel to a low ramp slanting thirty feet upward. Clearance was less than six inches.

"Step on it!" she explained. "Going cave here."

It irritated me that she wasn't digging back toward me. An image flashed in my motion picture mind: the brutal commander of a Greek war galley ordering the slave master to increase the beat to ramming speed.

"It would help if you dig back toward us," I suggested.

"Do you want to see this, or not?"

Jim and I dug through, and in an hour we traveled along one of the most beautiful passages in the cave. Light gray limestone walls undulated with scallop marks, and glistening gold highlights reflected from our lamps. Perfect ripple marks covered the undisturbed floor. A discovery like this was the visual equivalent of a sustained orgasm.

"Time to get to work," ordered Diana. "We'll survey from here back."

"Here" was a reduction in passage size to a tight crawl, with a strong breeze flowing from it. I was sure it would enlarge again and lead into Proctor Cave. Jim complained again of a growing headache and continuing nausea, but Diana silenced him with a lecture on how important it was to put this discovery on the map. Her appeal was sprinkled with phrases such as "They'll never forgive us" and "wimp-out." I had used intimidation of this kind myself, but hers lacked the subtlety I had cultivated.

We started an L Survey in the ten-foot-wide by nine-foot-high undulating passage and in a few hours tied back to the K Survey. Jim said his headache registered 9.5 on the Excedrin scale. What kind of party leader would let a party member get so trashed?

Well, I was forty-five years old, he was what—thirty? If people couldn't do this stuff, they shouldn't try. Diana was smiling as if in agreement. It was 4:20 A.M. when we left the cave. Jim Keith never came back.

We described Snail Trail as a four-star lead, which was how really hot openings were marked on the exploration lead list. From roughly scaling off the survey on the topographic map, we saw we were well away from Mammoth Cave Ridge, nearly halfway across Doyle Valley. Directly southwest of Snail Trail was Proctor Cave. "Thin people are needed!" Diana appended to the trip report.

If Diana was to connect caves, she'd need thinner cavers than Jim Keith or me. At 185 pounds, my girth was not the problem; it was chest and bone size that limited me. Or was it that I withdrew from confrontation with limitations? Diana escalated her recruiting, and one month later she fielded the first "All Woman Party," as it came to be known.

In the Mammoth Cave area, women cavers had made significant leadership contributions from the early days. One or two old-time CRF officials didn't like the idea of women leaders, but highly successful women explorers and leaders abounded in our field work. A physiologist said that women make good push cavers because they excel at endurance—something about better distribution of body fat. Exceptional upper-body strength was not the issue. In an almost legendary incident, Diana Daunt, working as a Park Service tour guide for the Wild Cave Tour in Mammoth Cave, was challenged by two burly football players: "*That* is our guide? We'll run her ass off. Not that she's got one!"

Toward the end of their tough trip, peppered by their derision, Diana quietly turned over the rest of their party and conclusion of the trip to the trailer guide. She said to the athletes, "You boys just follow me . . . if you can."

She led the pair of muscle men at breakneck speed through crawls, across ledges, down climbs, and through water and mud. They begged her to stop, gasping, swearing, and sweating. She said, "Now just keep up with this candy-ass guide. We don't want you lost." She took off like a shot. At the end of ninety nonstop minutes, the carnage was terrible and the mutilation of the male spirit complete.

When cavers, regardless of sex, heard her story of the football players, they howled. Women cavers are not all destroyers, but they can do anything the men cavers can and regularly lead push trips and major expeditions.

Diana's recruitment of Pat Crowther to lead the next trip to Snail Trail was a good choice. Pat was thin. She had discovered the way to the Flint Ridge connection with Mammoth Cave and had been a party member on that tough trip. She was as good as cavers come. Beth Grover and Cindi Smith were also competent, and with Diana, they expected great things.

At the Thanksgiving expedition, Pat's Snail Trail party got off to a slow start. Someone forgot a carbide bottle, so they returned to camp for it. The timing was unfortunate, and they waited two hours for a window between tourist trips going in the Frozen Niagara Entrance. Their speed in the cave was frustratingly slow, taking five hours to reach the end of the Snail Trail survey. But they were loaded with enthusiasm and tools—trenching shovel, crowbar, and trowel.

Pat and Diana squirmed through the tight crawlway beyond Station L1. It pinched to six inches of airspace with cold water and goopy mud below. It was a belly-crawl, so tight that Beth and Cindi could not fit through it.

According to Pat, it was lucky for Beth and Cindi that they couldn't make it.

Pat and Diana surveyed fifteen stations of N Survey starting on the far side of the tight swamp crawl. The gap in the survey would be closed later. It was relatively dry going, and they came to a passage trending northwest in one direction and northeast the other way. The passage to the northwest ended in a fill to the ceiling after a few feet. The other end, to the northeast, went five feet to a lead, a two-level passage that looked like it might drain a vertical shaft. The way into it was blocked by jutting rocks. Diana said, "We need a hammer to break off the projections. There's a good wind blowing. Remind me to bring a hammer next time."

They continued their N Survey, following a three-foot-diameter tube northwest about two hundred feet.

Abruptly, the passage broke upward. A two-foot-diameter tube went high one way; a foot-high by six-foot-wide lead went the other. Nobody larger than

Pat or Diana could fit, they thought. But the strong wind promised big cave ahead. Soaked to the necks, the two shivered with cold. Time to get out!

Beth and Cindi, left behind, had tried to keep warm in the chilly breeze by enlarging the swampy low passage between the end of the L Survey and the N Survey. They had made a crude groove in the heavy mud, but the ooze flowed back into the lowest spots.

Cindi was feeling sick. "Think it's intestinal flu," she said. She felt she had to retreat, but there was no escape from the wind blasting through the passageway.

Pat said they had to tie the gap in the surveys together or else a hanging, unconnected piece would be left. The prospect of surveying across the swamp from N7 out to L1 was grim. Pat guessed it could be done in a thirty-six-foot shot across the swamp, but it wasn't that simple. The digging had not broken through to the swamp to allow room for a person to hold a light on the survey station.

Pat attempted a one-person survey. She went back and forth through the center of the swamp, turned around each time, and attempted to read the compass at each turn. The wet cold was intense. Pat decided to sight to an imaginary point, crawl up to it, and improvise. A rotten way to survey, she thought, shaking with cold and fatigue. Those shots—N7 to M1 to M2 to L1—drew out the last reserves of the surveyor's energy.

The party was together at last, but the condition of the members seemed to be sinking toward a free fall of fatigue, discouragement, and illness. Now Beth was sick, too. Cindi took a Lomotil diarrhea pill, and Pat administered a Dramamine nausea tablet to both Cindi and Beth.

Pat said later, "I forgot how sleepy Dramamine makes people. We had no idea of the time. My watch gave up the ghost on its initial immersion at 4:30 P.M." Pat knew the only way to warm the party was to start moving toward the entrance, no matter how slowly. These are times when experienced party leaders recognize the endurance limit but know that encouragement and insistence on steady progress is the only salvation. The danger is real, but experience pays off. Pat reported later: "Take lots of wool clothing or shorty wet suits."

A month later, Denny Burns led a Snail Trail trip with Diana, another caver, and me in tow. We encountered high water at K49 and even higher water at K80. We aborted the trip, knowing the dugway would be filled with water.

Two years later, in November 1978, Diana Daunt led the Thanksgiving expedition. Lynn Weller, a tall and long-legged electronics engineering student at Ohio State University, had joined the caving. Recruited by John Bridge, who had a keen eye for the kind of engineers it took to explore these caves, Lynn seemed a natural caver. She had taken many weekend trips to small caves, and this year she had been introduced to the map factory at Scooter Hildebolt's house. Lynn could draw clear maps, survey, reduce the survey notes on her calculator, and write vivid trip reports. Diana had her eye on Lynn, as did Jenny Anderson, who was making progress spearheading the exploration of passages in Miller Trail that also headed across Doyle Valley toward Proctor Cave.

Lynn Weller's journal contains an intimate record of the details of that next trip to Snail Trail. Her account, plus the recollections of the other participants, characterizes the struggle and frustration of CRF cavers determined to find a way into Proctor Cave from Mammoth Cave.

On 22 November 1978, Lynn asked Diana about Snail Trail. "It goes!" said Diana. She drew a map of a T-junction. "This branch dead-ends. The other way is a tight lead. You can hear water noise, but you'll need a hammer to go beyond where we stopped."

Diana described the tight, going lead as two feet wide by three feet high. Lynn agreed it sounded better than anything else she'd heard.

"I'll go," Lynn volunteered. Diana instantly assigned Tom Brucker to go with her.

Tom was incredulous: "You *want* to go to Snail Trail?" He didn't do anything to change the assignments.

"Get to bed," Diana ordered. "To go out Snail Trail, you'll need all the sleep you can get." Lynn was happy. To her, Diana's tight-ship decisiveness was a sign this would be a good expedition.

Diana rolled them out of bed by half past seven and had breakfast over by quarter past eight. It was dark and rainy. Tom feared the water would make Snail Trail impassable. "It takes water directly every time a deer pees in the woods. It's been a couple of years since I was there, and the M Survey was filled with water."

Diana assigned Tom Alfred to their trip. He and his wife, Janet, were attending their first expedition, but Diana separated them.

They loaded their gear into Tom Brucker's Datsun and headed for the Frozen Niagara Entrance. Everyone ducked beneath the stairway at College Heights Avenue and doubled back under the tourist trail. They climbed down the breakdown to Fox Avenue, then Logan Avenue. After mostly walking canyons and a few minor climbs, they came to a rock placed in the middle of the passage with "SNAIL TRAIL" and an arrow smoked in carbide soot.

The arrow pointed to a low passage to their left, a flat-out chert belly crawl. Tom Brucker slid through the opening. Tom Alfred and Lynn followed.

"This is a neat passage," Tom Brucker said.

"Why?" asked Lynn.

"Because it's unusual. You just know a passage like this is going to do something."

They alternated between walking cave and flat-out belly crawls over coarse sand. "Big cave!" Tom shouted whenever the party reached a segment high enough to permit walking. He uttered a groan of disappointment as the belly crawl inevitably resumed.

Near the end of the K Survey or the beginning of the L Survey, one of the crawls had filled with sand. Two inches of air space remained with a gale blowing through. Tom removed his helmet to begin one of his favorite caving activities—digging. Alfred and Lynn began moving sand out of the way. They took turns; it was hard work.

On his second turn at the face, Tom managed to wiggle through the last fifteen feet and got his head through a narrow crack into a room. He literally used his head as a plow and finally moved enough sand to get in.

It looked like another short piece of walking cave. Alfred and Lynn took Tom's helmet and pack and wiggled through the hole. This crawl was a packs-off, helmets-off crawl. Sand sifted into their clothes, and their backs rubbed against the rock ceiling.

Tom was surprised to find the passage so dry. His hopes had risen considerably. They were in the nicest section of Snail Trail, the L Survey, and there ate a meal.

This was walking cave. Fantastic! Unfortunately, it ended too soon. The beginning of the M Survey was where Tom's party had been turned back by high water two years before.

"Wow!" said Tom, peering down the M belly crawl. "It's dry! I'll bet even the pool is gone!"

Diana had said they could keep dry until they reached the pool. Beyond it, the passage forked—the N Survey.

Diana had asked the party to resurvey the M Survey through the pool, but Tom decided to resurvey the whole thing. Alfred took the front point, Tom took the compass, and Lynn took the survey book. They started a T Survey, named for the two Toms. It was slow going. At T2, Alfred had to borrow Tom's putty knife to dig out the floor to get through a tight spot. It was still a chest compressor, and it took Lynn some effort to get her hips through because her coveralls stuck to the mud.

The passage had no standing water, but it wasn't dry, either. Lynn discovered that there was no way to move in a helmet-off belly crawl and keep her hands clean. Her hands and forearms were needed for locomotion. She couldn't carry the survey book in her teeth because it was impossible to keep from dragging it in the mud, and she couldn't lift her head high enough from the floor to clear the mud. She finally tucked the book under the suspension system of her helmet, put on her muddy gloves, and wormed her way to the next station, pushing her helmet and pack through the muck.

Lynn was awakened from her preoccupation with these matters by the sound of a distant splash. Alfred had found the pool. Apparently, the water level was about the same as when the pool had been discovered and first surveyed.

By the time Lynn reached the pool, Alfred and Tom had already crawled through it. Tom shouted back instructions: "Put on your pack and helmet. Push your knee-crawlers down your legs or they'll get stuck in the mud. If you sacrifice your left arm and both legs, you can keep your trunk dry. Stay high and keep to the right. Don't get your chest wet."

Lynn unhappily strapped on her wet pack and noticed that the bad waist strap on her pack had broken and the remaining strap was worn a quarter of the way through. Her helmet contained so much mud from pushing it through the ooze that she couldn't get the strap over her chin. The adjusting buckles were clogged with mud, also. Tom suggested that Lynn hook the strap under her nose. He was kidding, but Lynn decided it wasn't bad advice. In fact, it was the only thing she could do. The muddy strap barely reached her nose. Lynn's knee-crawlers refused to move, glued to the floor. The buckles were buried under great clods of sticky mud. She started into the pool.

There was one foot of air space, eight inches of water, and eight-plus inches of oozy mud. Lynn said she hadn't hit bottom yet.

She soon discovered the reason for removing the knee-crawlers, because she sank into eight inches of muck and stuck fast. She found it impossible to lift her knees clear of the stuff because of the limited headroom. And she couldn't push against anything to drag herself through the mud. Lynn turned and thrashed, and her foot hit something solid to push against. The total length of the pool was about fifteen feet, and eventually she emerged at the far end.

Tom laughed when he saw Lynn had followed his advice and put the helmet strap under her nose. She cleaned up the best she could in the clear trickle of water running into the pool.

Tom checked a lead. As Lynn and Alfred sat watching Tom's feet, Lynn noticed Alfred shivering; he was wetter than Tom and wasn't wearing enough wool.

Tom backed out and said, "There's an S-turn. I can't get through it. You want to try it?" Lynn crawled in. There was only a slight S-bend, but it was a tight passage. After some trial and error, she found a position to get by the first bend, but her knees were in the wrong position to make the second. The roar of the water was tantalizingly close, and the passage opened up ahead. She backed up and told the others of the problem.

The party returned to the junction room to eat candy bars. Tom Brucker decided they would look the other way before continuing the survey and started off. Lynn had to change carbide, but she told Alfred, "Go on, you need to get warmed up. I'll be along in a minute."

She finished changing carbide but could not get her lamp lighted with her wet hand. The flint was wet, too. Lynn waited for things to dry. By the time they did and she was able to light the lamp, she heard the sounds of the Toms returning.

"Goes five-hundred-plus feet," they reported without elaboration. They picked up their survey gear and started the slow process of measuring. Several times Lynn repeated back a distance or bearing but did not write it in the book. The measurements needed to be retaken because Lynn was dozing off and on.

Tom Alfred was extremely cold, but he ignored his discomfort as long as possible. Finally, he told Tom Brucker, gritting his teeth to keep from chattering, "I don't think I can go on. I'm frozen."

Tom said, "There's no harm in being cold as long as it doesn't affect your ability to leave the cave." Alfred visibly forced himself to stop shivering.

"I'm beginning to doubt my ability to leave the cave," he said.

If Tom was disappointed, he didn't show it. "Very well. We'll survey two more stations and then explore awhile. That will warm you up, and besides, we've got to get you some virgin cave!"

Was it the con? Tom Alfred had never been in virgin cave before.

Lynn put the book away after the last station and followed the two fast-moving cavers down the passage. They passed the end of their quick look, and Alfred led on into virgin cave. The passage varied from one and a half to three feet high and contained occasional shallow pools of water. Everything was muddy. The easiest way for them to move was to lie on their sides and shove forward with their feet. Lynn fell behind when she stopped to take off her pack, a move that later proved unnecessary. She delayed further to put the pack on again.

The missing waist strap shifted the full weight of the mud-laden pack onto her shoulder, which was already aching. It was a long way out.

Lynn moved faster to catch up. She came upon Tom and Alfred lying in the passage.

"I don't think I should go on," said Alfred.

Lynn admired Alfred for his courage in speaking up. It would be easy to tough it out and go on, but that could jeopardize the entire party. It was extremely important—lives could depend on it—for cavers to correctly assess their physical condition and let the party leader know before it was too late.

Tom paused. "Okay. You can rest or head back. I'll go on for ten minutes. Then I'll catch up with you." He took off ahead.

Alfred said to Lynn, "Don't you want to go on?"

"No, I'd just slow him down," said Lynn. She was feeling spent herself, but she had experienced that feeling before during her bicycling days. Always, that feeling had miraculously left after she ate. She was sure she'd be okay after a meal, and probably Alfred would feel better, too. They were long overdue for another meal.

They rested a bit, then headed back along the passage. They had come a long way, and it seemed even longer going back. Tom caught up with them about the time they reached the abandoned measuring tape. They dragged it behind them back to the junction room and began to clean and coil it as Tom told his story.

"I found the end."

Lynn didn't believe him; how was it possible that this passage, with its large airflow, ended? Wasn't its size increasing?

Tom continued, "Well, for practical purposes, it's the end. I guess it's about fifteen or sixteen hundred feet to where the passage seems blocked by a chert dike, but you can squeeze up through a slot in the ceiling over it and come back down into the passage. There, the passage changes to room-like walking segments, except that it's wet. The floor gradually rises to within two inches of the ceiling, and the cobblestone floor might be dug out. Not much breeze. I got stuck in the slot coming back . . ."

There was an odd note in his voice—fear? worry?—and something strange about the way he mentioned getting stuck. But Lynn thought she must have been mistaken. Tom Brucker would not be afraid. A ridiculous thought!

He continued, "It starts out heading southwest, but at its end it trends more south. I think we're about to intersect a major valley drain. The passage changes radically in character. Passages always do that before a junction. At least we've proved there's cave out under Doyle Valley. Probably someday some hot-shit caver will come out here and dig ten feet and get into a master drain. But we now know what happens to Snail Trail, and we won't send another party out here."

Tom grew quiet. "That's fifteen hundred feet of cave that won't be surveyed . . . by me."

The party started back out through the total immersion swamp pool. The chest compressor in the five-station M Survey was no longer remarkable; they were so well lubricated that they just slid through. At the beginning of the M and L Surveys, Tom fired up his carbide lamp and wrote "LUBE TUBE" over the low entrance to the crawl.

"It will be nice to get back to our dig and eat a meal," Lynn said. She was weary.

Tom frowned, "In my experience, when we're this cold, it's better to keep moving. If we stop now, we'll freeze." Lynn strongly disagreed but kept her mouth shut—Tom was the leader. But she began to doubt her ability to leave the cave without another meal.

Progress was slow. All of them were having lamp problems: too much mud had gotten into the water chamber of each lamp and had clogged the water dropper. Also, since no part of their clothing was clean or dry, they couldn't dry their hands—and wet flint won't spark. They stopped to change carbide in the room before their dig.

All three blew out their lamps at the same time. Then nobody could get a lamp going. In the dim illumination from a flashlight, Tom got out his match case, carefully tapping it until a match slid out far enough to grasp it without touching the end with his wet fingers. The other two watched anxiously. Tom scratched the match. Nothing. He struck it again. The tip crumbled. He shut the case on the damp matches.

"I think I've got a cigarette lighter," Tom Alfred said as he dug through his pack. Once found, he flicked it repeatedly. Nothing. "Oh no! My thumb was wet, and now the flint is wet. It won't work."

Lynn said, "I've got two cases of matches. One should be dry." She continued fiddling with her lamp.

Tom sounded annoyed with the delay. "Could you get them out now? I'm beginning to worry. It's a long trip out by flashlight." Lynn dug through her pack. She had stored the match cases in a plastic bag, along with a candle and heat tab stove.

She searched through the things she had taken out. The plastic bag wasn't there. "I must have dropped it last time we changed carbide."

"Aha!" shouted Alfred; his lighter had finally caught. The group fired up their lamps and packed. When Lynn stood up, her plastic bag with the matches and heat tab stove fell to the floor.

"Fantastic! I haven't lost it after all."

The trio started through the dig. Alfred's lamp went out in the middle of it, and a struggle ensued as Tom passed his lamp back to relight it. Various lamps went out at various times. The wind in the passage extinguished some of the feeble flames.

Tom's lamp was the first to die. Water would not come out of the dropper. No shaking, twisting, or reaming helped. Lynn handed him her spare lamp.

By this time, Lynn was miserable. She felt wiped out, her lamp kept conking out, her shoulders ached horribly from supporting the entire weight of her muddy pack, and her head throbbed because her helmet didn't fit with all the mud in it. She wondered when she had ever felt worse. Never, she decided.

Lynn was not hungry but realized she needed to eat. She knew Tom would not stop for anything more than a candy bar, and the very thought of candy made her nauseous.

Tom had been saying since her first expedition that he was out of shape. Lynn prayed she might never have the honor of caving with him when he *was* in shape. At one point, Lynn said she felt like sitting down in the middle of the passage and crying. She bit her lip—what a stupid thing to consider doing! "I had to laugh at myself. The whole thing was absurd enough to be funny," said Lynn later.

Suddenly, Tom Alfred turned around. His light had winked off. Lynn held her lamp out to relight his, but then hers went out too. Darkness. Lynn spun the spark wheel, which ignited the flame with a pop. She decided the lamp must be out of water. There was plenty of water in it, but it wasn't dripping. She tried all the tricks she knew, but none worked. Just too much dirt clogged the water chamber. Lynn screwed the darkened lamp back together and they hurried to catch up with Tom. "A can of fruit juice sounds good," said Alfred.

The party caught up with Tom at the chest compressor exit from the crawlway. It hadn't been bad coming in because they had been going downhill; however, it would be tougher going uphill. Lynn's chest went through, but her hips caught. Pushing with her feet just squeezed her in tighter. There was nothing to pull on, and her arms were weak from all the digging earlier in the day. She backed up and dug a little, then tried again. Not enough. She dug some more, then Tom Alfred extended a hand. She took it and he pulled her through.

Lynn's failed lamp was shot and Alfred's lamp was almost to that point. "Just pour a little water into the carbide chamber every ten minutes," was Tom's advice. Lynn poured in a slug of water and screwed the lamp together. With a spark, a giant flame leapt forward. They continued.

At the next lamp repair stop, Lynn said, "Tom, I'm going to mutiny and eat a can of fruit."

An unhappy sigh came from Tom. "Okay." He sat watching Alfred and Lynn down cans of fruit cocktail in a few gulps. He never reached for any food.

Energy returned almost immediately, but the spurt lasted only about forty-five minutes. Fruit is better than candy bars for a quick boost, but neither lasts.

They wiggled through the last chert crawl and finally emerged into Logan Avenue. Lynn was feeling bad again. She staggered unsteadily to her feet in the walking-height passage. She worried about some of the climbs ahead. Her arms were weak and she didn't know if she could walk without falling, let alone climb.

By the beam of her feeble flashlight, she put one foot ahead of the other in a semi-trance. Alfred also staggered along. It was a silent, stumbling journey—a forced march driven by necessity to get out of the cave. The outside air revived them somewhat as they closed the cave gate behind them.

On the drive back to camp, Tom commented that this had been a good group. "I never heard one complaint," he said. Lynn thought, What good would it have done if we had complained? But he was right. All of them had been miserable, and the problems had been many. Nobody had really complained—at least out loud. In fact, they had laughed at themselves many times to help relieve the pain, chill, and suffering.

Party after party returned from cave trips with survey notes and tales of discovery in all parts of the cave. Mammoth Cave was growing by miles and miles. The occasional small pushes toward Proctor Cave were a drop in the bucket compared to the productivity of cave-finding elsewhere in the system.

My focus was on Doyle Valley, even though most of my own trips were to other leads in the system. CRF expedition leaders followed lines of least resistance in dispatching survey parties. Leaders could send parties to a variety of leads, knowing only some would pay off. If they concentrated the expedition trips on narrow targets, they risked having nothing to show for hundreds of hours of work. When I led expeditions, I hedged the bet with a trip dispersion strategy. Sending trips to many different locations would yield more cave than dispatching trips to a concentrated area. When I wasn't leading expeditions, I enjoyed going wherever I was assigned.

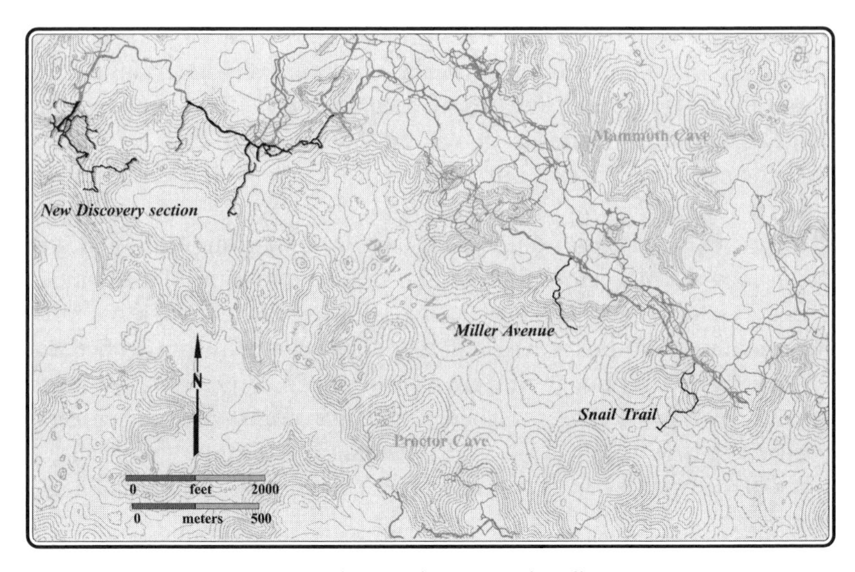

Southern Mammoth Cave, showing Doyle Valley extensions

When Tom Brucker's party returned, I scaled off their estimates of discovery on the topographic map. I speculated that the remaining distance to Proctor Cave was about seven hundred feet, but it may as well have been a mile. There had to be more promising ways to get beneath Joppa Ridge. My thoughts again turned toward Proctor Cave. Indeed, there were not many leads left, but most of the pits discovered off the trunk remained unexplored. Just maybe . . .

In 1978, Lynn Weller took numerous trips into Proctor Cave through the dreaded Proctor Crawl. Both Tom Brucker and Stan Sides had warned that I wouldn't fit through the tightest part and had discouraged me from even thinking of trying it. Lynn, on the other hand, had been talking with Jerry Davis, who was a good vertical caver. He had earlier led a trip into Proctor to a promising pit at Station P17 in the Proctor trunk passage. John Barnes had gone on the trip, too, and he agreed it was the place to go. John and Jerry convinced Lynn.

Lynn was leading the next trip to the P17 Pit on 21 April 1979 with John Barnes and Richard Zopf. She urged me to join them since I could do vertical work. Tom Brucker was horrified when he heard I was on the trip, but Lynn and Richard agreed I would do fine. John Wilcox told them I could go anywhere but might be slow coming out. Since Richard had dug out the tightest part of the Proctor Crawl, he could judge who would fit. Besides, I really wanted to go, because for some years I had been following the work in Proctor Cave in hopes that I could be on the party that connected it with Mammoth Cave.

We set off with a small coil of rope and some tools for digging. We wound the straps of our cave packs around our ankles so that we could drag them through the sand behind us. I found the Proctor belly crawl low and monotonously unending.

At the tightest spot, I had to exhale to squeeze through, but the trip in was otherwise uneventful. I concluded that there must have been a conspiracy to keep me out of Proctor Cave. Was it that they didn't want me to find something wonderful? Or were they genuinely concerned about my ability to make it?

At Station P17, we climbed down a short pitch into the bottom of a little dome to where a canyon led off. Richard Zopf canyon-straddled a deep void to find the anchor point for the rope. We tied the last ten feet of rope around a rock corner as a safety backup. Richard adjusted the rope pad to protect against abrasion where the rope went over a lip of rock. We put on harnesses and a carabiner brake bar and one after another clipped onto the rope where it dropped vertically into the pit. The first few feet down were a layback climb until we dangled free on the brake bars. The descent, about sixty feet, was a piece of cake. We found the previous C Survey station at the bottom of the pit.

The C Survey ran along the bottom of the pit, which was a chain of vertical shafts that made up a canyon about sixty feet high by twenty feet wide. A lot of wind blew through the place. The survey left the floor of the pit to angle up a sixty-foot climb. We knew there was a little more survey to finish, but our attention was on a slot in the floor on the far left side of the dome. Water drained out through it.

We had to dig to get through. We took turns prying rocks with a short wrecking bar. Lynn had packed a putty knife for digging; we also used our hands to scoop mud and loose rocks out of the plugged passage. Only one person could work at a time; the others froze in the stiff breeze flowing through the shaft.

There was plenty of time to talk, because the digging was extensive. We'd dig through into a comparatively wide place, then attack the next blockage. I used Richard's hammer to break up larger rocks. Rock chips flew in all directions from the blows, and I had no safety glasses.

John Barnes said he liked caving in Proctor, contrasting it with Roppel Cave. "They take a lot of risks in Roppel," he said. He reinforced the prejudice I felt, that the Central Kentucky Karst Coalition was made up of daredevils who believed in giving the cave a chance. Jim Borden had told me of handholds breaking off while free-climbing a deep pit. Were these tales of youthful machismo? Or just plain unsafe caving? Well, John was among safe friends here.

Chips continued to fly. All of us were accomplished diggers, making steady progress through the mud. Then using the hammer and rocks, we pounded off projections to make the slot larger.

Near the end of our allotted time, we broke through to the next room, a low-ceilinged alcove at the top of another pit. This one was about twenty-five feet deep. The only place to tie off our rope was back where we had first opened this passageway. Lynn went back to the bottom of the previous pit for more rope. We tied off our last 150-foot rope and threaded it through the tight passage, then lowered the end down the drop.

Richard Zopf rappelled down our second rope into virgin cave. He heard the sound of a distant waterfall, out of sight. He yelled there was still another pit below that one. We followed him into the second pit, where the walls glistened with descending water and mud was everywhere under foot.

"This place has been under water recently," Lynn said, examining a muddy high-water mark on the wall. "I'll bet it was back in December when the water was up fifty-three feet in the Green River."

At the far end of this room was the lip of the third pit. We lowered the end of our 150-foot rope into it, but with all the zigzagging from the anchor of the second pit, it did not reach the bottom.

Richard reached into his pack and pulled out a length of rope. "You never know when you might need a line." He uncoiled the thirty-foot handline. I made a note to myself that on our next trip here, we must bring more rope.

To test the depth of the drop, we searched for rocks to heave. Finding none, we packed mud into pretty good mud balls and hurled those into the pit. *Splash!*

Richard rigged the line, then clipped onto it with his descending rack, a steel loop containing an array of aluminum brake bars. He threaded the rope around alternate bars to provide friction during his rappel. He arranged his ascending Jumar clamps so he could get at them in case he had to stop on the rope and reverse his direction. Lynn, John, and I huddled as he backed over the edge of the mud slope.

"It's pretty dark," Richard said. "I see the water below me. I'm going down for a better look."

There was a pause, with appropriate twangs of the rope, grunts, and groans. He was about twenty feet below us.

"Hoooo-weee!" Richard exclaimed. "I didn't stop in time. Part of me is under water. I'm on top of some kind of lake . . . or rather, up to my middle in a lake or river or something. No telling how deep it is. I'm trying to re-rig to ascend the rope." He struggled. Minutes stretched out. We shivered.

Every sound of Richard's description reverberated. Next to the shaft was a single passage, maybe twenty feet wide, with only four feet of air between water and ceiling. Richard fished his Jumars from his cave pack, taking care not to drop them into the water. He clamped one Jumar on the line above his locked descending rack; its nylon leader streamed down from it. He fumbled to hook his chest carabiner into the leader. This was a difficult maneuver under dry conditions. Here, partly submerged, it was frustrating and cold. He had to transfer his weight from the rack to the Jumar leader. There was nothing to push against; he pulled himself up with arm strength alone. At last, his full weight dangled from the locked Jumar clamp.

The next task was to rig a second Jumar clamp that trailed a foot loop. It was located below the now loose rack. It clicked onto the rope. Now, he had to raise one foot high enough out of the water to pull the loop over it. This seemed impossible. Dangle, thrash, and splash. He launched out with one foot toward the loop held low in the water. It was a blind thrust. There was nothing to be seen in this watery hole. He had the impression his foot came nowhere near the loop. He tried the maneuver many more times before he was successful.

Richard thought of a way to plumb the depth of the water. He tied his hammer on some webbing from his pack and lowered it down the pit. When he felt slack, he pulled it up and looked for the water line. "It's about four feet deep," he said.

"How are you coming?" we asked from above. We couldn't see him.

"Keep your pants on." A testy response, I thought. It wasn't like Richard. He must be in trouble. I mentally ran through the drill to rappel down and rescue him if necessary.

He was climbing higher. He shouted: the echo amplified his voice. "It appears to be—and sounds like—a ponded river. I'm feeling pretty weak. Pull me up if I tell you." We braced as best we could in the muddy canyon. There wasn't any place to get a good purchase. "Pull!" We pulled him up the last few feet of the slippery slope.

Richard was caked with mud and drained of energy, although objectively I knew him to be the strongest caver on earth, tired or not. He said the pool was at base level. We had descended far enough by my reckoning, too. Echoes had made it nearly impossible for him to understand our words above him. Zopf and I agreed that we had heard an echo of such intensity only in Roaring River and Echo River in Mammoth Cave.

Next time, he'd prepare better for the transfer from descending to ascending gear. And he would make it a point to stop before getting his butt wet.

We surveyed from the newly found pit back up the second drop. We ate a meal and headed out. John Barnes was like a zombie, pulling prussik knots from his cave pack. Prussik knots to climb out, instead of Jumars? That's what cavers used in the fifties; now, thirty years later, only a cheapskate would think of using such primitive gear! The party zipped up the rope in short order, but John tediously moved each knot inch by inch up the rope. Prussik knot ascents can drain away as much energy as doing fifty pushups. I didn't think John could do ten pushups in his present state. Eventually, he struggled over the lip and we began the long trip out.

"This is the base level we've been trying to find for years. It's a prize because in base-level passages you can go for miles, crossing under sinkholes and valleys," I informed Lynn at a resting spot in the Bottle Room near the entrance. I proposed a return trip. The Green River water level should fall soon. Its flood stage had backed up water in all the lowest level passages, but when the Green River receded, the passages would empty quickly. Lynn was eager to go. She could piece together the evidence without my patronizing lecture.

Her trip report later indicated Richard Zopf's stream appeared to head the right direction and ought to open up all of Joppa Ridge. She also said it would be one hell of a difficult trip to traverse the Proctor Crawl and three vertical drops. I wondered from a practical standpoint how much cave could one explore from such a deep-cave jumping off point. Could anyone expect productive survey work from the cavers after such formidable obstacles?

John Barnes supplied his own answer. He never returned to Proctor Cave.

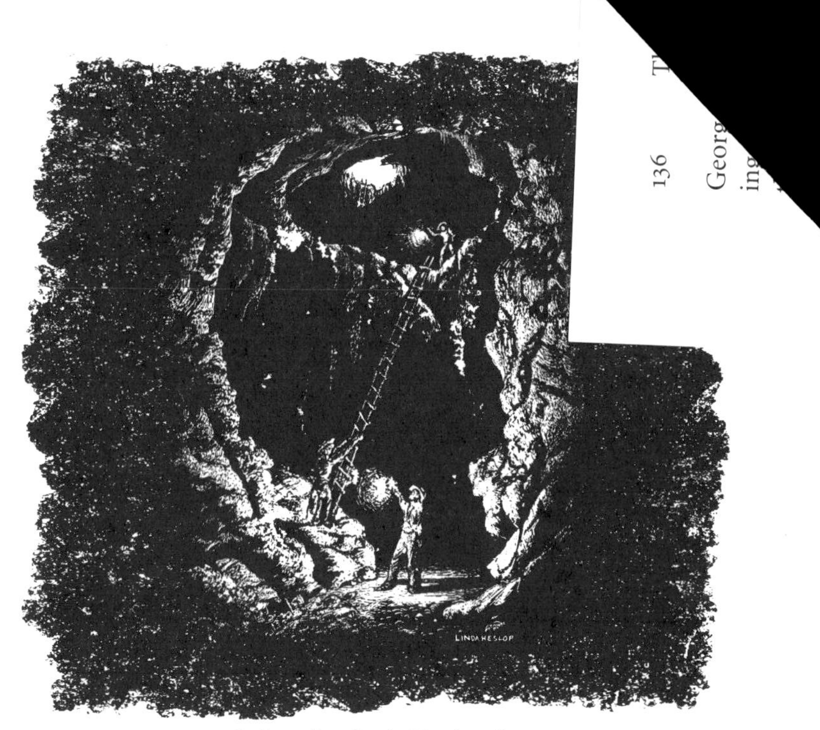

Early exploration in Morrison Cave

10

The Legacy of George Morrison

Cavers Follow in the Footsteps of the Old-Timers

Now the plot thickens. Let's go back eighty years and begin again. And we will not lose sight of Roppel Cave.

Between 1915 and 1925, the legendary cave wars reached their peak. The conflicts pitted landowner against landowner, each seeking to cash in on the fledgling cave tourist industry. Mammoth Cave was the big tourist magnet. The path to success during these turbulent times was to find a cave to commercialize or, better yet, an entrance to Mammoth Cave to siphon off tourists headed for the Historic Entrance on the northwest edge of Mammoth Cave Ridge. One of the more enterprising individuals during this period was e D. Morrison, a likable oil prospector who turned to cave prospect-. His determination, brute force, and ability to operate by stealth led to he discovery of many miles of passage in Mammoth Cave (to the dismay of the heirs of the Mammoth Cave Estate) and, more importantly, to the opening of the New Entrance and Frozen Niagara Entrance to Mammoth Cave. He was successful. He built and operated a hotel and conducted Mammoth Cave tours, making a lot of money operating the "New Entrance to Mammoth Cave" at the expense of the historic Mammoth Cave.

During his search for this back door to Mammoth Cave, Morrison probed countless holes along Mammoth Cave Ridge and other areas. His efforts were systematic and thorough. The Mammoth Cave Estate got most of the tourist money; everyone else had to scratch out a living off the land. A few operated souvenir rock shops. But every landowner knew that he might get rich if Morrison found an entrance on his property, so all the landowners cooperated with Morrison.

One of the holes Morrison looked into was perched on the side of an upland midway between Joppa Ridge and Mammoth Cave Ridge. This hole could accommodate a person—barely—and dropped immediately into a deep pit. This pit probably interested Morrison and his men because it exhaled large volumes of cold air in the summer—a sure sign of big cave. Although not on Mammoth Cave Ridge, the pit might intersect passages that were an eastern extension of Mammoth Cave. Elmore Borden, Pete Strange, and Will Warfe owned the lands surrounding the blowing hole and entered into an exploration agreement with Morrison. The four of them would explore the cave.

At the time, explorers descended pits on homemade ladders or, if the pit was deep, by being lowered on a hay rope. Morrison's hole was deep, over seventy feet, so they lowered Morrison on a rope threaded through a pulley that was lashed to a log placed across the entrance. The drop was narrow with two jagged offsets. The three on the surface had to lower the explorer slowly to avoid injuring Morrison or catching the rope behind a ledge. Despite these precautions, the tethered explorer had to continually avoid falling rock knocked from the many ledges by the rope.

Once at the bottom of the pit, Morrison untied the rope from his harness and moved into unknown cave. At the base of the pit, a low crawl over rocks wedged in a narrow and deep canyon led to a room. He scrambled down a slope covered with broken rock and arrived at the edge of a second pit. With a short line he always kept in his pocket while exploring, he lowered his kerosene lantern to see if the cave continued from the bottom. All he could see was a tiny canyon. Surely, a cave with such a wind as this had to continue a better way.

Across this second pit, however, there was a large opening in the wall—a window. He picked up rocks and tossed them up into it. A few seconds passed, then a clattering echoed from beyond the window. Morrison had not felt any air blowing up from the second pit, but now he could feel a cool breeze coming from the window. This must be the right passage!

Now that there was cave to explore, he needed an easier way to descend the entrance pit. The four labored for several days building ladders made of cedar poles. One by one, they lowered sections of this rickety ladder into the pit until they had assembled from the bottom up a zigzag route between ledges all the way to the surface. If the cave went, they could improve it later.

They continued the ladder building at the lip of the second pit. During many trips from the entrance pit, they had dragged the necessary additional materials for twenty-five feet of ladder, including extra-stout poles, since they were now building more of a bridge than a ladder. They extended the makeshift bridge across the gap into the window. Morrison hoped that the drop beyond the window would be climbable, but those hopes were quickly dashed. Back to work.

They assembled a third ladder to descend the pit beyond the window. The effort to place the ladder was both difficult and dangerous. Sections had to be lowered into the pit and assembled in place. Only one person could sit precariously at the window and steady the ladder while subsequent sections were nailed to its top end. When the project was complete, the foursome climbed into the large shaft beyond the window.

Over the next several trips, they explored more than a mile of complicated canyons, squeezes, and large domes. Unfortunately, the hoped-for breakthrough into the eastern extension of Mammoth Cave was not found. They had not penetrated beyond the extensive complex of shafts. Finally, at the end of one low, muddy crawlway, they found another deep pit ringed by an unstable pile of sandstone boulders. They were now far below the entrance. The sound of falling water hinted that the new pit might be the key to big cave. One of the explorers grabbed a loose rock above them and was greeted by a roar as the hillside slid into the abyss. Only by using one precarious ledge were they able to avoid disturbing the accumulation of loose rock.

By this point, dragging in ladders was impractical, so they later returned to the perch with a hay pulley wheel and a long section of half-inch hemp rope. They carefully avoided touching the loose rocks. Morrison cut a short length of the rope and looped it around a pillar in the wall above the ledge, tying it snugly with a square knot. With the hay pulley hooked onto the loop, they threaded the remainder of the rope through the pulley and lowered one explorer thirty feet to the bottom of the pit. It was then easy to lower the other explorers by feeding the rope through the pulley from the bottom. Large sandstone boulders littered the floor. A tall canyon led off to the north.

The tall canyon carried a cold wind that chilled the explorers as they followed the current in the foot-soaking stream. The passage was long, leading them far from the main cave. Ultimately they came to the top of a fourth pit. The stream they had followed was now quite large and crashed downward in a thundering roar. A lantern lowered on a rope revealed little due to the spray from the falls. Somebody would have to investigate the potential directly, but they were far from the entrance and at the edge of their capabilities and will.

They returned later with a stout plank, a second pulley, and an even longer section of rope. They wedged the plank above the pit to serve as a guide to keep the rope and the explorer out of the falling water. They lowered one man down the forty-foot pit, and once on the bottom, he untied and set off to explore.

The floor of the pit was sandy and flat; its walls were coated with a thin layer of dried mud. Obviously, this pit flooded at times, perhaps even backflooded from the Green River. The mud coating looked similar to areas in Mammoth Cave near the level of Echo River. The pretty waterfall impressed him, but the water flowed into a pool and disappeared under the wall. Looking around, the lone explorer found a canyon blowing air. He followed this down to a stream, probably the same stream that drained the waterfall splash pool. Unfortunately, just a few feet farther, the ceiling lowered to only a foot above the stream. This was no place to go with a lantern, and besides, belly-crawling in the water did not appeal to him. For Morrison's team, this was the end of the cave. It would not become the back entrance to Mammoth Cave that they were seeking. Morrison abandoned the effort, leaving the landowners disappointed.

But Morrison's exploration was not the end for the cave. Another adventurer, Old Man Hackett, took interest in the caves of the area. Hackett was an eccentric hippie forty years ahead of his time. He took up residence in Long Cave, an old saltpeter cave south of Mammoth Cave; he ate canned oyster stew and condensed milk. His hair cascaded down to his waist when not rolled up and stuffed under his ten-gallon hat. After a number of trips down Morrison's cedar ladders, he returned with tales of big cave and plans for commercialization. Understandably skeptical of Hackett, the owners didn't take him seriously. After all, why would he succeed where the competent Morrison had failed? In disgust at his failure, it was said, Hackett eventually sealed the entrance and left the area, returning to his native Texas, resolved that his secrets would be safe forever. Morrison's old cave was soon forgotten.

Don Coons's arrival in the Mammoth Cave area marked the beginning of a new era of modern off-park cave exploration. He began caving in the long, wet passages of Perry County, Missouri, exploring caves such as Mystery Cave and Rimstone River Cave. The caves of the Mammoth Cave region were perfect for his abilities and interests. Don first came to Mammoth Cave by way of the Cave Research Foundation in 1973. It wasn't long before he married fellow caver Diana Daunt and settled in a small, wood-framed house near the boundary of Mammoth Cave National Park. At that time, Jim Quinlan, the park's research geologist, coordinated hydrologic research in the caves and karst of the local region and was looking to recruit a field assistant. It was perfect for Don, a superb caver who was lucky enough to be able to live where he was surrounded by the big caves. He took the job. Many cavers would kill for a position like that. Fortunately, his job as an assistant was during the summers when the water levels were low, making cave field work conditions most favorable. In the spring and fall of each year, he returned to his family farm in north-central Illinois to plant and harvest corn and soybeans. During the intervening months in late fall and winter, when water levels in Kentucky were high and the fields of Illinois lay dormant, he joined caving expeditions to explore the deep pits of Mexico or the enormous caverns of Belize.

I—Jim Borden—first met Don Coons during the summer of 1974 on my third caving trip in the area. I was still wide-eyed and eager, full of intentions to discover a big cave with my partner Joe Saunders. Joe had arranged a meeting with Jim Quinlan, so one morning we headed to the park to discuss hydrology and dye tracing. I was impressed by Jim Quinlan and Don Coons. Wow, I thought, these guys were paid to do what I loved to do.

I liked Don from the first moments I spoke with him. Quinlan talked fast and long, totally hyper, and appeared to know everything about everything—and maybe he did, I thought. Don, quiet and unassuming, listened thoughtfully and added intelligent comments. It was apparent that he also loved caves. He was genuinely interested as Joe Saunders talked to Quinlan about the caves that Joe knew along the river. Don readily accepted Joe's invitation for a trip to Gradys Cave; any chance to see a new cave was not to be missed. We planned to go the following day.

Gradys Cave is a ten-mile-long river cave known for lots of water and mud and trips that last twenty-four hours. I was petrified at the prospect of a caving trip there; Don's acceptance of the invitation made me feel better. I knew I would not be suffering alone, although I knew I would indeed suffer. Joe was notorious for the killer trips he ran into this cave, and I felt that two of us against him would stand a better chance of survival than one would.

The following day, Don met us at the appointed time. Heavy rain filled the gullies, and fountains of water spurted from the sodden ground. A trip into Gradys Cave was out of the question. I did not show the relief I felt.

Don was sensitive to Joe Saunders's sour look about the reduced prospects of any Gradys Cave trips during the week. As an alternative, Don offered us a trip into Parker Cave, where he was leading a trip the following day. Parker is a stream cave, but it was far from the Green River and would not flood the way Gradys did.

We had a wonderful time. Although not a large cave, Parker was varied and interesting. Don and I talked about the many caves in the area, and I told him of my work on Toohey Ridge. We shared the dream of finding an enormous network of caves that ranged outside the park. He agreed with me—Toohey Ridge could be a key. At Parker Cave, we began a long and lasting friendship.

The year 1977 was an unusual time around Mammoth Cave National Park. The previous winter, park officials had become alarmed after surprisingly high levels of radon gas were detected in many of the passages of the commercial tours of Mammoth Cave. Knowledge of the risks of radon exposure was still in its infancy, although it was known that overexposure in uranium mines could cause lung cancer. Accordingly, the park adopted an aggressive regulatory position regarding radon exposure of its own personnel. Normal tourist visitation would not be affected, for tourists' exposure was minimal. But if an employee approached the strict limits of exposure, he or she would be temporarily reassigned to non-cave duties. Jim Quinlan's summer cave survey programs were curtailed because his employees were subjected to too much exposure. Low-level passages, the primary targets of Quinlan's research, had the highest known levels of radon. It was not unusual for his employees to log seventy hours of in-cave time per week. It did not matter that most of the effort was within off-park caves on private land; Quinlan was employed by the Park Service. Consequently, both he and his employees were subject to the strict radon exposure guidelines.

Roppel Entrance gate during the flood in 1981. Photo by David L. Black.

One of the Mammoth Cave area's large vertical shafts, which can range from ten to two hundred feet in height and which are often the keys to new discoveries. Photo by Ron Simmons.

Roberta Swicegood rappelling into a large dome near Elysian Way. Photo by David L. Black.

Yahoo Avenue, one of the spacious avenues under the broad sandstone cap of Toohey Ridge that enables explorers to cover large distances underground quickly and often leads to new, complex sections of cave. Photo by Ron Simmons.

Jim Borden with large gypsum flowers in Yahoo Avenue. Photo by Paul and Lee Stevens.

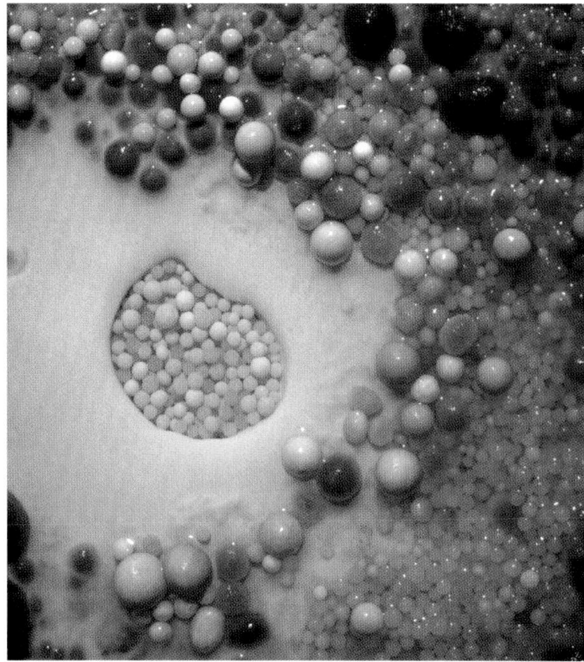

Cave pearls, up to one centimeter in diameter, found in a small pool in Yahoo Avenue. Photo by Paul and Lee Stevens.

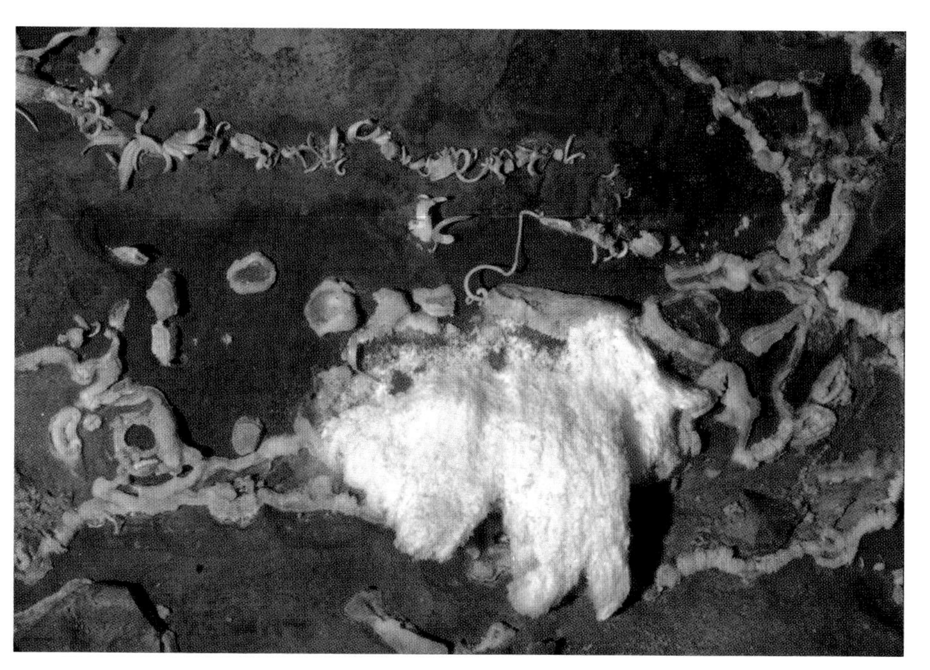

Gypsum cotton along the walls of Arlie Way. Photo by Paul and Lee Stevens.

Eight-inch gypsum flowers in Yahoo Avenue. Photo by Paul and Lee Stevens.

A dramatic large canyon of Lower Elysian Way that comprises part of one of the longest uninterrupted stretches (4.5 miles) of large passage in the Mammoth Cave system. Photo by Ron Simmons.

Large flowstone draperies and deep pools of water that characterize the far reaches of Lower Elysian Way. Photo by Ron Simmons.

Roberta Swicegood (foreground) *and John Zidian at the large rimstone dams guarding the route through the Watergate, far Lower Elysian Way. Photo by Ron Simmons.*

Ron Gariepy in the C Crawl, one of the passages in the "old cave" that gave Roppel Cave the reputation as one of the toughest in the country to explore. Photo by Bill Eidson.

One of the tight passages in the S Survey below the S64 Pit that almost defeated Bill Walter and Jim Currens on the trip during which Arlie Way in Toohey Ridge was discovered, which in turn paved the way for the connection with Mammoth Cave. Photo by Bill Eidson.

Caver rappelling from the ceiling at the point of the breakthrough discovery of Hawkins River in Proctor Cave in 1979. Photo by Pete Lindsley.

Logsdon River, far upstream near the point of its discovery from Arlie Way in Roppel Cave. Photo by Ron Simmons.

Deep water in Hawkins River near its downstream sump. Photo by Pete Lindsley.

Bill Walter (left) *and Pete Crecelius stopping for a lunch break in Freedom Trail. Photo by David L. Black.*

The Roppel-Mammoth connection team at the point where the two great caves were linked in 1983. Front row, left to right: *Dave Weller, Sheri Engler, Roberta Swicegood, Don Coons.* Second row: *John Branstetter.* Third Row: *Jim Borden, Roger Brucker, Lynn Brucker, Bill Walter.* Not shown: *David Black. Photo by David L. Black.*

Yahoo Avenue. Photo by Ron Simmons.

Lower Elysian Way. Photo by Ron Simmons.

Quinlan did not employ any cavers to work on his cave survey program that summer, and as a result, Don Coons lacked a job. No problem—Don hired on as a seasonal guide for the commercial cave tours in Mammoth Cave National Park.

The park employed many seasonal guides to handle the large numbers of tourists who visited the cave during the summer months. One of the fringe benefits was Mammoth Cave itself. During the evening and other off hours, the park allowed full-time guides to lead seasonal guides through wild parts of Mammoth Cave. Most seasonal guides looked forward to these trips, so when they reported for employment in June, they were disappointed to find that this practice had abruptly halted. The new radon exposure limits killed exploring by guides.

Seasonal guides Jeff Ulrich and Robert McDonald were extremely frustrated by the new set of rules. They itched for some wild caving, and this itch could no longer be scratched within Mammoth Cave. They asked around the Mammoth Cave guides' lounge for places to cave outside the park, places not subject to the park's supervision. Surely, they thought, some of the full-time guides would know of a few caves.

Ranger John Logsdon knew about a cave. The previous autumn, while deer hunting on his farm, he had found the blowing hole that Morrison had explored fifty years earlier. Recently, the cave had opened itself after being filled long ago by Hackett. The logs used to cover the hole had probably rotted and fallen in, allowing the wet Kentucky winter to finish the work of exposing the entrance. Logsdon had been intrigued by the tales of Morrison's earlier exploration, but the small entrance and seventy-foot pit prevented him from checking out the cave. Jeff and Robert wanted to explore the cave, but they were not vertical cavers. What to do?

While collecting tour tickets at the Historic Entrance to Mammoth Cave, Don Coons overheard Jeff and Robert talking about Logsdon's tip. In his typically ingratiating fashion, Don talked the situation over with them, modestly offered his expertise in vertical caving and the loan of some gear.

The trip was a go.

The next day, Don, Robert, and Jeff parked at John Logsdon's farm. They hiked across the wide field to the opposite edge of the ridge and walked down the gully Logsdon had described. There was the hole, just above the floor of the gully, with an old fence circling it. You could miss it if you weren't looking for it. As they uncoiled the rope, Don noticed that they were standing less than fifty feet from one of the many prominent concrete boundary monuments of Mammoth Cave National Park.

Interesting, he thought as he gazed down the hill into the park.

The National Park Service had always been sensitive about passages of Mammoth Cave that extended outside park boundaries. Inevitably, some Mammoth Cave passages would be found to extend under private property. In fact, the CRF knew about some of them already. Exploration of a cave with an entrance as close to the park boundary as this one would be bound to generate Park Service interest.

Don outfitted Jeff and Robert for the descent, then each rappelled to the bottom of the seventy-foot pit. The top was narrow but soon widened to ten feet. The cold wind refreshed them as they passed through the narrowest section. At the bottom, they stood on rocks, dirt, and rotting leaves and branches. Logsdon had told them the stories from the days of Morrison, but cavers view such tales with skepticism because they are often wild exaggerations. However, the part of the story about the cave entrance being filled and recently reopening itself rang true—the bottom of the pit was littered with cut logs and planks.

A strong breeze blew from a small crawlway under one wall. The three pulled out loose rocks and scraped away dirt. After fifteen minutes of easy digging, they were through. They soon stood at the edge of another pit. Here was the first real evidence of explorers from periods long past. Leaning across the pit to the far wall was a rotting wooden ladder. At the ladder's far end, they could see a window into a third, parallel pit. Don reached down to test the ladder, giving it a gentle but firm tug. The whole structure groaned and slowly collapsed, crashing to the floor thirty feet below.

Don cringed. "Jesus! Good thing we didn't try to cross the pit on that!"

Jeff and Robert stared into the pit, awestruck, as the last of the rotting timbers let go and cartwheeled into the darkness.

Don was curious. Someone had gone to a lot of trouble to construct this ladder. Had they found something worthwhile that lay beyond? Climbing to the window was now out of the question. The only remaining route was down. But for today, they were at the end. Their only rope was tied to the tree at the top of the seventy-foot entrance pit.

They would be back.

A week later, Don Coons, Robert McDonald, Jeff Ulrich, and Sue McGill returned with a second rope so they could descend the pit below the window. Don had recruited Sue and had given her quickie lessons in rappelling and ascending cave rope. (Sue said later she feared that vertical work was a lot more dangerous than Don had let on.)

This time, Jeff and Robert were more experienced and descended the entrance pit with little problem. Don again evaluated the prospects of crossing the second pit to the window on the far side. It looked possible, but today was not the day to try.

Concentrating on the task of going down, Don threaded the rope through a hole in the wall and secured the rope with a bowline. Then he grabbed the coiled rope and tossed the remaining line into the pit. As the most experienced, Don led the way down. Below, a narrow, popcorn-encrusted canyon led off. A slight breeze cooled his face as he looked into it. After six feet, the passage turned a corner. Don called up for the others to come down. There was cave here to push.

The canyon was narrow. They struggled through fifty feet of painful passage, eventually emerging at the base of a large, impressive vertical shaft. After a few minutes of exploring, they realized that the shaft through which they had entered was just one of several that formed an extensive complex. It was like visiting a museum. Evidence of ancient exploration lay everywhere: metal buckets, chunks of cedar poles, rusty nails, and a few bottles. In one dome, they found another rickety ladder that led up forty feet to what was probably the window across the second pit. The old-timers had perhaps reached this part of the cave from that window. Certainly, none of them had followed the route through the narrow canyon. Don knew better than to test the ancient ladder rising above them; it was surely as rotten as the one that had crumbled beneath his touch earlier. They would go back the way they had come.

There were leads everywhere. They had found a cave! But with the narrow, fifty-foot-long canyon between them and the entrance, it was time to head out. They took one last, longing look up the ladder, knowing that the entrance pit was probably just beyond, before heading back through the canyon. The tight stretch of horrible squeezing seemed infinitely longer going out. Besides being tired, they had to force their bodies uphill and against the grain of the scallops; Sue's hips proved to be a particular disadvantage here. The party took forty-five minutes to get to the base of the second pit. Don Coons named the canyon Sue's Sorrow.

Several hours later, the party of four tired cavers reached the surface after dark. It had been a difficult trip. Robert McDonald and Sue McGill would not return; the cave was too demanding. But Don Coons smelled big cave. He would be back.

One evening, Don and I were sitting around drinking beer in his living room, as we often did on my many visits. As usual, we were chatting about nothing in particular until Don began to talk about the new cave on John Logsdon's farm. He described the pit traverse that would bypass Sue's Sorrow.

"You know, Jim, I think you could climb across that pit over to the window where the old ladder is," Don said nonchalantly between swallows of beer.

I knew that he was trying to suck me in; I smirked to let him know that I was on to his ploy.

Don said nothing, drinking his beer with a glimmer in his eye.

I grabbed the bait. "Tell me about it."

He began his inch-by-inch description of the obstacle. I tried to imagine it as he spoke. Despite his best animated efforts, I could not visualize the climb. "What kind of holds are there?" I asked.

"Well, I don't remember, really." He looked embarrassed. Seeing my growing reluctance, he continued smoothly. "You're so tall and lanky, you can probably just stretch your way across." He stood up and extended his arms for emphasis. "Besides, I'll belay you."

"Great!" Like that made a difference.

I frowned as I imagined clinging to the wall for dear life. Finally, with no more strength to hold on, I would let go. Don's rope would catch me, but because the climb was a traverse, I would pendulum and crash into the wall below.

"It'll be easy," Don said.

"Sure, why not." I surrendered. There was really no face-saving way not to try. I could at least look at it. If it was too dangerous, I would decline.

I guess I was the natural choice for this project. For the last several years, I had been climbing extensively at various rock-climbing meccas across the United States. I was a fair climber, certainly better than average, and was probably technically suited for the task. Cave climbing is very different from traditional rock climbing, however. In caves, one cannot depend on the holds, the rock is often rotten, and mud and water afford slick, unsure footing. Moreover, wall crack systems tend to be poorly developed, making the placement of pitons or nuts a chancy or impossible undertaking. I preferred to climb on the sun-warmed cliffs of Tuolumne Meadows in Yosemite National Park. This would be no Yosemite.

A couple weeks later, Mark Stock and I followed Don Coons across the large field to the wooded gully leading down to the cave. Each of us carried rope and vertical gear. In addition, I had brought several runners, carabiners, and nuts in the hope that I could protect the climb and prevent rib-smashing wall crashes. As we headed down the last few feet of the gully before the cave, something piqued my memory. Vague recollections of the big field, the crossing of the fence into the steep gully, and the rock outcrops lining the hillside all precipitated at once from the haze of things past. The image suddenly became clear.

"Hey! I know where we are."

Don looked at me quizzically.

"A couple years ago, I came down this gully looking for a cave someone told me about. But I couldn't find it."

He was now standing at the entrance, uncoiling the first rope down into the hole.

"And I can see why. You'd walk right by and not see it." I motioned down the gully to the signs marking the park boundary just a few feet ahead.

I thought about it, looking first up at the cave entrance, then up and down the gully. I shook my head in disgust. "I can't believe I missed it."

"Well," Don snickered, "tough break."

Thirty minutes later, I stood at the edge of the second pit studying a series of holds on the right wall leading up to the window. I had a rope tied around my middle, and Don was seated firmly behind a boulder ready to belay me, waiting for my word that I was ready. The traverse across the wall really didn't frighten me. A couple moves and I would be over.

I took a deep breath. "On belay," I said, barely loud enough for Don and Mark to hear.

"Belay on."

This was routine for both of us. My experience came from the cliffs of the Shawangunks in New York, and Don's came from the deep cave systems of Mexico. We both knew what we were doing.

I sunk my right hand deep into a crack out over the second pit and smeared my right foot onto a sloping ledge. I leaned out to my left, using my right foot for support while reaching over with my left hand to a projecting rock horn just a few feet from the window.

"Looking good," Don said encouragingly.

I was pleased that the rock was solid. I was now brimming with confidence. With two good handholds, I stretched my left foot over to the wall below the window and leaned back. Cranking with my left hand, I moved my right hand to a second horn just above my left hand. In one swift motion, I reached up with my left hand, grabbed the edge of the window, and then swung my right hand over next to it.

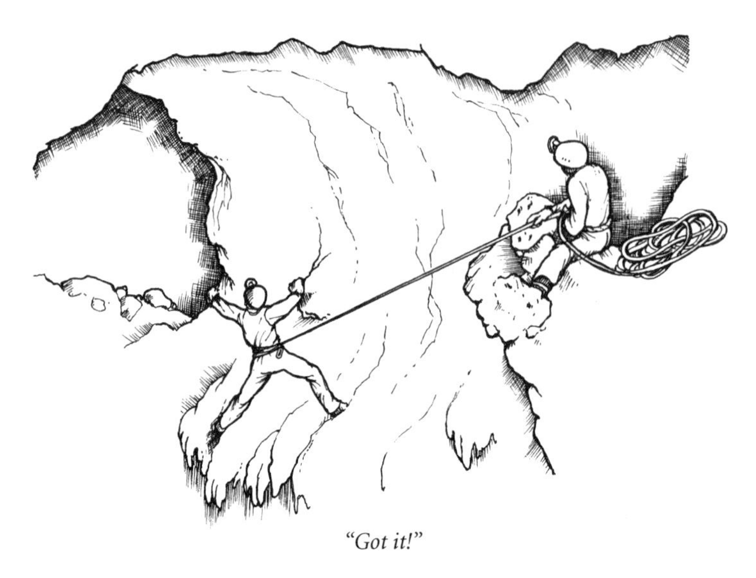

"Got it!"

"Got it!" I shouted.

I pulled myself up and swung my butt around, sitting as I would sit on a fence.

Behind me and below was the darkness of a large room. A wooden ladder led down into it. I described what I saw to an expectant Don.

"That's it! You did it," he said. "You've bypassed Sue's Sorrow!"

"Okay," I answered. "Off belay."

I was secure on the perch where I sat, so I untied the rope from around my waist as Don tied his end in a loop through a hole in the wall. I lowered my end of the rope down the pit beside the old ladder. We would use this line to rappel down.

I then tied into the standing rope while Mark Stock tied into a third rope. He tossed the loose end over to me, and I set up the belay. Mark and Don would repeat the traverse clipped into the standing line with a free carabiner while I belayed them. This was a secure rig, as they would each be protected in two ways from falling. One at a time, they came across and rappelled down into the third pit next to the ancient cedar ladder. Once Mark and Don were down, I untied and rappelled down to join them.

Don gave us a tour of the impressive shaft complex, showing us places where relics of previous explorers littered the floor. He carried the last of our ropes over his shoulder. Our objective lay at the far end of this chain of shafts.

It seemed to me that a large number of people had spent a great deal of time exploring this cave. Most of the more easily reached holes contained rotting remnants of cedar poles that these old-time explorers had used for short climbs.

Don turned off into a lower canyon, chimneying to the floor eight feet below. I could feel the ever-present breeze as I slid the last few feet to the floor.

Fifteen minutes later, we were at our objective, the farthest place Don had explored. We stood at the brink of what might have been the lowest point in this complicated network, the master drain. A steep slope of sandstone boulders led up to the right into blackness. Ahead, the floor dropped away thirty feet to the bottom where a canyon led off—the drain. The climb down the pit looked formidable and, besides, there might be other things to check first. Mark and Don were intrigued by the void above the sandstone slope.

There was no point in all of us checking the same area. I recalled a place fifty feet back where the pit might be narrow enough to chimney down. I knew the best lead would be out the bottom of this big pit, so I focused on that objective.

"Don, I'm going to see if I can climb down to the bottom of this thing while you guys look around up there."

A distant answer. "Fine. Have fun."

I lowered my body down into the crack, flailing my legs for purchase. By pressing the sides of my feet hard against the smooth walls, I was able to brake my slide.

Chimneying down a narrow canyon

I stood at the bottom. Far above my head, I saw two bobbing lights near what had to be the ceiling of the dome. The whistling sound of a falling rock sped my departure into the canyon.

I announced my success. "I'm down! I'll see you later."

Muffled, indecipherable shouts came from above.

I wondered what they said. I shrugged and stepped into the canyon.

A decent lead. I was moving in a canyon about twenty feet high that was a bit too narrow for easy walking but spacious enough. I was surprised to see that I was still following the footsteps of some unknown and determined explorers. I moved swiftly through the canyon, heading downhill, following a small stream. This was definitely the deepest lead in the cave. Down was the direction to the big base-level streams.

I was sweating as I heard the sound of a waterfall in the distance. That quiet roar invigorated me. I picked up my pace, eager to see what lay ahead.

I looked down into a large pit. A thundering waterfall crashed into it. Wonderful! I chimneyed out over the pit for a better view. I spotted a handy ledge that spanned the canyon—a rest stop! I sat down, finally able to catch my breath. While relaxing, I noticed something unusually dark, almost organic. I moved closer and found the remains of a long rope neatly coiled on the ledge. Beyond the rope across the ledge, back into open space, a large, cave-weathered board was wedged between the two walls, ends jammed under ledges on each side. I thought about this awhile. The board was used to guide the rope. Some unlucky caver probably was tied in and lowered to the bottom of the pit, using the board as a jumping-off point. The descent was probably controlled by human strength. I pondered my surroundings, wondering what lay below at the base of the waterfall. What did those old-timers find? What were they looking for to have gone to so much effort?

The canyon continued beyond the drop, but I wanted to report my discovery to Don and Mark, who were probably still at the dome. I had half expected them to catch up with me since I had written off the dome as unpromising. Perhaps they had found something.

Barely a hundred feet from the waterfall, as I headed back to the others, I froze. In the distance, I heard a loud, deep rumble that gave way to a muffled but very distinct roar. A cave-in! My heart raced as I flashed through probable scenarios. We could be trapped. Or, perhaps worse, my companions could have been crushed in a rockfall. Whatever it was, it sounded serious.

Adrenaline pumped through my body. I ran as fast as I could toward the source of the noise.

"Don! Can you hear me?" I shouted.

I stopped to listen. No answer. I ran on.

Another much louder rumble.

"Shit! What's happening!" I yelled.

Wham! My helmet hit an unseen ceiling ledge, and I fell backwards. Darkness. I was dazed but shook it off, fumbling with my light to get it relit. I got up and continued my run.

I yelled.

Again, no answer.

A few minutes later, I reached the base of the pit. I looked up in hope.

At that precise moment, I heard an earsplitting roar above me. Instinctively and nearly without looking, I dove under a ledge just in front of me as a landslide of rocks poured over the lip of wall that I was standing below.

For at least five seconds, the avalanche continued. Shit, they're dead!

Slowly, the rockfall diminished until only separate stones bounced off the wall.

Cautiously, I stuck my head outside the shelter so my shout might be heard above—if there was anyone to hear it.

"Don!" Silence.

Finally, faintly: "Borden! You okay?"

"Yeah! Are you okay?" I had not expected that question from them. But I was relieved to hear any response.

"We're fine up here, just a little preoccupied at the moment. The whole ceiling seems to be coming down. It's just a bunch of loose sandstone—"

A low rumble.

"Shit!"

The deafening roar again filled the room. I ducked back to safety. Tons of rock fell from the ceiling, splintering against the slope and thundering to the bottom, just a few feet from where I hid. Seconds passed before it again became silent. My lamp had been snuffed out by the wind of the avalanche.

Surely that was the last of it, I hoped. I poked my head up around the corner. Satisfied that the way was clear, I scooted back up the slot to the base of the sandstone rubble slope and scrambled along a wall to where Don and Mark should have been.

"Don!"

"Stay clear!" was all I heard before the resounding groan of slowly shifting rock began again. I hunkered down below an overhang as another stone avalanche passed over me.

This one was by far the biggest rockfall yet. It sounded like the entire slope was moving over me, the sound painfully loud. The slide lasted a full ten seconds.

Again, silence.

"We're coming down!"

Five minutes later, the three of us stood at the bottom of the slope, around the corner from any potential danger. We told each other our versions of the events of the last three-quarters of an hour. Once we had separated, Mark and Don had picked their way up the sandstone rubble slope looking for leads in the large shaft. After forty feet, the slope met a ceiling comprised of interlocked sandstone rocks in a mud matrix. It was not encouraging. However, they saw a low crawl that appeared to lead to passage with a solid, limestone ceiling. Don was in the lead, so he climbed into this crawlway, being cautious not to touch the menacing ceiling of loose rock. A wisp of air came from the crawl, so Don called for Mark to follow. As Mark was levering his body into the low space, his head bumped into a rock jutting from the ceiling. The ceiling sagged and the big rock slid slightly and, with a *thump,* settled an inch onto his helmet, knocking his carbide lamp off and sending it clanging down the slope. Mark knew that this rock might be supporting the loose ceiling.

Now in the dark, he managed to work the twenty-pound rock onto his shoulders where he eventually was able to jam it fast between the narrow walls, suspending it over the drop, hoping that the ceiling would not fall. Then, he gingerly crawled from below it towards Don, who had turned around and was lighting his way. For an instant, the rock teetered, then fell, tumbling into the dark below. At that moment, a sickening groan began from the ceiling outside the crawl in the main shaft. Mark rushed in to join Don in his tiny shelter under a solid roof. Small rocks began to rattle out of the ceiling above the entrance of the low crawl. Suddenly, they saw a steady rain of falling rock, like a waterfall, pouring from the unseen ceiling.

After several seconds, the rockfall stopped, but the groaning continued as the ceiling seemed to gather all its energy to spit out its next wave of boulders. It was not until four of these awesome rockfalls had passed that the audible groaning finally subsided. With fear and great care, they had crept out of the crawl and down the slope to safety.

As I described what I had found, Don and Mark turned their attention from the horrors of our recent adventure. When I reached the part of my story about finding the waterfall with the ancient rope, any thoughts of leaving were immediately set aside.

There was the one small problem, Mark's carbide lamp. An ugly gash on his helmet showed where the rock had hit it. The bracket was mangled. I unfastened my pack and fished out a spare lamp and handed it to Mark. The lamp was battered, having had a rough ride in my pack for many trips, but it worked fine. Mark jammed my lamp into his damaged helmet bracket. He had light!

I proudly showed Mark and Don the waterfall I had found, the board wedged between the walls, and the neatly coiled rope on the ledge. I described how the unknown explorers had probably lowered one of their comrades down this pit. Unfortunately, even with our rope, there was no way for us to get down on this day. Most of our vertical gear was hanging on projections at the base of the window. Anyway, the rope was too short. We could continue following the canyon beyond the drop. Maybe we could find an alternate way to the lower level, or the canyon itself might lead to more cave. We chimneyed over the widest part of the pit and continued in the canyon, following the footprints of those determined explorers.

Two hours later, we could follow the canyon no farther.

It had gradually become smaller until we could no longer squeeze through. We had followed a quarter mile of the canyon, never leaving explored cave, never finding a possibility of going down except at the waterfall. We apparently were as resolute as the old-timers, but no luckier.

Somewhat disappointed in our failure to find major going cave, but still hoping that the waterfall pit might provide a surprise on a later trip, we retraced our steps toward the entrance. We stopped at the base of the pit where so many tons of rock had fallen, eight hundred cubic feet by our conservative estimate, in a short set of catastrophic collapses. We warily glanced up into the darkness of what we christened "Avalanche Pit" for signs of more falling rock. Certainly older, and possibly wiser, we emerged into the sultry, summer evening.

Satisfied that there was sufficient cave, Don Coons put together a survey crew to begin a map. On 26 November 1977, with Jeff Ulrich and John Branstetter, he began the survey, choosing the initial survey point some distance into the cave from the entrance. Already, Don had been quizzed by several Park Service officials about the extent of the cave. Not wanting to create an issue, his rationale for a hanging survey was that if the entrance was not marked on the survey, there would be no way to prove that the cave crossed the national park boundary. Don would just "assume" that the cave was totally contained within private lands; there would be no need to involve the Park Service.

They surveyed nine hundred feet through the confusing array of domes and canyons. It was a fine trip. Although they found nothing new, John Branstetter was captured by the cave, and he too shared Don's confidence that this could be the beginning of a significant underground discovery. The cave felt right. With the scent of big cave ahead, it did not take much for John to commit himself to the cave. From then on, it was his project too.

John Branstetter was born and raised in the cave country, so he had been around caves all his life. Far from being a "good old boy," John had studied hard to become a dentist and would shortly return to the town of Horse Cave to set up practice. He was articulate, had a good sense of humor, and was a pleasure to be with. He pursued everything to which he was committed with great persistence and deep passion. In caving, he always intended to make a big discovery. Living in the heart of the Mammoth Cave region gave him the advantage of knowing nearly everything that went on related to caving in the area. He and Don became close friends and caved regularly together.

Given the cave's proximity to the Mammoth Park boundary and the interest in it by the Park Service, Don and John concluded that a low profile—if not a totally secretive one—was the right approach. They would keep the cave to themselves. From then on, all work in the cave was done by Don, John, and Sheri Engler, to the full exclusion of everyone else. They liked it that way.

Sheri Engler was a recent arrival to caving in Kentucky. A transplant from Illinois, Sheri had met Don, recently divorced from Diana Daunt, at a caving club meeting at the University of Illinois at Urbana in 1977. Sheri had begun her caving career around Bloomington, Indiana, and easily became friends with Don. Her bubbly enthusiasm and personal intensity invigorated him. After a few months, Sheri came back to Kentucky with Don, and they were married. Sheri was five feet tall and a strong and earnest caver. She had a special aptitude for talking nonstop about anything and everything, at a loss for words only when tired. However, above all, Sheri could be counted on to put in a hard day's work.

Over the next twelve months, Don, John, and Sheri painstakingly surveyed their way toward the waterfall drop beyond Avalanche Pit. On each trip, they followed the footsteps and scuff marks of the old explorers. They checked and mapped all the leads thoroughly, monitoring their exploration and survey progress relative to the national park boundary just a few feet away. When they were satisfied that everything they had so far surveyed trended away from park property, they finally tied the entrance into the survey net only to discover that at one point they had indeed penetrated beneath the park. However, interest in the cave by the park personnel had waned after months of their low-key, vague reports of narrow canyons, tiny crawlways, and rubble-filled domes. The trio volunteered reports only when pressed. Nobody suspected that the cave might be anything special.

I called Don on several occasions to try to arrange a trip back to this cave that, as far as I knew, had not been visited since the Avalanche Pit episode. Not wanting to usurp what I viewed as Don's cave, I was satisfied by his reply that he was too busy and his promises of some future trip. I was not suspicious but was perplexed by his lack of interest in the cave. Don, Sheri, and John kept their secret well.

Finally, on 2 December 1978, the survey reached the top of the waterfall pit. This had been a goal from the beginning. They began a T Survey—"T for 'thrill,'" Sheri said as John rappelled into the mist of the waterfall spray. Their enthusiasm peaked.

However, Thrill Shaft disappointed the three cavers. Instead of the enormous walking passages they had expected to find, all they discovered was the stream flowing into a small pool—a sump. More poking turned up a canyon lead that intersected the stream beyond the sump. Unfortunately, this passage immediately degenerated into a low, wet belly crawl. A cool breeze hinted of cave beyond. Since this was the only lead out of Thrill Shaft, they resigned themselves to the cold and splashed into the miserable crawl.

The fifty-four-degree water penetrated. Expecting dry cave, they had not dressed for such chilling conditions. After a few feet, the belly crawl mercifully opened to a hands-and-knees crawl, but as they progressed, the water deepened and they were again wet up to their necks.

For over a thousand feet they continued in the grim, wet passage. The mudbanks progressively grew higher and wider, forcing the cavers to squeeze over them to get around the many sharp bends. John's energy and enthusiasm were quickly ebbing as prospects for finding anything down this miserable hole seemed bleak. Don, too, felt the temptation to bag it, but he was determined to push the passage to a definitive end. Don knew that if they quit now, they, or anyone else for that matter, might never return. They pressed on. Sheri was enthusiastic and right on Don's heels. John dutifully zombied along some distance behind them.

The water continued to be neck-deep as Don crawled around one more corner, almost giving in to defeat. He looked up as he had at each bend to evaluate what lay ahead. Ten feet down the crawl, he could see the blackness of vast open space. He surged on in one tremendous wave of brown, muddy water to emerge in an enormous borehole!

Don's echoing whoops were enough to raise John from his stupor. In no time, the three of them jubilantly splashed in a ten-foot-high and thirty-five-foot-wide river passage extending into darkness in both directions. They probed both upstream and downstream, but deep water barred the way after just a few hundred feet in each direction. This river passage was no place to be without wetsuits; Sheri's chattering teeth attested to that fact.

The river discovery was most unexpected. The colossal size of the passage and the amount of flowing water left no other conclusion: this must be the backbone of the Turnhole Basin drainage that Jim Quinlan had predicted to be nearby. Although the existence of such a water flow was no surprise, it had been supposed that it would not be a river with a free surface but a water-filled borehole below the water table. But here it was, a true underground river, open to exploration. A stiff breeze blew downstream as the air from Morrison Cave added to make a wind tunnel.

Reluctantly, they left their new discovery and made the long crawl back to the Morrison Entrance. The trip out of the cave was difficult after the half-mile water crawl without wetsuits. They were exhausted, but their spirits soared. The possibilities of that river were enormous.

What had been the trio's private playground had suddenly become too big for them to handle. The new underground river was extremely significant and would likely take them far underneath Mammoth Cave National Park, as the water apparently flowed to Turnhole Spring, five miles to the west. The surveys showed that most of the cave was remarkably close to the park boundary. The river that Don Coons, Sheri Engler, and John Branstetter had found almost certainly swung beneath park lands. If that were so, they were trespassing. They had to be cautious, but it was clear that they needed help to decipher the vast puzzle they had stumbled upon. They would keep this secret for as long as they possibly could, but they knew they would eventually have to notify both the Cave Research Foundation and the National Park Service of their discovery.

But not just yet.

The temptation to keep it all for themselves was too great to resist. They had to learn exactly where the river lay and how far it extended. After all, perhaps they had found only an isolated segment. It would be bad science to neglect further personal investigations. There was no need to stir up speculation until they were sure.

John Branstetter brooded over their cave's potential while Don and Sheri went to Mexico for their annual caving junket. So far, they had kept their secret and figured that the river would be safe from unscrupulous and rapacious cave poachers until their return in mid-April. Besides, water levels in Kentucky during the winter months would probably be too high for them to explore the low-level river.

On 21 April 1979, the threesome wrestled vertical gear and wetsuits into the cave and headed to Thrill Shaft. One of the last things they had done before Don and Sheri had gone to Mexico was to replace the tenuous tyrolean traverse across the second pit with a prefabricated steel pipe bridge they had built in Don's garage. The tyrolean traverse, where cavers hang on a horizontal rope and push across with legs and feet, was slow and unsafe for tired explorers. The bridge would make the route to Thrill Shaft practical.

They resumed the survey from the top of Thrill Shaft and set stations in the wet and cold flowing drain. This deep water was debilitating. John envied Don and Sheri, who wore wetsuits. He, too, wanted a wetsuit, but his tight budget would not allow the investment. Always resourceful, John planned to make his own wetsuit in the cave. On this trip, he carried a roll of Saran Wrap and a roll of duct tape. At the bottom of Thrill Shaft, he wound himself like an Egyptian mummy in the Saran Wrap. He then wrapped the transparent suit with duct tape to hold it together and to keep the water out. But when he crawled into the river, the water penetrated the clinging film wrap, causing it to relax its cling. Festoons of soggy Saran Wrap loosened and floated away. Water seeped, then poured in to saturate his clothing. Sheri and Don howled at his distress. Which was worse—death from cold water, or dishonor from Saran Wrap? John recovered to tell the tale of his failed experiment.

It took them eight long and chilling hours to chart the twelve hundred feet through the crawl and into the river. Finally, they set Station T42 on a mudbank at the mouth of the crawl. Logsdon River, the name they had all agreed upon, was now finally on their map.

Pleased with their efforts, they abandoned the survey and turned downstream to explore. John was freezing, but they had to see where this passage went.

During the next two hours, the three ran through almost twice as much cave as they had seen in the twenty-one months since their work began. Long waist- to chest-deep pools alternated with extensive stretches of babbling rapids and tall yellow sandbanks. As they continued down this incredible passage, a growing rumble lured them onward. Around a corner, the entire river crashed through a crater in the floor, gurgling away into a low drain. The volume of water nearly filled the hole with a vertically plunging wave moving continuously downward. No easy or safe route down the falls could be found without risking being swept away in the torrent. Don and Sheri had seen similar falls in Mexican caves, but this was a first for Kentucky.

The main trunk loomed darkly ahead. Drawn forward by the vastness, they walked through the now silent passage, sloshing through long pools left by receding floodwaters that occasionally surged through this immense tunnel. They shivered at the thought of being in this low-level cave during times of high water. They walked for over three thousand feet before finally giving in to exhaustion at a spot where a pure white stalagmite and stalactite provided a stunning contrast to the jet black chert of the passage walls and floor. The effect was mesmerizing. They called the formation the Sentinel, as it seemed to guard the cave beyond. The passage continued as large as before, but it was time to turn back. It would be a long haul out. They turned around and slogged back towards T42 and surveyed cave, peeking upstream for a thousand feet. Like the downstream portion, it showed no signs of ending.

Dazed from exhaustion, the tired cavers began the long crawl through cold water to Thrill Shaft. One at a time, they climbed up the thin rope to the top of the waterfall.

Three hours after reaching the top of Thrill Shaft, the wet and tired but very happy cavers squinted in the bright morning sunshine. They were invigorated, basking in the glory of their grand discovery as they walked across the wide field back to their waiting car.

Every caver dreams of the big discovery, the indescribable feeling of being the first to see never-before observed grandeurs of the underground. The urge to share the knowledge of this discovery was great, but the desire to keep the find for themselves was more compelling. They would keep the knowledge of Logsdon River a secret, for now. But surely a little subtle gloating would not hurt. And their resolve to keep their own find secret did not prevent them from wanting to know what others were doing.

Reenergized by the warmth and the memories of the trip, Don, Sheri, and John cleaned up, dressed, and drove the fifteen minutes over to the Austin House on Flint Ridge to visit one of the scheduled CRF expeditions. They did not know that the previous day, Richard Zopf had dangled in the deep pool below the P17 Pit in Proctor Cave.

Don, Sheri, and John were intrigued by the news of Richard's discovery. That river in Proctor had eluded diligent search for a long time, and the fact that the discovery was more or less at the same time as their own was quite amazing. Careful not to let their cat out of the bag, they probed Richard for information.

Which way did the water flow? "Couldn't tell—the water was too deep. It looked ponded."

Was there any airflow? "None that could be detected."

Was it base level? "Yes, it looked like base level, but the open space might just be a migrating shaft."

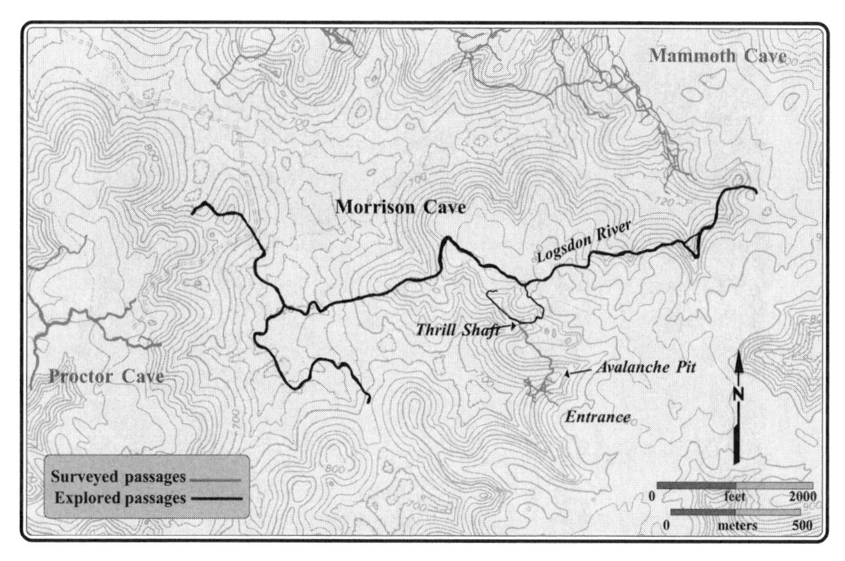

Logsdon River discoveries in Morrison Cave

Richard looked at the trio, perplexed, and finally asked the inevitable. "Why are you guys so interested?"

"Just curious," Don said in his most relaxed tone. Don and John avoided looking at each other. "Quinlan's dye tracing suggests that there are major base-level passages beneath Joppa Ridge just waiting for someone to get lucky. Maybe you finally found one."

"Maybe. But those passages are probably siphoned. If they were there, we probably would have found them by now. We've looked pretty hard for a long time."

"Maybe not," Don said. "Maybe they're not all sumps."

Richard Zopf gave no hint that he sensed that the three were playing with him. "The Green River was pretty high this weekend; we'll have to give it a try next month after the water drops," he said.

"Sounds interesting; we'll try to be here," Don replied.

Richard watched as the three got into John's truck and drove off. Mysterious.

An exhausted Lynn Weller pulling herself over the mud lip of a pit

The Abyss

Roger Brucker and Lynn Weller Drop into Richard Zopf's River

For the next several weeks, I—Roger Brucker—spent sleepless nights thinking about our trip to the P17 Pit in Proctor Cave where Richard Zopf had dangled in deep water. On our next trip, 26 May 1979, I led Lynn Weller and Blu Picard through the Proctor Crawl and up into Proctor Trunk. We had a list of leads to check. Richard, the expedition leader, had warned, "Survey or else! You'll be restricted to one minute of showering for every hundred feet of survey you bring back." We brought along an additional fifty-foot rope to augment the ropes we had left in the cave back in April.

We used the rigging we had left in the pit at P17 on the last trip. On the bottom, we followed the C Survey up the slope to a twelve-foot chimney. A second climb above that led to a horizontal passage about ten feet wide by four feet high. We started a Z Survey and carried it along the passageway for fourteen stations, then arrived at the brink of a forty-foot pit. With a belay we could have traversed a slanting ledge on the left side of the pit to reach an extension of the passage we were in. It looked like a great lead that could go a long way, but I wanted to check the big body of water below instead. I'd trade a four-minute shower for that.

We packed the survey gear and retreated, rappelling to the bottom of the P17 Pit once again. Blu Picard said he would stop here and snooze a little. His vertical gear was not in top shape, and it had slowed him down on the last drop.

Lynn and I squeezed through the low spot and crawled along the narrow slot. We rigged onto the second rope we had left in place, noting Richard Zopf's spare shirt that he had put there as a pad for the rope hanging over the sharp lip of rock. The rope sheath already was frayed from earlier sawing over the lip. Maybe that was why John Barnes didn't come with us this time. Would John tell his friends that *all* Kentucky cavers are unsafe, and masochistic as well?

At the bottom of the twenty-five-foot drop, we walked seventy feet to Richard's mud funnel, the pit leading down to the water. We talked over the plan. Neither Lynn nor I had ever changed from rappel to ascending gear in mid-rope. We decided that Lynn should go first, making sure her Jumars were at the ready in case she, too, needed to yo-yo back up the rope. I could rappel down and untangle her if she got stuck, or maybe pull her up on the rope if necessary.

Lynn leaned back on rappel, walking backwards down the mud slope. There was less water on the floor than before, so things were less slippery. At the lip of the free drop, she looked down into deep blackness. There was no watery pool, but she could see the faint trace of the rope disappearing into the gloom. A strong breeze picked up as she maneuvered over the lip.

She dropped into what seemed to be midnight outdoors on a moonless, starless night. It was a vast place where she had no immediate sense of walls. She stopped and twirled slowly on the rope to get the best look. Below her was a wide river flowing swiftly.

Suddenly, Lynn's heart nearly stopped. She had caught sight of the knot marking the end of the rope six feet below her—and about ten feet from the bottom of the drop. Her heart began to pound.

"Roger!"

"Yes?"

"The rope is too short."

"How much?" I asked.

"Maybe ten feet. I'm coming back." She had to yell to make herself heard above the reverberations of the water.

Lynn engaged in the fight of her life. She unclipped her foot Jumar from its parking place on her seat-sling Jumar and wrapped the loose standing rope around her waist for a safety. Holding the rope with her left hand, she snapped the foot Jumar onto the line, then stood up in the stirrups connected to the clamped Jumar. Next she reached for her chest Jumar. She had to clamp that Jumar over the standing rope so she could set up the inchworm climbing system. Properly rigged, a caver could stand up in the stirrups of the foot Jumar. The upward movement of his trunk would carry the chest Jumar up the rope. He would then sag his weight onto the chest Jumar, which clamps the rope, then draw up his legs to unload weight from the foot Jumar. He would raise the foot Jumar up to the chest Jumar, then stand up again . . . if the rigging was proper. Lynn's wasn't.

Lynn could not get the chest Jumar gate over the rope. With repeated tries, her hand began to shake. She felt the onset of panic. Her thoughts raced in all directions. Could she continue the rappel? She lurched upright once again. The Jumar still would not go onto the rope.

"How are you doing?" I asked. She said it was good to hear a human voice. She was breathing hard.

"I got the bottom Jumar on. Having trouble with the top," she yelled.

She rested momentarily. Then, the Jumar went on properly. Her legs and arms were weak.

"I'm starting up."

Lynn appeared at the lip. She slipped and groped her way up the mud slope. She fell forward on the floor, then picked herself up. She derigged and lay down.

I untied the bottom rope, added about fifteen feet to it, and retied the knot. I hooked my rappel brake bars onto the rope, backed over the lip, and slid to the bottom and into the water.

Briefly I imagined sinking into hip-deep mud, but the footing was firm sand, a characteristic floor of passages in the lowest level of the caves.

"It's big passage with a big river in it," I shouted. The echo was strong, but not as intense as the last time when water had filled the passage. "I'm getting off to look around. Lynn, you come down." The trunk passage was thirty feet wide by twenty-five to thirty feet high; the stream, fifteen feet wide and about a foot deep.

When Lynn reached the bottom, we floated bits of paper and timed them to estimate the stream flow at twenty-five cubic feet per second. We figured that was about sixteen million gallons per day. A big river!

We knew that Blu Picard would not go anywhere until he heard from us, so we set off to see what we had found. We had come down in the south side of the passage, which was rectangular in cross section with smooth walls.

The upstream passage headed to the northeast. It was like wading in a cold trout stream. About forty feet upstream from our landing spot, a side passage veered north, which we followed for two hundred feet. It kept going, but we turned back because we had overstayed our prudent turnaround time. At the main passage, we turned upstream, wading about three hundred feet mostly up to our knees, sometimes up to our waists, seeing blind crayfish and cave blindfish. We stopped where the water widened to fifty feet and extended from wall to wall, becoming much deeper. The passage continued. I guessed it would go a long way.

Downstream, to the southwest, the stream flowed fifty or sixty feet and ducked under a low arch. The stream may have siphoned; we could hear the gurgle and echo of water slapping on the ceiling. The swiftness of the current rushing downstream made it difficult to stand in the streambed. The bottom of the stream was gravel and sandy silt, but I did encounter a couple of surprising knee-deep quicksand pockets. We wanted to look some more, but Lynn reminded me we had signed out for a 2:00 A.M. return time, and it was now midnight.

We named it Hawkins River after park superintendent Amos Hawkins.

Later, Lynn said she felt happy that I was the first to ever see this beautiful river. Of course, she was wrong. Richard Zopf had seen the top of it on the previous trip. And, unknown to any of us, others had seen the same river in another location, at exactly the same time Richard had dunked his butt in it.

Where the water disappears in Morrison Cave

12

Run for the River!

River Adventures and the Brass Ring

Roger Brucker, Lynn Weller, and Blu Picard emerged into the bright morning sunshine, slow-moving but smiling, and trudged their way back to their parked car. They were four hours past their planned return time. On the drive back to Flint Ridge, an approaching driver flagged them down.

"Who the hell is that?" Roger mumbled. He looked at the distant vehicle in the rearview mirror as he slowed down to pull off onto the shoulder.

The backup lights came on and the car drew closer. Roger saw the bearded face of Richard Zopf.

Roger glanced at his wristwatch. The reason for Richard's intercept was evident. "Boy, are we late," Roger confessed.

Roger and Richard climbed out into the fresh air and met on the side of the road. Roger was red-eyed, his faced was caked with mud, and his slouch gave away his exhaustion. The two exchanged a few words about the important river discovery, then they returned to their cars. Before Roger could even start the engine, Richard had turned around in the grass on the side of the road and sped off back to camp.

By the time Roger had pulled his car into camp, Richard was standing in front of a crowd of expedition attendees. Cavers, plates in their hands, were still shoveling breakfast into their mouths. Richard had already announced the hot news of Roger's trip to the expedition: the deep pool Richard had found the previous month was not just a cul-de-sac but a vast underground river.

Richard had always steadfastly maintained that he did not believe in secrets. Big discoveries were something for all to share, and the new underground river was no exception. After all, discoveries were sequentially built on the accomplishments of all cavers who had come before. To keep discoveries secret slapped the face of everyone and achieved nothing. It promoted elitism—those in the know versus everyone else.

Richard's announcement had electrified everyone in camp, and Richard had already assembled two parties to follow up the new leads in Proctor Cave.The plan was bold—as it turned out, too bold. A total of eight people dragged wetsuit bottoms and vertical gear through the Proctor Crawl. It was a strong crew, but the large number of cavers clogged up the route as they all tried to thread their way through the succession of crawls, drops, and canyons to the newly discovered river. Hours were lost in waiting.

Once gathered in the eerie gloom of the large subway tunnel, they were calmed by the gentle murmur of the flowing river. The full weight of the discovery sank in. Jerry Davis, Gail Wagner, Bill Holland, and Walter Mayne made their way between the sandbanks in knee-deep water.

The other survey party of John Branstetter, Don Coons, Sheri Engler, and Tom Gracanin stood watching as moving headlamps revealed the full majesty of the passage. Hundreds of feet away, they could still see the lead party. Reflections of their carbide lamps sent shimmers of light glinting and dancing off the wide river back toward the waiting team at the base of the rope.

Don, Sheri, and John were desperate to know whether this river and their cloaked discovery in Morrison Cave were one and the same.

"Oh, no! The water is too deep!" Gail's disappointment echoed down the passage.

Just past where Roger Brucker's party had halted their exploration the previous day, Gail's party was stopped where the sandbanks were replaced by the inky blackness of wall-to-wall deep water. There was no end in sight as the water stretched into the darkness ahead. Although mentally prepared for almost anything, neither party could risk an indefinitely long swim this far into the cave without the proper equipment. It would take full wetsuits and inner tubes to continue. With just six hundred feet of survey—ridiculously small survey output for two parties—they began their struggle out of the cave. The drops and crawls leading to the dry upper levels of Proctor Cave took several hours for the party of eight. Four hours later, they had completed the Proctor Crawl and headed back to camp.

Back at camp, Richard lamented the waste of so much effort, chiding Roger for what he thought was inadequate reconnaissance. Richard noted in his expedition report summary that, in the future, survey parties should make an effort to provide better information for the planning of follow-up trips. The zeal of discovery was no substitute for well-thought-out party management, and in a cave such as Proctor, this was doubly important.

The Proctor Crawl had always been a formidable obstacle for bringing equipment into the cave. Trips through it were difficult, but the prospect of dragging along full wetsuits and inner tubes in addition to vertical gear and other normal caving gear was daunting, to say the least. But a lot of people were ready to go. Tom Gracanin said it first: the river could lead to the back door of Roppel Cave, which was miles away!

During the week following the Memorial Day expedition, Don Coons and Sheri Engler discussed the situation. They had still not made the CRF aware of their secret river discovery. In their estimation, the two rivers were probably the same. Some questions still remained, however. The water flow they saw in what had been named Hawkins River seemed substantially greater than that in Morrison's Logsdon River. And where they had turned around in the passage that carried Logsdon River, there was no stream. Did Logsdon River disappear and then reappear as Hawkins River? Could there be two rivers? Their minds raced with the possibilities.

The private knowledge of Morrison Cave and its river that Don, Sheri, and John had could not be kept for long. If they delayed revealing the secret, the CRF might swallow up Morrison Cave in an orgy of discovery without even realizing the threesome's prior claim. The CRF had no certain knowledge that there were caves to the east with which Proctor might connect. It was time for the Morrison explorers to act.

The next CRF expedition was scheduled for a full week in July. This was the traditional Independence Day expedition, and many cavers would attend. Pete Lindsley, from Dallas, Texas, the expedition leader, had already announced his intention to push hard in the new river. There was big cave to be found, and he would throw everything he had at it. Clearly, if Don and Sheri were to do anything in Morrison Cave, it had to be done in the few weeks remaining before Pete Lindsley's invasion.

John Branstetter was off for his two-year dental stint with the U.S. Indian Health Service and, to his dismay, could not join the effort in the coming weeks. But the large passage in Morrison Cave now had to be pushed past the Sentinel westward to the presumed connection with Proctor Cave. In the minds of Don, Sheri, and John, that would establish their claim on the eastern extension of Logsdon River, toward Roppel Cave.

Morrison Cave was dangerous because of its potential for flooding and the relatively great amount of rope work. It was too risky for Don and Sheri to go alone. So, Don recruited his old friend Thom Fehrmann as the necessary third caver.

On 28 June, Don, Sheri, and Thom dragged their wetsuits into Morrison Cave en route to Thrill Shaft. The timing would be close; in just two days Pete Lindsley's CRF expedition would begin the attack on the river. This would be the only chance from the Morrison side to connect Logsdon River with Hawkins River. If they failed, they would try to be assigned to the CRF party that might make the connection through Proctor.

The three cavers quickly put on their wetsuits at the base of Thrill Shaft in Morrison Cave and began the long, crawling slog through the T Survey to Logsdon River. This was the third trip through the hog wallow for Don and Sheri. The crawl seemed longer today than it had the first time. Thom thought the crawling would never end.

The walking and wading through Logsdon River to the hole where the river disappeared with a roar invigorated them, and soon they were whooping and hollering, enjoying the echo of the brilliantly resonant passage. Just one hour after leaving the T Survey behind, they reached the Sentinel and were in virgin cave. The thirty-five-foot ceiling swallowed the yellow glow of their carbide lamps, and the immense, black passage was eerie. For the next thousand feet, they walked wide-eyed through a dazzling array of orange flowstone draperies and large rimstone dams that glowed in stark contrast to the black chert. This was caving!

Don, a few feet ahead of Sheri and Thom, ducked his head beneath one especially large flowstone drapery, expecting nothing more than continuing large passage.

"Damn!" he said.

"What's wrong?" Sheri asked, stopped in her tracks.

"Breakdown!" Don stood at the foot of an enormous pile of black, jumbled boulders that extended up into the darkness of the ceiling. Breakdown was the only thing that could prevent a connection to Proctor Cave, and here it was.

The wind continued to blow past them at high velocity. The boulders were clean-washed with no signs of river ponding. A small stream of water was flowing through the pile. Maybe cavers could go through too.

The trio checked hole after hole in the enormous pile of boulders. The breakdown blocks were large and the rockpile was full of voids. They moved through a succession of holes farther into the cave toward Proctor Cave. But, the pile continued. They occasionally popped out into passage fragments, but they either ended abruptly or led back into the main breakdown zone.

They spent hours poking out leads, but there was no way on to the main passage that by all rights should lie below them. They finally called it quits in a large breakdown room where a small waterfall fell down an adjacent dome. High leads continued among the boulders, but the sought-for way down continued to be blocked. As they sat cooling themselves in the large room, the air drifted lazily by, hinting of continuing cave . . . somewhere. But where? The opportunity for their non-CRF connection was ebbing. As each minute passed, their exclusive playground seemed more threatened. The CRF was coming.

Defeated in their main objective and exhausted, they retreated. They would have to take their chances with the CRF expedition the coming week. If they could not make the connection themselves, at least maybe they could join the CRF connection trip. The elation from their discovery of the huge trunk passage was now overcome by their disappointment.

Around noon on Friday, 29 June, the three cavers dragged themselves out of the last pit of Morrison Cave, drove back to their cabin, and crashed into a long, deep sleep.

Saturday dawned. The Austin House at Flint Ridge bustled with activity as crowded cavers ate breakfast. It was nearly July, and it was going to be hot. Everyone would be going underground today, if only to escape the searing midday heat. A special excitement filled the air. Most of the arriving cavers knew about the recent discovery of the river below the P17 Pit in Proctor Cave.

Bleary-eyed and still tired, Don and Sheri wandered into the house. Five seconds after they walked in, a smiling Pete Lindsley marched up to Don and grabbed his right hand in a firm handshake.

"Hey, Don, welcome! I'm glad you could make it." Pete led Don into the expedition leader's office adjoining the main dining room in the Austin House. "I'm putting two parties into Proctor Cave today. I have you and Sheri on one trip and Zopf and Lynn Weller on the other. I'll fill out the rest of the parties later. What do you say?"

Don looked towards Sheri with concern. Pete beamed at them for an answer. There was no way either of them could go into Proctor today; they were too exhausted from their trip into Morrison Cave, and their gear was a sodden mass. But they couldn't admit they were shot, because that would lead to questions about what they had been doing.

Finally, Sheri blurted, "Pete, I sorta have a bad cold and think I should stay on the surface today."

Pete looked at Sheri, assessing this unexpected excuse. "Oh . . . okay, then." He then looked expectantly at Don, who was now on the spot.

"It's been a tough week for me too. I'd like to go caving with Sheri in a day or so. If that's okay, I'd like to pass."

Pete looked confused but didn't push it. "I don't know how you can miss this trip. Suit yourselves."

Twenty minutes later, Pete stood in the door to the kitchen and looked over the mass of cavers. After a pause, he cleared his throat and began his expedition welcome speech.

"Welcome to the July expedition! I want to thank each of you for coming. It's going to be a great week, and we'll have lots of cave for you guys to find. You just have to go look for it . . ."

He covered a standard list of things to do and to avoid, chores that needed to be done around the facilities, and new operational directives. With such a crowd, the need to project brisk efficiency was critical. The disciplined organization style of the CRF gave rise to accusations of bureaucracy and arrogance by offended independent cavers, but unarguably, it got the job done. Long-time attendees usually tuned out this part of the meeting.

Don perked up when Pete's monologue returned to the expedition's objectives at hand.

"This week, we have something special for you. As you know, last month a large underground river was found deep in Proctor Cave." All the cavers now tuned in. "The trips to the river will be difficult, but the discoveries will be magnificent. I think we can accommodate everyone who is able and wants to go. Trips will likely be twenty-four hours, but we have found a way to make things easier for you. We have prepared a map." He pointed to the cork bulletin board where a new map had been posted.

Don strained to see the squiggles from across the room but couldn't make it out.

"This is a map of what you *will find.*"

Laughter filled the room.

"Don't laugh! I'm serious!"

Pete grinned and waited while the chuckling subsided. "Each party can choose ahead of time which discovery they want to make. All you have to do is just mark it on the map, and it's yours. And be careful! We'll be checking to make sure that you explored the correct lead! And, if you are lucky, you will be on the party that makes the connection between Proctor Cave and Mammoth Cave! Now, won't that be something!"

Pete began to announce the trip assignments. Only one trip would go to Hawkins River today. Richard Zopf would lead Roger Brucker, Lynn Weller, and Scooter Hildebolt to continue the survey upstream into and beyond the deep water. They would carry inner tubes, wetsuits, vertical equipment, and other caving gear. They would be loaded going in but could leave the rope and inner tubes for the parties that would travel into Proctor the remainder of the week.

After the meeting broke up and the cavers scattered to get ready for their assigned trips, Don began to study the fanciful map on the wall. From the base of the P17 Pit, the river was drawn leading gracefully to the east, passing waterfalls and large side leads. One lead to the north pointed to Pete's predicted connection with Mammoth Cave. What a dreamer!

Don shook his head at the irony of the bogus map. If Pete only knew how prophetic his joke map probably was!

The trip leaders picked up their cave gate keys, passes, and last-minute instructions from Pete Lindsley. Then they drove off, fanning out across the park. Last in camp was the Hawkins River party. With all the multiple packs and heavy gear that they had to move, they spent a long time preparing.

Roger Brucker pulled out some peculiarly bent metal wires from his pack to show off to everyone.

A jab was hurled from some caver across the parking lot. "Hey Brucker! Whatcha gonna do with those things? Cook hot dogs?"

Roger ignored the insult, explaining that the pieces of bent coat hangers would serve as excellent survey stations in the featureless river passage lined with mudbanks. "You see, you just stick these into the mud and then someone else will be able to find the stations." Roger stuck the end of the wire into the ground beside the car to demonstrate his point. The wire looked like a little golf flag. "I call them 'Highly Visible Flags.'"

By that time, a small crowd had gathered to hear Roger's pseudoscientific lecture about odd little pieces of coat hanger that more closely resembled some lunatic's weapon.

The irreverence continued to issue from the crowd. "Aren't you afraid you're going to stick yourself with those things?"

Roger scowled at their laughter, then replaced the skewers, carefully wrapping his spare shirt around the wires so they did not poke through the sides of the pack.

At last, they were ready, the car loaded with gear and people. As Richard Zopf's party began eight-mile drive to Joppa Ridge, Don Coons stood in the front yard watching them disappear over the hill, leaving a plume of dust behind. Would they find the connection to Morrison Cave?

At this point, Don decided that if there was any chance of putting a lid on the explosive work that had begun on Pete's expedition, he had better let Pete know a little about the goings-on in Morrison Cave.

Don and Sheri found Pete typing a memo for the park superintendent in the expedition leader's office.

Quietly, they told Pete enough about Morrison Cave to make it clear that a connection could be imminent. Since no survey had been completed in Logsdon River, there was no way to be sure. Don's only request was that he and Sheri be allowed to make the connection attempt; Pete could select the rest of the party. Pete immediately suggested Art and Peg Palmer, who were to arrive on Monday, 2 July. The quiet deal made, Don and Sheri headed back to their cabin to get some sleep.

But the CRF assault had begun.

Roger Brucker, Scooter Hildebolt, Lynn Weller, and Richard Zopf slithered through the Proctor Crawl in a long four hours, twice the usual time. The heavy packs tied to their ankles were like boat anchors. Despite the effort, their spirits were high, buoyed by the anticipation of discovery.

Several more hours of fumbling around—descending ropes, moving equipment, putting on wetsuits, inflating the inner tubes—brought them to the pile of rocks marking the end of the survey line in Proctor Cave. They pulled out their brand-new Brunton compass and a one-hundred-foot tape and began the survey, paddling on inner tubes up the underground river. Roger had brought a Mylar notebook in which to record the numbers and make sketches of the survey in the wet passage.

The trip exceeded everyone's expectations. A thousand feet from the beginning of their survey, a major side passage bringing in half the flow of

Hawkins River issued from a low arch twenty-four feet wide. Floating in inner tubes, the party explored fifteen hundred feet of this nearly flooded passage.

After this, back in the main passage, they continued the survey for another fourteen hundred feet to the boulders of a giant breakdown complex blocking the big cross section of trunk passage. Holes led off everywhere, but there was no main route. However, this was not surface breakdown—the limestone rocks were massive and clean. Surely, a way on would be found. But it was late, and the party packed away the survey gear. They walked and floated downstream to the ropes and the climbs back into the upper levels of Proctor Cave. At the top of the first rope, the foursome stashed their wetsuits and inner tubes for use by other explorers later in the week.

The following day, Sunday, Pete Lindsley led the second assault on the river. Pete was fascinated with the potential of the large river tributary Brucker's party had discovered leading to the south, so he signed out for that objective. Don Coons and Sheri Engler, who were lingering around camp to see what was happening, were relieved to learn that Pete did not plan to return to the breakdown pile—by their estimation, the best chance for the connection to Morrison Cave. But Pete did not know that; Don and Sheri had disclosed as little as possible. Why would anyone with such a large lead beckoning to the south go anywhere else?

The connection was still safe!

After Pete and his team of Bob Buecher, Jim Goodbar, and Ron Bridgemon made the long, hot trip through the crawlway, descended the ropes, and floated past the deep water section, they paused to rest at the beginning of the Right Hand Fork. So far, they had made great time and deserved a break.

A gasp from Pete shattered the silence.

"Shit! Where's the damn Mylar?" He dug through his pack, looking for the Mylar notebook that they would need to record survey data in the wet passage. Mylar plastic, which had a rough surface, was about the only material that would stand up to all the water—ordinary notepaper would disintegrate. In desperation, Pete dumped the entire contents of his pack on the sandbank where they were sitting.

"I know I put it in there. Where is it?"

After a few more minutes of fruitless searching, he finally gave up.

"Well, damn," said Pete in final resignation. "We can't very well survey this thing." He gestured towards the low, ominous watery arch of the Right Hand Fork. "We'll have to find something else to do."

After considering the situation briefly, he had a solution. "Zopf said that there were plenty of leads in the breakdown where they quit and that there should be a way through. Let's go there and survey what we can. See if we can get through the collapse."

After Pete returned all his equipment into his emptied pack, they turned to the east, to the end of the Zopf party's survey, and began their work. Pete always carried a spare paper notebook in his pack.

The breakdown was more extensive than they had guessed from Richard Zopf's description. Holes went everywhere! There was no way to be sure of the best way to go. All day, they painstakingly surveyed through what looked to be the best route, never finding the end of the extensive breakdown zone.

Near the end of their allotted time, Pete checked a low crawlway. After a few feet, he emerged into yet another breakdown room. Totally disoriented, he had no idea if he had been heading farther into the cave or back towards Proctor Cave. As he sat down to cool off, he spotted fresh footprints on the floor leading from a hole on the opposite side of the room where he was sitting.

Puzzled, he studied the footprints closely. "Hmm," he said to himself, "I wonder whose footprints those are?" He contemplated following them. Maybe their owner took an easier route than the one he had followed getting here. He shrugged and turned back the way he had come. It was too late in the day to press onward.

After returning, Ron Bridgemon asked Pete, "You've been gone awhile. You find anything?"

"No, but I found some footprints in an odd place." He frowned. "I can't figure out how they got there."

Ron replied, "That's strange; I saw footprints too. I was in virgin cave, then there they were. They must have come from out of the breakdown somewhere."

"They must somehow be from Zopf's party from yesterday. They said they had poked around a lot." Pete looked thoughtful, then continued, "This area is so confusing, how can you even tell which way to go?"

Frustrated at their lack of any big success in getting through the collapse zone, the party packed up and started out. There had been solid accomplishment but no thrilling breakthrough.

On Monday morning, there was no sign of Pete Lindsley's party, but then they were not due back for several more hours. Don Coons was curious about what they had found, but he and Sheri and Art and Peg Palmer were heading to a different objective. As the foursome reached the Proctor Cave En-

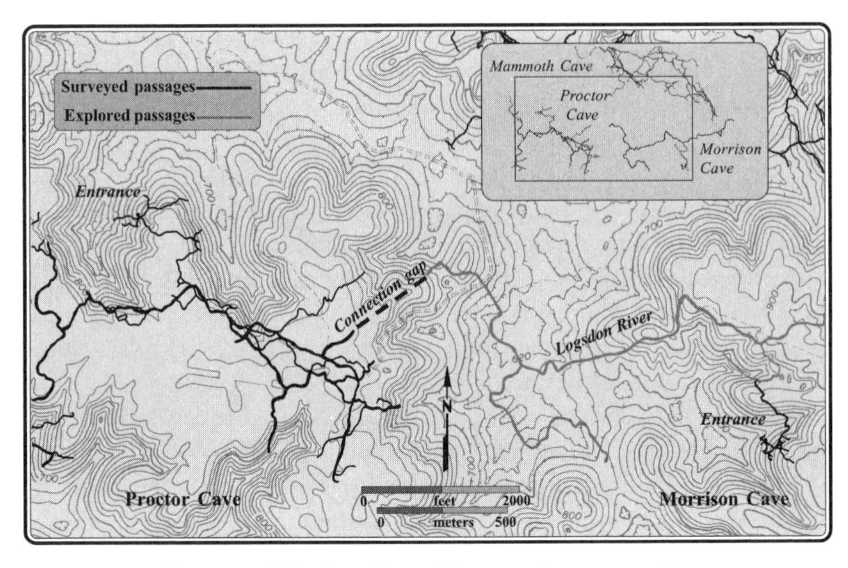

Proctor and Morrison Caves, July 1979, prior to connection

trance after the mile walk through the woods on Joppa Ridge, they found Pete's party emerging from the cave entrance to begin their hot walk back to the car.

Don smiled. "How was your trip? Find much?"

Pete looked exhausted. "We didn't make it to the wet lead—forgot the Mylar notebook."

"So where did you go?"

"Where else? We ended up pushing the breakdown pile that Zopf found."

Don's heart sank. Did they make the connection? He bit his lower lip and asked as calmly as he could, "Find anything?"

"Surveyed about twelve hundred feet. More breakdown, but it still goes, and it's really complicated."

"Sorry about the notebook. I think we're going to continue your survey into the breakdown. It's gotta go somewhere," Don mused.

Pete's party had started up the hill. "Good luck," Pete shouted back. Then they were gone.

The connection was still to be made!

Later, as they were walking through the woods, Pete recalled the mysterious footprints, a little detail he had forgotten to tell Don. He shrugged his shoulders. It probably was not important.

Proctor Cave blew a lot of air in the summer months. The four cavers cooled off in the breeze as they fired up their lamps and cinched their packs.

Clang! The ancient entrance gate slammed shut behind them. This was the trip. Don and Sheri smiled in anticipation of what the day might bring.

Many hours later, the elation of their anticipated connection had crumbled away under the weight of the breakdown pile. Throughout the day, they had surveyed about fifteen hundred feet but, like Pete's party, had failed to find the other end of the breakdown. No connection was apparent. Although the unsurveyed leads seemed abundant, none seemed promising. The day was winding down. They had run their survey into a low side passage that looked like it would be just another of many cutarounds. After a few feet, several small holes led off.

The party split to check for leads. Art Palmer entered a virgin passage that grew smaller. After squeezing past some rocks, he saw a footprint, then another. Then he spotted a large turd! Art puzzled over its origin. Could it have been left by one of the other three in his party? Or by another party? Art's cave sense told him that he had penetrated far into the breakdown and this should be virgin cave. Maybe Pete's party had explored farther than he thought, reaching this spot from some other route. No explanation was apparent, and to ask the others if anyone had taken a crap would be in bad taste, so he resolved to keep silent when he returned.

Don, meanwhile, was crawling in another small passage, trying to follow the elusive breeze. Shortly, he joined another, larger passage that led farther into the breakdown. On the mud floor, he found footprints leading left and right. Footprints! His mind numbed with disorientation. Where did they come from? The set of prints to the right finally ended at an obscure hole that Don followed back to surveyed cave. To the left, the passage was clogged with rocks, the muddy footprints leading upward. Don began to squeeze upward into the rocks. An idea struck. Up! Maybe this was the answer. Their explorations in Morrison Cave several days ago had ended in a high-level room that seemed to be above the breakdown; maybe he should try to go up rather than through the pile.

Renewed by this possibility, he squeezed up between piano-sized boulders and shortly popped into a small chamber. At the far end, a low crawl between boulders led off. Don was still following the trail of footprints. Ahead, he could hear water falling. Connection?

Don raced through the low crawl and emerged high along the wall of a large breakdown room. As his eyes adjusted to the blackness, a wave of elation overcame him. This was the breakdown room containing the waterfall that just a week earlier he had seen in Morrison Cave!

Ten minutes later, Don was looking over Art's shoulder as he sketched in the survey book.

"Find anything?" Don was as deadpan as he could be.

Art looked up. "No. How 'bout you? Any luck?"

"Just a minute. I've got to talk to Sheri." Don huddled and whispered to Sheri. They both broke out in laughter. Sheri jumped up and down.

The pair immediately shared the news that they had found the route to Morrison Cave! The Palmers joined in their revelry, although the importance of the breakthrough was not immediately apparent.

"Good show, you guys!" Peg Palmer exulted, pumping their hands.

"What should we do with the survey?" Art asked.

Don and Sheri exchanged glances, then Don said, "Let's just end it. Come on! We'll give you a little tour." The Palmers didn't know what to expect.

Don and Sheri showed off the sights of the route back toward Morrison Cave in the best running, rinky-dink manner: the Sentinel, the hole where the thundering water of Logsdon River disappeared, the crawl leading to Thrill Shaft, and the wonderful rapids and grand passage that led to points unknown to the east.

Back at the breakdown after this blazing, whirlwind tour, they discussed the significance of the connection.

Prior to this, passages had led beyond the park boundaries, but this information was closely guarded, even within the CRF. Now, here was a passage from a major park cave that not only blasted way beyond the boundary but was also accessible from an entrance outside the park. Although the link couldn't be kept secret, it was important to release the news carefully. Next to the sketch of the breakdown, Art replaced the enthusiastic "GOES!" with "NO GO." All agreed to keep quiet about the link until Don and Sheri could talk privately with Pete Lindsley.

Questions raced through their minds as they made the long slog out of Proctor Cave. How would the National Park Service handle an entrance off the park? Should the news of it be made public at all? It would be up to Pete Lindsley to control the situation.

After emerging from the cave, Don and Sheri approached Pete Lindsley, who was eating a breakfast prepared by Richard Zopf. The pair described the connection and the details of discoveries in Morrison Cave.

"Aha!" shouted Pete, slamming his fist on the table. "I knew something was strange. On my trip the other day, I found footprints. Couldn't figure out where they came from. They were yours. Damn! My trip made the connection and we didn't even know it!"

"Pete, they could've been anyone's," pooh-poohed Don.

Art never mentioned the turd.

"Maybe," Pete admitted, waving Don off. "But tell me more about Morrison Cave!"

Pete was highly enthusiastic as he listened to their tale. Don described the numerous opportunities for a connection to Mammoth Cave.

By Don's reckoning, parts of Logsdon River could be less than a thousand feet from where he supposed the limits of John Wilcox's closely guarded surveys in Cocklebur Avenue lay in Mammoth Cave. He ended by saying that efforts in this area would be more practically served through the Morrison Cave Entrance rather than through the long, arduous route in Proctor Cave.

"Besides," Don added, "what would the park say if they knew the CRF was pushing cave outside the park from Proctor Cave? Politically, I think it would be better for all parties concerned if we worked the river from the Morrison Entrance."

Pete agreed. "Let's do it! Don, I want you to take a party tomorrow. Go to Morrison Cave. Survey towards Mammoth Cave. I'll call Wilcox."

Don's smile disappeared. He was deeply disturbed, as well as exhausted from lack of sleep.

Pete had missed the point about keeping things quiet. "No . . . I don't think so." He paused. "Not right now, anyway. In fact, I think we need to keep everything quiet until we figure out what the next step is. At that point, we can talk."

The sketch in the survey book showed a terminal breakdown where the connection was, and Don wrote a trip report declaring there was no reason for anyone to go back: "All the leads end!"

Pete Lindsley brooded. His grand plan for surveying miles of river had been threatened by Don Coons and Morrison Cave. Pete considered his options: he could defer to Don, but that would effectively shut down operations in the river for the remainder of the week; or he could ignore Don and continue the surveys in the river to the east via the newly discovered connection with Morrison Cave. Pete had high hopes for this expedition. Over the years, he had grown weary of setting up big discoveries during his expeditions only to have all the goodies gobbled up by others before he could return to Kentucky.

Once, on the last trip of an earlier expedition he had run, he peered through a breakdown in the back of New Discovery in Mammoth Cave. The air was rushing through and open blackness lay ahead—a big discovery loomed. As he began to crawl through the hole, trip leader John Wilcox stopped him: "Loose rocks . . . it may be too dangerous." Reluctantly, Pete had backed out.

Wilcox found another hole and squeezed in while the others waited. He found a walking lead that he followed to a room so vast that his flashlight beam would not touch the walls or ceiling.

Their party surveyed the first gooey part but quit at a pit. As they were packing to leave, Pete understood Wilcox to say he didn't know how big the room was. The party left the cave and Pete headed back to Texas.

The next month, the ambiguous size miraculously became clear and Wilcox's party easily moved past the pit to discover the largest passage cross section in Mammoth Cave! They explored and surveyed several thousand feet of it. Pete had felt cheated. Now, for once, the big discovery had at last been made before his expedition. He had at his command the strong cavers and resources, and he was finally free to show everyone how big caving could be done for an entire week.

In what seemed a continuation of his bad luck, the most recent turn of events was now *too* good and his plan to survey miles of cave was at risk.

Pete was certain of one thing: he should proceed with the exploration. Don had seemed unyielding in his determination to keep the river in Morrison Cave to himself. After all, Don had turned down Pete's offer of the CRF's help and did not want anyone to continue exploring through the connection unless he was involved. Pete made phone calls for counsel.

First, he called Roger Brucker. As chief instigator of the plan to push the river, Roger could certainly make an informed recommendation. Roger, expectedly, gave his support to Pete. "Go for it now. I would." The sensitivities of a couple of secret cavers was no reason to stop exploration, Roger declared. Pete then called John Wilcox and quizzed him about the lead in Cocklebur Avenue in Mammoth Cave. The two of them decided that the lead probably did not connect to Logsdon River—Cocklebur was at a higher level than any base-level river. Was Wilcox available to join the expedition to help connect Morrison-Proctor Cave with Mammoth Cave? No, Wilcox had other plans.

Dauntlessly now, Pete decided to press ahead in the river with the full resources of a mighty expedition. Don and Sheri would be pissed, but they had had their chance.

On 4 July, Pete Lindsley assembled a strong party led by Richard Zopf to survey through the Morrison-Proctor Cave connection. Pete ordered Richard to extend the survey line east as far as they could possibly could. They had to know where this river went! Richard, always a team player, was going to follow these instructions to the letter—the line was the most important, the complex sketching would come later. In what was forever to be labeled the "Wall-less Sketch Trip," Richard Zopf, Tom Brucker, Lynn Weller, and Walter Mayne surveyed over a mile of cave.

Upon viewing the survey book when the party returned the next day, Pete expressed shock to find that Richard had sketched only passage detail that he could readily observe—in fact, there was an entire room with no walls. "What have you done? This sketch is terrible!"

Richard protested vehemently, "I couldn't see the walls! How could you expect me to draw the walls?"

Pete rolled his eyes.

"Besides, you said you wanted us to survey as far east as possible!"

Howls of disgust roared through the Austin House when the cavers heard what they regarded as a self-serving alibi to justify scooping a mile of cave.

Realizing the futility of his argument, Richard gave up. His protests died, drowned in the laughter and derision.

Indeed, Richard's survey was one big chunk of cave. The CRF had effectively annexed Morrison Cave. Already, the CRF had done more survey in Morrison Cave than Don Coons and his companions had. No matter that their effort in Morrison had been a heroic struggle. They had pushed sur-

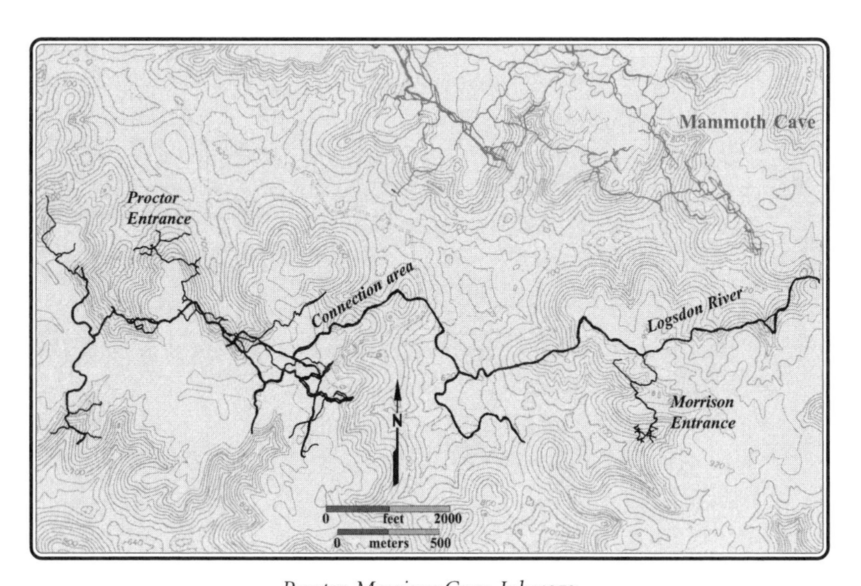

Proctor-Morrison Cave, July 1979

veys through tight and twisting passages while the CRF's survey had been an easy glide accomplished in just one trip. The deed was done. Pete thought about calling Don and letting him know what was going on, but Don and Sheri's small cabin had no phone.

Later that day, in a chance encounter at the local grocery store in Cave City, Don and Sheri ran into a CRF caver who told them of trips to the river and the discoveries of gigantic trunk passages. The caver was fuzzy on the details, but the fundamental fact was obvious: Pete had disregarded their wishes and had continued the survey into their cave.

Enraged, they sped to Flint Ridge to find Pete Lindsley. They fumed and sputtered about broken promises, deceit, and lost faith. But the damage was done. Pete gave a diplomatic shrug, sympathetic to their position to a point, but defended his actions of the last few days. "I would do the same thing again," was his last word on the subject.

The CRF's march east had begun.

A low, wet lead requiring checking by a nude dunk

13

The French Connection

A Nude Swim and a Farewell Send-Off

One thing was clear to me—Roger Brucker—in early August 1979 after Pete Lindsley's expedition: Logsdon River in Proctor-Morrison Cave was flowing generally from east to west less than two thousand feet south of Mammoth Cave, which is oriented from southeast to northwest. The two caves formed a giant upside-down T on the map with only the junction missing. Tom Gracanin, back in Ohio at Scooter Hildebolt's CRF map factory, had pointedly said, "Roger, we can connect Mammoth Cave and Proctor anytime right here." He was pointing to a map. John Wilcox had come to the same conclusion. I had too. Since I was familiar with Logsdon River from participating in the rapid succession of surveys, I had already narrowed the probable reach in the river where the caves might intersect. I was so sure of my dead reckoning that a few weeks before I had wired a CRF permanent station poker chip at survey station Z4 on the Logsdon River passage wall near some ceiling domes where I thought the connection was.

Would anyone try to scoop the connection—just go into the cave, find the right hole, and connect the caves? Not likely. Don Coons, who had proved himself capable of secret work, had been forthcoming since the July CRF expedition. He and Sheri Engler had graciously swallowed most of their disappointment and had cooperated in the ongoing work. Also, Don knew little or nothing of the lay of the cave south of the Frozen Niagara Entrance to Mammoth Cave. The Frozen Niagara Entrance to the closest point in the river measured three thousand feet.

Wilcox, who spearheaded the exploration and survey in the East Bransford area of Mammoth Cave, had concealed the compass and pace survey results and the few newly drafted maps. With the CRF board of directors' approval, he had never commented publicly about cave passages extending outside the park. We further believed the National Park Service would not be interested in publishing the location of off-park caves. At best, it could lead to an enterprise like George Morrison's New Entrance to Mammoth Cave; at worst, to another Floyd Collins disaster. There was ample historic precedent for withholding the information.

As for the Roppel cavers, whose paranoid insecurity seemed boundless, what could they do about it? They already were highly upset about our explorations in upstream Logsdon River. A connection to Mammoth would probably upset them even more, because that would make the connection to Roppel Cave all the more likely. They wanted to keep Roppel Cave separate from Mammoth Cave.

On the other hand, from the CRF's point of view, a connection of Morrison with Mammoth Cave would have no practical value in shortening the trip to Logsdon River. To get to the end of Logsdon River from the Morrison Cave Entrance took about four hours and thirty minutes; it could only take longer starting from the Frozen Niagara Entrance to Mammoth Cave (assuming a connection were found), and the trip would likely be six or eight hours. Anyway, hadn't we said to the Park Service, the CKKC, and anyone else who asked that the CRF will follow the cave passages wherever they go?

Then there were the practical problems of probability and realism. I had been involved in more work leading to more Kentucky cave connections than anyone. They are extraordinarily difficult to find and are usually the result of the work of hundreds of cavers over many years. Nevertheless, there were some guidelines that could be followed: First, the closer the passages are to the river or base level, the greater the likelihood of finding a connection route unobstructed by breakdown. Second, the stronger the airflow, the better. Third, a short distance between caves is utterly unreliable as an indicator of connection probability. Wilcox knew this as well as or better than I did. Tom Gracanin had not lived long enough to know it, but he was plenty smart and probably would figure it out on his own.

We would try a connection between Mammoth Cave and Morrison-Proctor Cave on 11 August.

Small party, short notice. John Wilcox, Tom Gracanin, and Lynn Weller said they'd go. I telephoned my son, Tom Brucker, and without telling him too much, I promised him a great trip. He said he might or might not come. With no expedition scheduled that weekend, we'd have to plan our own food; in addition, we'd have no backup in case we had an accident.

The arrangement was for Wilcox and Lynn to meet me in a Lebanon, Ohio, supermarket parking lot at 8:30 P.M., Friday, 10 August. We'd buy food and drive to Cincinnati to pick up Tom Gracanin at his parents' house.

Gracanin had made a small-scale map of where we were headed. It was an overlay of the topographic map, so we could assess the likelihood of encountering breakdowns. Wilcox studied the map. "It's one thousand feet to eight hundred feet away," he guessed.

I was surprised when Tom Brucker arrived. "You didn't tell me not to come," he said brightly. He had not returned my call when I had four people in the party, so I assumed he would not join us. But we were all pleased to have him along.

After breakfast, we piled into John Wilcox's car and drove to Jim Wiggins's house in the park housing compound. Wiggins was the assistant superintendent, and he gave us the key to Mammoth Cave. We told him if we weren't back by 2:00 P.M. Sunday to initiate the CRF emergency callout procedure.

Inside the Frozen Niagara Entrance to Mammoth Cave, we walked to College Heights Avenue, ducked under the tourist trail metal stairway, and started the journey through a succession of canyons and crawlways off the visitors' route. We went through Fox Avenue through the A, F, N, O, S, and A Surveys, scrambling for a couple of hours through these small passages that Wilcox, Lynn, and others had mapped. We came to the far end of the Third A Survey to a place where a sandbank sloped down to a wall-to-wall pool of water that extended off into the dark distance under a low arched ceiling.

We retreated and headed for the first side lead to the south. We continued a B Survey as far as a pit in the floor at the end of it. Tom Brucker went to the bottom of the pit. The drains were too small and he could not fit into them. (Rats! That was the one I was *sure* would go.) We reserved the second pit for later.

We went back to survey station B5 for lunch and changed carbide. Tom's pack had developed a rip. Lynn produced a needle and heavy thread from somewhere, and Tom sewed his pack.

Wilcox and Gracanin finished eating before the rest of us and said they'd go back and start exploring the pool at the end of the Third A Survey, which was now the best lead we had left. We would follow them.

The two viewed the pool of water with some dismay. No one had brought wetsuits; this was a lightning-fast commando trip. If they got their clothes wet, it would make for an energy-sapping trip out of the cave. So they stripped off all their clothes, buckled on their knee-crawlers and hard hats, stuffed one baby bottle of carbide and one water bottle in a pack, and set off.

They waded, then crouched, then crawled several hundred feet on hands and knees in water. The bottom was sandy mud, firm, not the barnyard consistency they had feared. They came to where the ceiling dipped to within four inches of the water surface, with twelve inches of water beneath. They were prone, struggling to keep from slipping into deeper water as they held their mouths against the wet ceiling. Keep the lamps dry. Don't slip on the slope. They were moving sideways. "Don't make waves!"

Their watery squeeze ended when the ceiling lifted a little. Now they crawled on hands and knees again. Since they were excited and moving fast, they kept warm enough not to complain.

The pair reached a T-intersection. A fresh stream flowed from north to south. Wilcox said later he knew at that moment that they would connect to Logsdon River through the south-flowing stream.

They raced downstream. Enjoying greater headroom now, they could move fast, almost at a run. Farther along they heard a roaring sound. The ceiling raised a little. They climbed out of the stream onto sandbanks and went up one bank to a low squeeze where they faced blowing air and vastness. They scrambled down the other side of the sandbank. They had found it!

Logsdon River flowed from left to right, northeast to southwest, in a magnificent trunk passageway twenty-five feet wide by fifteen feet high. They had connected Mammoth Cave with Proctor-Morrison Cave, coming in at the ceiling through an inconspicuous hole that would have been hidden in shadow from speedy explorers wading up the river. However, just the previous weekend, the tiny lead had been noted in a survey book for checking later by one sharp-eyed note-taker.

After a quick reconnoiter, they located survey station Z6, then Z7. It was 4:00 P.M. In marked contrast with the route from Mammoth Cave, the Logsdon River passage reeked of sewage, and they were growing cold despite their excitement.

On the way back through low water passage, which they were now calling the U-Tube, Wilcox smoked a carbide arrow on the ceiling at the T-intersection. Coming from the Logsdon River end, an explorer would probably miss the right way and continue crawling in the water. Wilcox made a note that the upstream watercourse should be surveyed; he thought it might lead to a vertical shaft. He counted body lengths to keep track of the estimated distance between passage bends. He tallied 975 feet between Z7 and the end of the A survey station at the edge of the U-Tube. When I led a trip to survey the connection the following weekend, we measured the passage at 850 feet.

Lynn, Tom, and I waited on the comfortable sandbank at the last A Survey station. The piled dry clothes told us Wilcox and Gracanin must have proceeded in the buff, an unconventional mode of caving in Mammoth Cave. We reasoned that if they came to an end, they would be back soon. If they came to Logsdon River, they'd also return soon. No matter what, without clothes they'd freeze their butts and come back out of necessity. None of us really wanted to get wet.

After thirty minutes with no sound, we speculated, "Why get wet if we don't need to? They may find a dry way around all this water."

"They must have found something big; they've been away forty minutes," Lynn said. She had mentally prepared herself to enter the water. She had worn wool underwear just in case the U-Tube required immersion. I wore polypropylene underwear, so I would have gone ahead, too. Tom counseled waiting a few more minutes.

"Lights!" The faint word floated toward us over the water. We looked down low in the water and saw two headlamps reflected there, then we heard their splashing. Wilcox and Gracanin slithered toward us, pink bodies glistening wet in the mud-colored cave, then they stood up, dripping wet, and waded toward us. The shock of all that flesh in a cave made me blink. Most shocking was the colorful red bandanna tied around Tom Gracanin's neck. How French! The passage was called the French Connection from that moment on.

"It's 975 feet to Logsdon River," John Wilcox said. As they dressed, they described the connection. We made them go over it again.

"It's all one cave now—212 miles of it," Tom Gracanin said.

The day was still young. We dry cavers briefly considered going through the water to have a look at the connection, but Tom Brucker said: "No point going through the water if we can find another way in."

We returned to the unexplored pit, and Tom rappelled into it. Its drain was two inches high.

Next, we went to survey station B5 to survey some passages Wilcox knew about. We carried a C Survey up several climbs and canyons into a horizontal passage at the top, about forty feet above the lower passage, and followed it south, surveying to Station C34. There we found a breakdown, but there were upper-level leads and an extension back to the north.

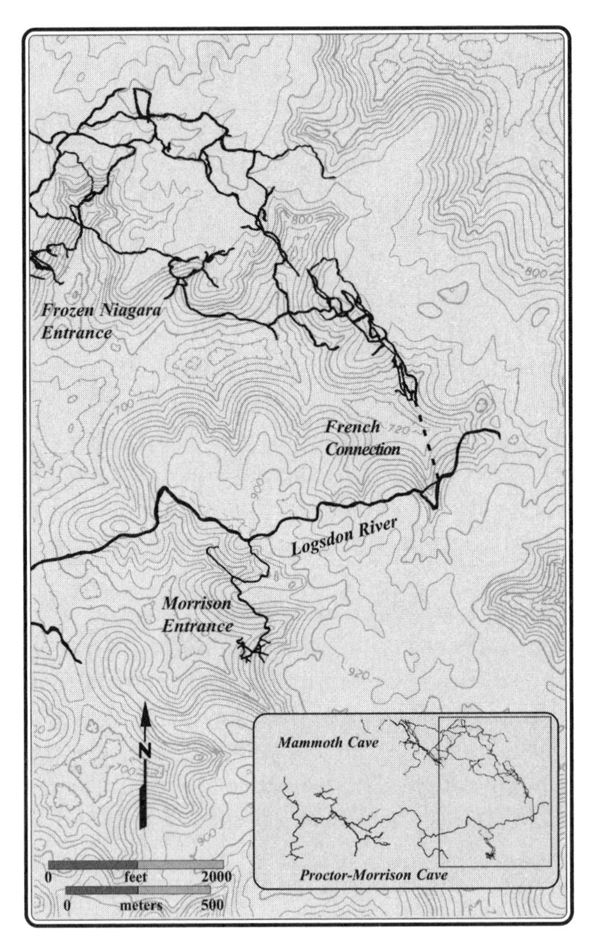

Proctor-Mammoth Cave connection, August 1979

We packed the survey gear, with a major cave connection and twelve hundred feet of new survey to show for our efforts. Wilcox said he'd show us some sights to "make our trip worthwhile." Back at Station C15, an unsurveyed passageway to the north led to some beautiful crystal needles and gypsum snow. There was plenty of passage yet to survey.

Since it was a long way out of the cave, we started back toward the entrance, moving almost automatically. Occasionally, we rested and snoozed. We left the cave through the Frozen Niagara Entrance at quarter past seven on Sunday morning, 12 August. I told everyone to stop and sit down on the road so that we could get a self-timed photo of the connection party. Gracanin asked if I wanted photos dressed and undressed.

We agreed on the way back to our Flint Ridge camp to call Amos Hawkins, the park superintendent. By protocol, he should be the first to hear the news. Inadvertently, I dialed Jim Quinlan instead.

"Did you connect?" he asked.

"No, no," I lied.

I dialed again, this time the right number.

The phone rang many times before a sleepy Amos Hawkins answered.

"This is Roger. We have something extremely important to tell you. Can you and Ethyl come over to Flint Ridge for breakfast?" Hawkins and his wife said they'd come right over.

We started cooking breakfast, including the steaks I had brought for the occasion. Joe Kulesza, the former superintendent of Mammoth Cave National Park, phoned us to discuss other matters. "Come on out!" I said.

Soon the Hawkinses arrived. We told them we had made the connection between Mammoth Cave and Proctor Cave.

John Wilcox and Tom Gracanin told their immersion-connection story several times. There was the suggestion that I should have taken blackmail photos of the nude returning heroes.

The discussion turned serious. We talked about the need and timing for telling Robert Deskins, the new superintendent who soon would be coming to the park to replace Hawkins. We proposed keeping the discovery quiet until we had surveyed the connection route and prepared some background materials. We wanted to be sure of what we had before announcing it. Amos Hawkins agreed and extended his cooperation.

Big passages in Roppel Cave

14

Fortuitous Intersections

Roppel Cave Explodes North—Again

Meanwhile, at the far reaches of Roppel Cave, four miles from Mammoth Cave, Don Coons and I—Jim Borden—stood looking across a pool that stretched out of sight around the corner. Infinity. That was the feeling we had as we considered where this passage would lead. We were far below the floor of the deepest surface valleys, standing in a corridor with walls twenty-five feet apart that stretched sixty feet up to a dark ceiling. The black walls swallowed the light of our carbide lamps. The floor was smooth and flat, strangely unlittered with breakdown. In a passage of this size, there usually are large fallen rocks to traverse, but not here. Don and I had sprinted the last two thousand feet, feeling out of place and out of time.

Don pulled out the compass he had tucked inside his shirt. He sighted through the eyepiece, measuring the trend of this enormous canyon passage.

"Three hundred degrees. Straight northwest, right toward the Green River."

I considered the possibilities. We were deeper than we had ever been in Roppel Cave. No longer under the sprawling sandstone-covered mass of Toohey Ridge, we had penetrated below the floor level of the deep karst valleys that separate the major ridges of the Mammoth Cave Plateau. We were at least one hundred feet below the deepest valley. There was no possibility of intersecting the surface, the usual termination of the dry upper-level passages that we had been exploring for the last year beneath Toohey Ridge. The opportunities at such a low level in the cave were boundless.

"Don, the Green River is five miles away. Do you think this passage goes all the way there?"

"I don't know about that. But Northtown Cave isn't too far away. I bet we can get pretty close to connecting with it."

Northtown Cave's passages had been explored over the last few years by Don Coons and others. The cave was horrible but blew a windstorm. They had mapped just under three miles. I calculated the distance. The image of the topography was clear in my head.

"I figure it's about a mile to Northtown Cave," I said. "There's no way this passage can stop."

I turned back toward the pool, taking in a large breath and bellowing into the unknown. We heard a long succession of echoes that trailed away into an eerie silence. In movies, caves always echo like a large tiled bathroom. Real caves seldom produce echoes unless conditions are just right. We repeated the echo experiment with different toned hoots. A staccato burst of three tones would return a sustained musical chord. Such perfect acoustical conditions were almost always caused by extensive areas of ponded water that formed a reflective surface for sound waves. We resisted the temptation to continue in this wonderful passage. In fact, it almost seemed anticlimactic. We knew that the ultimate length of this passage would exceed anything we could explore today. It was very late and a long way to the surface. We turned heels and reached the top after twenty-four hours underground.

A few months earlier, in August 1979, two survey parties had retraced the explorations of Ron Gariepy and Bill Eidson in the North Crouchway. With me leading one party and Jim Currens the other, we orchestrated a leapfrog survey north into the newly found passages. The purpose was to maximize surveying productivity. One party commences a survey while the other advances an arbitrary number of stations, such as twenty bends in the passage. At that point, the second party starts its survey. When the first party joins its survey to the second party's initial point, the first party leapfrogs the second party by counting twenty bends to start a third segment of survey.

Both Jim and I knew how to count during a leapfrog survey so as to get to survey the most desirable cave, so we tried to position ourselves most favorably for the best segment. I won this one. Pete Crecelius, Geary Schindel, and I surveyed a continuous string of Z Survey stations through over a mile of cave. Jim Currens, Bill Walter, and Chris Welsh cleared just under three thousand feet. They ran out of cave at the breakdown that had halted the original explorers. Still, not bad. Our two parties had logged over eight thousand feet of passage, all of it the low, wide elliptical tubes that were characteristic of this northern area of the cave. These passages were dry, often of crawling height, and more often than not floored with sharp-edged rimstone that wears out cavers. Tough cave, but nevertheless a fine discovery. Unfortunately, however, all these passages ended in ugly, wet sandstone collapse under the flanks of Toohey Ridge. Leads were scarce, and what leads we did find did not encourage us.

When Jim's party reached the breakdown that had halted Bill Eidson and Ron Gariepy, they were dismayed by having run out of cave. Although Bill Walter had been able to squeeze through an unbelievably tiny hole to find a continuation of the main passage, no one else in the party could get through. The lead would have to wait. On the way back, they shoved into every likely hole, looking for the discovery that eluded them. Vertical shafts with falling water hinted of lower levels, but the shafts' unreachable floors were full of rock, blocking any drains. The group was soon back in the main Z Survey line where a low parallel crouchway led off to the east, blowing air. Since it was close and paralleled the passage they had been surveying earlier in the day, it did not seem promising; it would likely lead to the same breakdown that had interrupted their previous passage.

Resolute, they began a D Survey into this side passage, drumming up their reserve energies. Within a few stations, they were crawling along on their bellies, wheezing in the dry dust kicked up from the floor. Every now and then they could rise onto their hands and knees, but only briefly. Soon, Bill Walter smoked "D10" on the wall of a larger canyon that cut across the crawl. The tired cavers ended their survey at this point; it was a good place to quit and begin the long slog out to the surface. Bill, never satisfied, looked briefly in each direction. Lots of sand fill one way, dripping water the other; both ways went on.

My party was sitting near the entrance of the D Survey side passage when Jim's party emerged.

"Anything?" I asked.

"Well, sorta," Jim answered. "We followed this passage for ten stations to an intersection with a large canyon."

I perked up. "A canyon?"

"Looks like it is heading to shafts."

"Oh." I shrugged back down.

Shafts often spelled doom for horizontal passages like these. From our experience, shafts that provided a route to more cave tended to be the exception rather than the rule, contrary to local caving lore.

Disappointment. We discussed what we had found, agreeing that although nothing stopped, nothing was very exciting, either. We had explored nearly two miles of cave out here, and it seemed to be just a big bust. A route to nowhere. We sat there lamenting our bad fortune. Poisoned in spirit by our negativity now and dead tired, we began the long crawl back to the surface.

A few weeks later, Pete Crecelius, Bill Walter, and I returned with a newly recruited horde of Canadian cavers who had traveled south from McMaster University in Hamilton, Ontario. They had rented an enormous van, taken out most of the seats, and crammed eight people and all their gear inside—geology students. We were also joined by Miles Drake, one of the D.C. cavers I had recruited. We were determined to make this unpromising part of the cave go somewhere.

Pete Crecelius was the newest hard-core caver addition to the CKKC. He was a blessing to our efforts in Toohey Ridge. Bill Walter had introduced him to the caves of Toohey Ridge, having met him while surveying in the caves north of the Green River in Mammoth Cave National Park. Pete was driven by the same greed for discovery that drove the rest of us. Probably it was this obsession that also made him one of our most productive contributors at Toohey Ridge.

Soon we were back in the cave again, at the end of the North Crouchway through which only Bill Walter had previously been able to pass. An energetic bout of breakdown hammering by Pete enlarged the tight spot so the rest of us could get through. Our large parties measured out fifteen hundred feet of cave survey, but once again nothing went. There was just more mud and breakdown as the upper-level tube ran to an end in breakdown and fill beneath the surface of a deep valley.

Several hours later, we arrived at the entrance of the D Survey, sweating after the strenuous three-quarter-mile duck-walk in the North Crouchway.

I was spent, and the Canadians were drooping. Two-thirds of the party were routed and aching to get back to the surface and home. Miles, Bill, and Pete, still energetic, decided they might as well look at the end of the D Survey. Bill recalled that it looked as though shafts were just ahead, but he knew it would be a long time before anyone would return to what seemed to be marginal leads at best.

I gathered my newly formed party of the weary souls and departed. Remaining behind, Miles, Bill, and Pete refilled their carbide lamps and set aside cave packs and flashlights. They would be minimally equipped for maximum speed, and their short supply of light would almost guarantee discovery of something, according to superstition.

With a wry grin, Pete queried his party. "Well, do we have our three sources of light?"

"Of course," Bill answered in his most serious tone. "We have yours, that's one; Miles's, that's two; and mine—that makes three."

"Yup, three sources," Pete agreed.

Miles, wide-eyed in mock horror, said, "Boy, you guys are really unsafe. I might have to report you to the NSS!"

All three of them laughed.

A primary caving rule is that each explorer shall carry three sources of light. No, they were not following the rule; but they would all stay together (most of the time), and they were seasoned veterans who (usually) left little to chance. They were taking a risk, but a risk was sometimes what it took to make that big discovery.

Well-rested now, Pete followed the D Survey—Kris Krawl, Jim Currens had christened it. Ten stations, to reach the end of the survey, should not be far.

Pete crawled through the low, wide tube, slowly passing the sooted station numbers. Ten stations *was* far. Finally he saw D10 labeled on the distant wall. However, between him and the wall lay a deep canyon that cut across the passage they had been traveling.

"Hey guys! Take a look at this!" he yelled.

The canyon was far larger than he had imagined. He climbed down into it, feet pressed against each wall for support. There was no need to climb all the way down; there were plenty of ledges to move along at the widest level. The canyon was still quite deep below them. Miles struggled to keep up, but Bill was right on Pete's heels.

Pete paused at one particularly wide spot to make sure no one needed assistance. Rapidly dripping water—which Bill had heard on the previous trip—had dissolved away the ledge along one wall. Getting past the obstacle required a balancing move against gravity as the slippery walls afforded practically no purchase. Pete reached out a hand to the others at just the right moment. This was no place to have an accident by letting pride get in the way of common sense. Bill and Miles made the scary leap across the void to the ledge on the far side.

They moved east down the open canyon. For the last several hundred feet, both the ledges they were following, as well as the ceiling, were gradually descending. The canyon was actually a series of bell-shaped rooms connected by narrower areas. On the map, it looked like beads on a necklace.

After one particularly difficult climb down where the bell-shaped rooms descended even more steeply, they found a large room. There was no apparent way on; this seemed to be the last bead in the chain. The main way filled around the next corner. However, this was not the end; a number of small holes led down through the floor. But reality began to catch up with the party. They were very far from the entrance—and from their packs.

They sat down and discussed their options. Was it better to go on, or should they save these leads for a later trip? They had already traversed the equivalent of a full day's survey to this point from Station D10 at the end of Kris Krawl. Only greedy cave rapers who would steal caving fruits without paying the survey price would go farther. But how could they stop? Was the way on worthwhile? Was it even possible?

"Hey, quiet!" Pete said.

They froze.

"Listen. Do you hear that?"

They strained their ears. After a few seconds, they heard a distant sound. Far below-and above the sound of their own beating hearts—they heard the telltale rumble of running water.

"Its a stream! Let's go!"

They climbed rapidly down into the narrow, sinuous canyons beneath the large room and squeezed around tight corners, each following a different route. In short order and more or less at the same time, they met on the same ledge, their feet dangling over open space.

"Wow! Take a look at this!" Miles said, his mouth agape.

A wide passage with black walls led off below. A glistening stream flowed through a broad and deep canyon in the floor.

Pete had already climbed most of the way to the floor twenty feet below. Miles was close behind. They moved downstream.

After just two hundred feet, an immense wall of breakdown reached from floor to ceiling, apparently blocking this wonderful passage.

"Shit!" Pete yelled.

The last thing expected this low in the system was breakdown. They had descended at least a hundred feet since D10, far below the lowest reaches of any valley wall. They stood in disappointment and disbelief.

Finally, Bill said, "Pete look at this." He pointed at the wall of rocks. "This isn't sandstone; it's limestone."

Pete inspected the pile. "Hey, you're right. I don't know what we have, but it isn't a surface break! We can find our way though this."

He picked his way up the pile, spirits lifted, looking for openings. Bill and Miles sat exhausted at the foot of the pile.

Pete paused, looking back at the party hopefully.

"You go on. We'll wait here," Bill said. "See if you can make this thing go."

Pete climbed up to the ceiling where he saw a likely looking gap behind a large, flat piece of rock leaning against the left wall. He slid in on his side, his chest facing the wall, his left shoulder against the slab. He dragged his body along by grabbing doorknob-shaped chunks of chert on the main wall. Sharp rocks on the floor poked into his ribs. After twenty feet, he reached a closet-sized room with a small, triangular-shaped hole at the far end. Through it was blackness. He crammed his body through the narrow slot, nearly falling headfirst into the void.

As his eyes adjusted to the black gloom, Pete examined his surroundings. Enormous black slabs of rock lay everywhere. The ceiling, floor, and both walls consisted of these rocks—a breakdown chamber. This was a large void, but he was obviously still in the main breakdown zone.

The only way on was through a low, horizontal squeeze between two flat slabs of rocks. He belly-crawled for fifteen feet to where he could climb down onto a wide ledge.

Blackness!

He had made it through. The passage resumed as before.

Just for good measure, Pete went about two hundred feet down this fine virgin passage, walking on wide ledges eight feet above the swiftly moving stream. At a large bend to the right, he was at last satisfied that this passage was not going to end anytime soon. He collected about a dozen encyclopedia-sized rocks and built a cairn, then removed his carbide lamp and shook it to get a longer flame. Response was slow—not much carbide remained—but the flame grew longer. With the two-inch yellow flame, he smoked his initials on top of the cairn: PWC—10/79. One rock below, he added an arrow that pointed the way back toward known cave. Some future explorer might one day come from the opposite direction. Then he hurried back through the breakdown to join the rest of his party.

The trio returned to where they had left their packs, their tiny flames in their lamps barely lighting the way. They changed carbide and started back toward the surface.

Pete Crecelius, Bill Walter, and Miles Drake emerged at dawn, and soon everyone was talking about this new discovery. Wow! A new river flowing away from Roppel Cave. I quizzed the three of them, debriefing each about every facet of the trip. I yearned to relive all of their experiences.

This was the biggest discovery in many months, maybe the biggest discovery since Bill Walter and Jim Currens had discovered Arlie Way. But where did it go? As with any major open-ended lead, the inevitable grand predictions came next, everybody contributing his own. It was not so much a matter of what was discovered but what *might* be discovered.

Such predictions were further enhanced by the aura of a remote, mysterious place, and Pete's river discovery was certainly remote. Many miles and many hours of difficult caving lay between it and the Roppel Entrance. The average travel time from the Roppel Entrance through the S Series along Bill Walter and Jim Currens's discovery route to Arlie Way was about two hours. The traverse to the beginning of Kris Krawl, about a mile up the knee-crushing rimstone of the North Crouchway, required another two hours. To reach Pete's initials at the limit of his penetration beyond the big breakdown took an hour and a half from the North Crouchway. Six hours from the surface—six hours if one was fresh. On the return trip to the surface, when one or more members of the party hovered at the threshold of exhaustion, travel time would stretch to eight hours. The final obstacle after the rigors of the S Series to within just a few hundred feet of the entrance was an ascent of 140 feet by way of two rope climbs and a chimney up through a narrow slot. In the winter, a cold wind howled in from the surface, adding chilling misery to an already often-broken spirit and body. Nearly always, the first of the cavers who made it out would quickly fall asleep on the ground, no matter what the temperature.

After everyone was finally topside came the obligatory untying and coiling of the rope used for the forty-foot entrance drop. Cold hands then fumbled with the padlock, hidden behind a steel plate inside the gate. A lock located where only one hand can reach it through a small armhole helps prevent unwanted break-ins but gives rise to wailing curses when one tries to unlock or, worse, lock it.

Finally came the mile walk back to the field station. It was usually pitch dark and often foggy in the predawn hours. The weary procession of tattered cavers often would be scattered on the route, which sometimes resulted in their getting lost. Cavers unfamiliar with the mind-numbing confusion of sinks and valleys between the cave and the field station might wander around for hours trying to find their way home to warmth and bed. Trips were hell.

Part of the attraction and challenge of the caves in Toohey Ridge, and specifically of Roppel Cave, was this very hellishness. With a single entrance and the intrinsic hardships of the routes to reach unexplored passageways, Roppel Cave was most difficult. But at the farthest extremes of the system, we were turning up promising walking passage. The D.C. cavers had a passion for this difficulty. When the news of the river discovery reached them, they immediately devised plans of how to attack this new and obviously promising lead.

Coincidentally, Chris Welsh, Bob Anderson, and Linda Baker had been planning a small underground camp in Arlie Way to overcome the rigors of the S Survey. This bivouac could be used as a base for exploration of the many leads reachable via Arlie Way. It would save the long travel time and energy lost going to and from the entrance. On the trip during which Pete Crecelius had discovered the new river, I had carried one of Chris's supply packs overstuffed with camp provisions. From the Roppel Entrance, I delivered it to the junction of Arlie Way in preparation for setting up the advance camp.

The timing could not have been better. The camp was now half stockpiled, and a wonderful lead pointed into the unknown. It was far in and the journey would be difficult, but fabulous discoveries likely awaited the explorers. The great prospects were sufficient enough to dissipate all doubt about whether the camp was worth it: this camp plan was perfect.

Camping in a cave is serious business and a stout test of spirit. One can never carry in enough supplies. Some major item will always be inadvertently omitted. Everything needed for the camp had to be hauled through a stretch of cave that one could barely squirm through. And even the absolute certainty of discovery is sometimes not enough to offset the depression that usually accompanies in-cave camping. European and Mexican cavers agree that cavers should not camp in a cave unless there is no alternative. Everything is damp in caves; just keeping warm is a real effort.

That is why I did not want any part of this camp despite being pressed strongly to participate. From my perspective, it was not worth it if I had to camp. I resisted the peer pressure and the prospects of being called a wimp.

I would rather make long trips traveling at a high rate of speed than stay underground so long that all the extra gear had to be carried in. In my view, if cavers carried camp gear, they were doomed to require it. Light and fast trips, that was the answer. Besides, underground camping in Kentucky has resulted in very limited success. It was not much fun. After a hard day exploring, nothing could replace a shower, hot food, and a warm bunk back on the surface.

But the planning continued for the camp. The planners, of course, had my most sincere support and good wishes. I continually bolstered their enthusiasm with reminders of how productive their camp could be. I boldly predicted they would survey a mile of cave in Pete Crecelius's new river. How could it end? I asked.

With four packs each, the die-hard cave campers entered the Roppel Entrance at three o'clock in the afternoon on Thanksgiving Day, 1979, after spending the entire morning packing and repacking. I was not at the cave, purposely insulating myself from any last-minute pressure by staying home in Maryland to attend my first family Thanksgiving in many years. Seven people—Chris Welsh, Linda Baker, Bob Anderson, Miles Drake, Bob Johnson, Randy Rumer, and Peter Keys—and twenty-eight packs made for a slow trip to Arlie Way. Packs had to be passed laboriously from caver to caver to overcome the countless obstacles; with the burdensome load, there was no way any one person could manage alone. A rope had to be used at the climbs to lower clumps of packs to the bottom. Climbing with such loads would have invited an accident.

Four hours after entering the cave, the energetic group finally stumbled into Arlie Way. With sighs of relief, they dropped their camp packs and took a long, deserved rest. The soft sand of Arlie Way would be an ideal campsite. In fact, a few members of the crew were tempted to stretch out right then and there.

Thirty minutes later, they continued their trek farther into the cave. Five of the seven party members were newcomers to Kentucky caves, having been sucked in by the continuous and unrelenting salesmanship of Chris Welsh, Miles Drake, and myself. As the last of the group walked north, they marveled at the sandy-floored tubes characteristic of passages in the Mammoth Cave area. They loved Arlie Way.

Delight soon turned to despair, however, as spacious Arlie Way yielded to the low-ceilinged, rimstone-encrusted, horribly long North Crouchway. These passages seemed to go on forever. They were miles from the entrance, hours into the trip. So much crawling! When would it end?

Finally, the cavers made a right turn into Kris Krawl, leaving the North Crouchway behind. At its beginning, the Kris Krawl is a crouching-height passage through long-abandoned water-carved potholes. The passage soon becomes broad and low. Leading the way through the crawl, Miles and Chris snickered as the others greeted the low ceiling with groans. It was a wide tube—but only twelve inches high. This was worse!

After fifteen more minutes of crawling on their hands and knees, the group came to Station D10, the end of the map and where they were to begin. They planned to divide into two survey teams and conduct a leapfrog survey. One team would begin at D10 and continue the survey toward the new river; the second would proceed to the river and start surveying there. The first party would then pass the other party and continue on into unknown cave.

As Chris uncoiled the tape, Bob Anderson, Linda Baker, and Bob Johnson followed Miles, who was still leading in cave familiar to him. The long straddles and steep climbs challenged the short-legged Linda, but with little assistance she climbed down into the river.

Anderson smoked "P1" two inches high on the wall of the large room above the climb.

"They should have no trouble finding that," he said.

The worst things that can happen in a leapfrog survey are having the trailing party go past the beginning of the other survey, not noticing where it began; or stopping the survey before connecting to the next segment, which leaves it as an unattached survey, floating with no connection to the map. Besides annoying the hell out of the victimized party, this wastes valuable time and effort. Anderson made sure that it would not happen today.

It took some time to thread the survey down into the floor among the boulders to where they could plumb the tape into the river passage. Then the four were standing in the passage, which had been named the River Lethe. They sat silently on a ledge, ten feet above the stream. Below them the water ran across the bedrock floor in a three-foot-wide canyon. A strong breeze headed downstream.

Linda broke the silence. "It's all so black," she said.

Every piece of exposed rock was as dark as coal—the walls, the floor, the ceiling, even the boulders. Where the water had washed the exposed surfaces clean, the blackness glistened.

Miles shrugged and said, "I'm not sure, but I think we're in an enormous chert layer, probably the Lost River Chert Formation."

"What's the Lost River Chert?" Linda muttered quietly, not wanting to sound dumb by asking.

"It's a relatively insoluble chert layer. It prevents most passages from downcutting below it and often indicates you are as low as you can go."

The passage was spectacular, especially after the miles of back-breaking cave they had followed to get there. Miles had, however, neglected to inform his party of the extent of the breakdown they were about to survey through. Distances between survey stations suddenly were less than ten feet.

"Hey! What's this shit!" Bob Anderson exclaimed. "You didn't tell us about this."

"Don't you remember me telling you about the breakdown?" Miles asked.

Anderson did not remember.

Surveying was slow, painful, and maddening. It took them two hours to work their way the tortuous eighty feet to the other side of the rock pile.

"Well, is there anything else you haven't told us?" Anderson asked.

"No, just going cave. Walking stream passage."

As they resumed surveying in the river passage, Anderson said, "We'll see."

Time melted away as they continued northward in the large canyon along a boulder-littered ledge system. Overall, this was pretty decent cave passage for surveying—roomy and comfortable, easy to walk in. Somewhere around Station P30, Chris Welsh's party moved past them to begin the third leg of the day's line.

Miles's party worked on into the early hours of Saturday morning.

After setting Station P50, Miles's party still saw no sign of the beginning of the other party's survey.

"Where in the hell is the beginning of Welsh's line?" Miles said. "Have we passed it?" All four of them had been looking carefully for a survey station, but no one had seen one.

"Maybe we're at the wrong level," Linda suggested. The large passage had many levels, any of which could have been surveyed.

"I don't think so," Miles said. "They said they would stay at this level. Besides, it's the most obvious."

One of the last instructions the two survey teams had agreed upon was that they would leave an unmistakable and prominent marker at the beginning of each leg to avoid overlapping.

"I don't think we've passed it, either," Anderson said. "But we've come a long way since we've seen them."

The group was exhausted. Compass readings had to be repeated because Miles, who was keeping book, kept falling asleep.

Anderson declared that enough was enough. They had not and would not be able to connect to the third survey. They would just have to violate the leapfrog rule that states survey segments must be connected. It was a long way back to their camp in Arlie Way, they had done a lot of work, and it was late. Their party had surveyed over fourteen hundred feet. They wondered how the other party was doing.

As Bob Johnson, Miles Drake, and Linda Baker put away the survey gear, Bob Anderson disappeared down the passage, stating he wanted to see if he could find the beginning of the next survey. Actually, his curiosity about the destination of the passage they had been surveying was too much to resist. This passage looked promising!

A few minutes later, he returned to the rest of his party.

"It's not time to go yet. You have to see what's ahead."

Anderson turned and headed back into the cave, knowing the others would follow.

Four hundred feet past P50, Anderson stopped to let the others catch up. They were met by blackness—an enormous passage twenty feet high and forty feet wide. The River Lethe flowed into the passage and joined a larger stream that continued to the north. None of them said a word.

Anderson pointed to the right. "That's upstream. I think Chris, Randy, and Peter must have surveyed up that way."

They tried to see through the gloom. The passage was big, although not as big as where they were standing. On the far wall, they saw "O1" labeled in large letters. This had to be the other party's starting station.

"Let's go see where *this* goes." Anderson motioned straight ahead into a large breakdown-floored passage. They set off, away from the other party, following the large virgin stream flowing underneath enormous boulders.

They walked three hundred feet down a steep slope of breakdown back into the stream, the ceiling now over thirty feet above their heads.

"God!" Miles exclaimed.

Although tempted mightily, they resisted the urge to explore onward into this huge borehole passage. That could wait for the next survey party. After all, no one would appreciate a rape of virgin passage.

They began the long trip back to the dry passages of Arlie Way, weary from sixteen hours of continuous work.

Along the way through the River Lethe, Chris Welsh's party caught up with them, and the combined group continued back toward camp. The thousands of feet of cave bore on endlessly, especially for the new Roppel cavers unfamiliar with the cave's long succession of obstacles. Twenty-one hours after entering the cave, they reached the camp packs stashed in Arlie Way. They were more than ready for much-needed rest.

As they unpacked their camp, problem after problem arose from a combination of poor planning and plain bad luck. The watering hole I had suggested they use turned out to be a Niagara Falls for anyone who tried to fill water bottles. Chris's new mini-stove refused to function, despite diligent coaxing, and most of the group's food required hot water to prepare. Nearly everyone had underestimated how much clothing was required for sleeping. Linda had neglected to bring dry socks. The shivering group, painfully cold, struggled to sleep for six hours.

The final insult came when it was Bob Anderson's turn to retrieve water. He walked down the trail, stopped, and stared at a murky pool of brown water. Eight hours earlier, only dry rocks had been there. A deluge of rain had apparently fallen over the last several hours, flooding into the cave.

After he returned to camp, the group discussed possibilities. A flooding cave, especially one in which the dynamics were not understood by the group, was a caver's worst nightmare. Nearly every year, cavers are trapped and drowned in caves. Visions of being trapped in the cave or, worse, drowning in a low-level passage dominated their thoughts. Realistically, they recognized that most of the passages between them and the entrance were at levels too high to flood; nevertheless, concern hung over the group like a dark cloud.

Randy Rumer and Peter Keys finally announced that they were heading to the surface. The others watched as the pair swiftly packed and disappeared into the darkness. The remaining five discussed what to do, finally deciding to head out of the cave also. They packed their gear, leaving what might be useful for a later camping effort. Surely, the remoteness of the new trunk passage would require additional camping trips to explore it.

With their load lightened, their travel through the S Survey was easier than it had been on the way in. However, as they passed the Boundary Dome into the smallest-sized section of the route between them and the surface, they realized that the advantage of their reduced load was offset by the difficulty of traveling uphill in this passage. All the sharp flutes on the wall surfaces pointed downstream, making their travel against the grain. Squeezes that they had easily slid through on the way in were now friction-ridden struggles. All their packs had to be lifted up rather than dropped down.

Four miserable hours later, they stood in Coalition Chasm. The small waterfall beside the rope now thundered as a frothing torrent. Everything was soaked, including the vertical gear they had stashed on ledges hours before. To their relief, Randy and Peter's gear was gone, so they must have made it out safely. Now very cold, they struggled into their vertical gear, and one at a time they made the seventy-foot climb to the top. They passed through the squeezes and finally stood below the forty-foot entrance pit. Sunlight streamed in from above.

Forty-two hours after entering the cave, the last of the group of seven reached the surface, eyes squinting in the bright morning sunshine.

The underground camp was a noble try but a failure. Significant time and effort had been invested for accomplishments that could have been made on a far shorter trip without camping.

This was the last camp trip for many years in Roppel Cave, and the cache of camp gear left by this party on a ledge in Arlie Way still waits for someone to use it.

The new discovery was incredible. It exceeded my most optimistic expectations—an enormous borehole heading north away from the main cave. This was what we had been dreaming about and looking for. Although the campers said they had learned much and were convinced that the next time the kinks would be ironed out, I remained skeptical. I was smugly confident that I could do more without camping on a trip of what I considered to be a reasonable duration: anything less than twenty-four hours.

The following month, I led a strong party of Don Coons, Sheri Engler, and Tom Gracanin into the cave to continue the efforts in this new north section. Wanting to see where the new river drained, we wrestled a five-pound bucket of fluorescein dye through the cave to trace the water. Jim Quinlan, the park geologist, would monitor the many springs in the area with dye traps—bags of activated charcoal that would pick up even the most diluted fluorescein dye trace. Our travel was also slowed by having to resurvey some of the route through the difficult canyon beyond the Kris Krawl, called Death Canyon because of the terrifying leaps required to traverse it. We dropped the dye and began to survey beyond P50 toward the new trunk passage now called Elysian Way. It took us over six hours to travel from the entrance to P50. By the time we had begun the survey, we were already fatigued.

In just a few hours of fast surveying, we placed Station P65 at the junction of the stream passage with Elysian Way, then swung the tape to tie into O1 of the hanging survey. Moving onward, we continued the survey north. We passed the end of Bob Anderson's exploration and headed into virgin cave, leaving the first scrape marks on the black rocks littering the floor.

The passage cross section was enormous! We surveyed up and over many large piles of breakdown, the passage growing larger after each pile. Finally, a pile of rock extended from floor to ceiling. Sheri and Tom Gracanin sat down, by this time almost spent, the long trip back to the surface looming in their minds. Don and I were still excited about this fabulous passage and immediately started investigating the pile.

We saw what was ahead: lots of rocks and several black holes.

"Do you want to check above or below the pile?" Don asked.

I looked up. At the top of the pile, a black hole led into the ceiling. Down below, the pile was just more rocks with doubtful looking holes.

"I'll go up," I said.

Without moving an inch, we carefully studied each other's faces. Suddenly, we both sprinted recklessly up the pile of rocks towards the most inviting black hole. Pride was at stake here; I was not going to let him beat me to that hole! Don cleverly headed me off at a tricky spot, reaching the top of the pile first. Laughing, we climbed up through the hole, finding ourselves in a large walking canyon that cut across the ceiling of the passage we had been surveying below. Big leads. The best way on seemed to be down, back into the blackness of enormous cave. Together—for we declared a truce—we picked our way down the far slope back into large passage.

Looking forward into the vastness, we unwittingly stepped on a particularly precarious slab, which began to move. Struggling to retain our balance, we fell backwards onto our asses on the rock as it began to slide, gain speed, and toboggan down the long slope.

"Oh, shit!" I yelled.

It was a wild ride. We clung to the rock as it ricocheted down the steep slope. I may have been screaming; Don was holding on for dear life. It was right out of the Road Runner and Wile E. Coyote cartoons.

After an eternity—probably three seconds—the rock ground to a stop with our bodies, if not our poise, intact.

"Whooee! Holy shit, what a ride!" Don exclaimed.

We stood on the now stable rock, gathering our wits.

"Guess we should be more careful," I said.

In the distance, the sound of a large waterfall drew us ahead to the largest room yet. A pulsating stream of water fell from the eighty-foot ceiling, crashing onto car-sized rocks on the floor. What a place! Past the waterfall, we could see that the passage continued as before.

We retraced our steps back to where Sheri and Gracanin were snoozing. We coaxed and intimidated them to survey on for just a few more stations. They *had* to see the grand waterfall we had found.

An hour later, having reached that goal, they summarily quit after we set P88 on a large boulder. They yawned when they saw the waterfall, annoyed by the spray in their faces. Being transported directly to the surface would

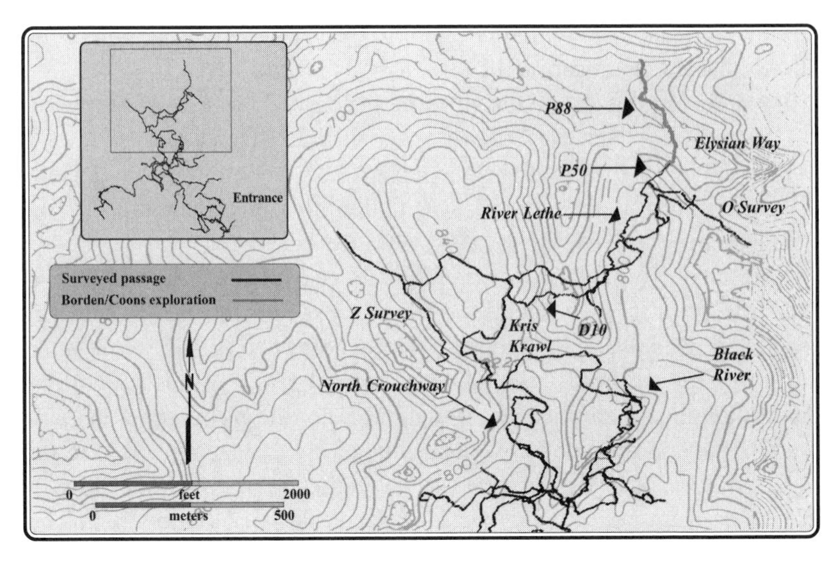

Elysian Way breakout in Roppel Cave

have been the only thing to impress those two. No matter. Don and I could not resist pressing forward. We fired up the flames on our carbide lamps and struck out to the north into virgin cave.

Just beyond the waterfall, the passage split into multiple levels, all walking. We each selected one and continued, reemerging in a large breakdown-filled trunk extending for hundreds of feet. We climbed up and down over enormous blocks for several minutes before sliding down into a fine, flat-floored walking canyon passage that led off into the gloom. We continued, our pace now nearly a run. We went around bend after bend, the walls rarely less than twenty feet apart, disappearing above us into the inky blackness. As Don and I moved along the wonderful passage, we both paused to peer at two solitary stalagmites jutting upward on a ledge far up the wall. Two sentries in the night, they seemed to be guarding what lay ahead. The effect was spooky. After nearly half a mile, we stopped abruptly at the edge of a pool that stretched out of sight around the next corner.

"Well, we've seen enough," I said. "Let's save something for the survey team."

Now this was real Roppel caving.

Negotiations between the CKKC and the CRF

15

Stress Fractures

Roppel Cave Spreads Out and Tempers Wear Thin

I unrolled the large piece of Mylar drafting film and carefully plotted the points established by the new survey Don Coons and I had made beyond the River Lethe. The map included the hanging O Survey to which we had tied. The passage took shape as I began to connect the dots. It looked like an athlete's javelin poised to be thrown. The passage—Elysian Way—was a long, slender line that jutted across the top of the tangled mass of squiggles that represented the passages making up most of the map of Roppel Cave. The tip of the spear was aimed toward the northwest, ready to fly forward to the eastern reaches of Salts Cave in Flint Ridge.

This was great going cave, but Jim Currens resisted pushing it from the very beginning. I spoke to him on the phone shortly after the trip. I tried to infect him with my enthusiasm, but as we talked it was apparent that he was not catching my disease.

"I don't understand why you think this cave is so promising," Jim said. "All that cave to the north won't really go anywhere."

I listened in disbelief to his words. In my mind, I could feel wind in our faces as we stood at the edge of the long pool and could sense the daunting vastness of the passage continuing ahead into darkness, a palpable black atmosphere pierced only by the occasional sound of a distant drip of water.

"Not promising? How can you say that? It's so big, and blows so much air!"

Jim droned on, "We all know the real cave is to the south. That's where we have to concentrate our efforts. So far, we've barely cracked the main body of Toohey Ridge. Look at how much ridge cave we've already found. We can go south and find many more miles. To the north, there are no major ridges, just a bunch of dissected knobs and chopped-up valleys. There are no . . ."

I was totally unprepared for his hostile attack. I tuned him out, his words receding into a babble of meaningless sounds as I waited for him to end his diatribe regarding our folly in exploring to the north.

"That's ridiculous!" was all I could think to say.

Silence.

Jim tried again, "That cave to the north can wait. It will always be there; Toohey Ridge may not. The CRF could push their way into it any day. I think we need to make it a CKKC priority, maybe even a policy, to push into south Toohey Ridge. We have to stake our claim to it."

"Claim? To hold off the CRF? They told us they wouldn't invade our territory, and besides, it's too far. It's six hours from their nearest entrance. Their hands are full elsewhere."

"We have to move now; we can't wait."

I was not buying it. The CRF was not a worry—yet.

"Well, do what you want, but don't expect us not to explore a lead just because you say so."

"Because I say so? Hey, I'm just trying to do what is best for the coalition. We have to think strategically. We just can't indulge ourselves by simply scooping cave."

"Jim," I said, "we will follow the cave wherever it leads us." I shoved that statement home hard and twisted it so it would hurt more. That was our policy.

In fact, it would do little good to legislate anything more restrictive. It never worked.

The conversation was over.

"You guys are making a big mistake," Jim said.

"Fine."

I hung up the phone.

I considered the situation. What about Jim's fear of losing south Toohey Ridge to the Cave Research Foundation? It was true that the CRF had been on our minds quite a bit recently. We had been able to ignore the group for many years, but now the organization loomed as a big (and bad) player in the complex political game that was caving in the Mammoth Cave Region.

Cave politics revolves around who has access to what caves and who does not. It isn't as simple as ownership of the cave. Landowners with caves on their property are seldom cavers, so access to caves involves relations with landowners, who can permit or prohibit exploration. Sneaky cavers can bypass prohibitions by stealth and by night. The politics can become intense and hostile. Some cavers say that secrecy about cave locations and what has been discovered is the only political equalizer. In this region, we had all of the above, plus rival factions of cavers, the rapidly changing dynamics of expanding cave, and the institutional presence of the National Park Service, the CKKC, and the CRF. The political means included secrecy, publicity, leases, intimidation, clique-building, sabotage of caver-landowner relations, and letters to the editor. The only elements missing in Kentucky cave politics were death squads and car bombs, and at times we even seemed to be close to that.

As vast as the region was, we of the CKKC were suddenly feeling quite intimidated by the CRF presence. There used to be thousands of feet between our cave turf and theirs; it may as well have been miles with all the sumps and low-level passageways that had separated Mammoth Cave and Roppel Cave at their closest point. But recently, the CRF had been discovering and surveying cave heading straight for Roppel Cave at somewhere between a few hundred feet a month and light speed. The killer was that we did not know exactly what the CRF was doing; they kept their secrets. In the last several months since the announced and highly publicized connection of Proctor-Morrison Cave with Mammoth Cave, I had heard vague rumors of secret trips through the Morrison Entrance to push upstream in Logsdon River. Miles of spectacular river passage were explored, it was said, that led to the western flanks of the main body of Toohey Ridge, but details had been sketchy. Nothing had been released, but I had stitched the story together from various bits and pieces. Apparently, the stealthy CRF cavers were confidently expecting to follow the river all the way to Cave City on the other side of Toohey Ridge but became astounded when, at Station Z129, the ceiling abruptly dipped beneath the surface of the water, terminating their eastward march at a sump. They were at the doorstep of Toohey Ridge but were blocked! Leads led upward and there was wind, I heard. But they were miles from the entrance. Bigger and better things awaited that were much closer.

My conclusions were based on conjecture, although I was fairly confident of them. I was also concerned about the implications of my assumptions. When I broke this news to Jim Currens, he became livid. To him, the situation was clear: the CRF's intransigent actions were a hostile territorial incursion, and they had to be stopped. Toohey Ridge belonged to the CKKC and no one else! Our entire organization had been built around the presumed cave beneath the vast Toohey Ridge. Our reason for existence and our vitality depended on it being a separate cave. I, too, felt that the CRF's actions were definitely an incursion, but little could be done now after the fact. All we could hope to accomplish now was damage control. The appropriate next steps were clear to both Jim and me; but, unfortunately, for each of us these steps would be very different.

Jim began a vigorous letter-writing campaign that attacked the threat on two fronts. First, he wrote directly to the CRF leadership, using a combination of direct pleading and a bit of intimidation to encourage them to back off. Second, he attempted to draw in outside forces to lean on the CRF to force them to abandon Toohey Ridge. This approach climaxed with his letter to the superintendent of Mammoth Cave National Park inquiring about Park Service policy concerning exploration of cave outside the boundaries of the park. Jim thought that if the park administration was unaware of the CRF's activities beyond the park boundary, the administration might start asking questions, which in turn might lead to an insistence that the CRF rein in excursions beyond the park boundaries. Although dubious of the campaign's effectiveness, the rest of us could do little to dissuade Jim from this approach. The strategy seemed reckless and damaging. Official CKKC support was withdrawn, forcing Jim to continue alone in his crusade.

CRF members were totally unprepared for Jim Currens's barrage. They were appalled to learn he had actually written to the park superintendent. In any event, neither prong of his strategy had any effect: the CRF did not capitulate to his demands, and the park superintendent did not care.

In fact, Jim's actions strengthened the CRF. His letters put the organization on notice that there were concerns outside the park that they needed to think about and act upon; they would not be surprised again. No longer was the longest cave the CRF's private playground. The group had to be prepared to cooperate with equally vested groups such as the CKKC. No, the two caves were not connected—not yet—but the possibility now required real action.

For the last couple of years, I had maintained a more or less cordial dialogue with the CRF. I was the perpetual fence-sitter, trying to play both sides and make everyone happy. In this case, I wanted to keep on top of what was happening at Mammoth Cave, but I also did not want the CRF to forget who we were and that we were significant. After Jim had hurled his volley of letters, something needed to be done before the situation escalated out of control. Tempers were hot and threats were made.

Roger Brucker and I agreed that the best approach would be for our groups to meet face-to-face. We brokered a joint meeting between the CRF and the CKKC to resolve the debate.

On 25 November 1979, Pete Crecelius, Bill Walter, Jim Currens, and I drove from Toohey Ridge to meet with the representatives from the CRF. The agreed-upon meeting place was the back bedroom at the old Austin House at Floyd Collins' Crystal Cave on Flint Ridge.

As we walked into the house, greeted by smiling faces and outstretched arms, I could not help but reflect on how caving in the Mammoth Cave area had come full circle. Over thirty years ago, the Austin House had been the birthplace of modern-day caving, starting with Floyd Collins' Crystal Cave. Years had passed and hundreds of miles of cave had been discovered. Now in 1979, the renegade cavers who had split off from the family to seek their own fortune in Toohey Ridge were returning to confront their progenitors. This meeting was going to be a landmark event in caving politics, whatever happened.

Tensions were high at this conference. Initially, polite discussion centered on uncontroversial issues of general interest such as cooperation in rescues, exchanges of data, and the like. These were resolved without significant debate but still failed to crack the mounting anxiety. At last we began to hammer on the pivotal issues that focused on CRF activity in the vicinity of south Toohey Ridge. Jim Currens echoed his earlier letters, insisting upon a full halt of CRF exploration efforts in this area. The CRF refused. The reason was simple: cave exploration could not and should not be legislated; exploration should always be able to proceed unfettered by any political baggage. The position was unassailable. Pete and I could only agree, emphasizing that territorial claims were empty if there was no cave to back them up. Jim looked resigned, but this was the only equalizer we had. Jim later said that we were selling out to the CRF.

The exploration issue closed, the CRF members then presented us with several policy statements that had been agreed upon at their board of directors meeting the previous week. The document was typed and well written, obviously prepared with much thought. I was flabbergasted! They had taken the initiative away from us and were presenting what was essentially de facto policy with little for us to say or do. Ingenious! I only wished that I had thought of it first.

In silence, we four CKKC members at the meeting read over the policy statements. There were many points of an indisputable nature. After distilling the two-page document, the bottom line was clear. The CRF had acquiesced to only one of our key demands: they agreed not to seek a connection with Roppel Cave. But there was nothing in the policy about CRF cavers not pursuing cave in south Toohey Ridge. They could still find a connection by accident.

This new CRF policy was significant. It was the first time that a formal position had ever been drafted that specifically prohibited an attempt to connect to a cave in deference to another group. Nevertheless, the CRF affirmation seemed hollow. Roppel Cave approached to within four thousand feet of Station Z29 in Logsdon River. CRF explorers could continue their work in the vicinity of the sump unimpeded.

After a few more polite discussions, the meeting was over. Despite the frustration and tension, the meeting had been a good first step in improving our relations. Both groups had talked rationally and had agreed upon a number of things. Pete Crecelius, Bill Walter, and I concurred that we had come away with about everything we could have reasonably expected, but Jim Currens was still unsatisfied since no agreement had been struck concerning the prohibition of CRF cave exploration in Toohey Ridge. That was true, but I was confident that the CRF would leave well enough alone. Agreement or no agreement, the group had to know that further transgressions near Toohey Ridge would result in further political explosions. In any case, Jim intensified his determination to beat the CRF to the cave he was so sure existed under the main body of Toohey Ridge.

My spirits were restored when Bill Walter and Pete Crecelius were suitably impressed with my glowing reports of northern discoveries, even if Jim Currens had turned into a sourpuss. Pete, Bill, and Jim had been planning a trip the following weekend to push a southward-trending breakdown passage near Yahoo Avenue, but after hearing my tale, both Pete and Bill summarily rejected that objective. They opted in favor of continuing the push northward in Elysian Way. I told them that Jim might not approve of this change of plans. No problem, they assured me. I wished them well on the trip, smarting with envy since I would not be able to join them.

Saturday morning, 28 December 1979, Jim arrived from Lexington armed with sledgehammers, crowbars, and other tools necessary to remove breakdown. As far as he knew, his objective was still the one and only. He had been looking forward all week to pushing this dry, localized collapse with the cool breeze blowing from it. Surely, he had been telling just about everyone, big cave awaits just a few feet on the other side. A push from Bill Walter and Pete Crecelius was exactly what was needed to make this lead go.

Bill and Pete had decided that it would be best to wait until they saw Jim in person before trying to divert him from his planned objective. A dirty trick, perhaps, but it would have been futile to try to persuade him long distance; once Jim had made a decision, little could change his mind. So far as Pete and Bill were concerned, they were going to Elysian Way. Why bother to cajole Jim?

Pete and Bill were sorting gear beside Bill's blue pickup truck as Jim bounced his way up the rutted driveway to the Toohey Ridge fieldhouse.He pulled up beside the other two vehicles and shut off the engine, opened the door, and stepped into the December sunshine with a beaming smile.

"Pete! Bill!" Jim nodded as he enthusiastically shook hands. "How've you guys been? Have a Merry Christmas?"

They exchanged pleasantries for a few minutes, then Bill broke the news. "Jim, we want to head out to the end of the P Survey in Elysian Way today. Borden told us all about it. Sounds like a great lead!"

Jim's smile slowly turned to a black scowl.

"I was counting on you guys to help me push this lead. If we can work our way through that collapse, we might break out into south Toohey Ridge today." He looked toward Bill. "Bill, you remember how promising it was, don't you?" His eyes pleaded.

"Yeah, I remember," Bill said. "But that lead will still be there next time. This lead Borden and Coons pushed sounds pretty significant. I want to take a look."

Although the CRF had a policy of non-connection with Roppel Cave, Jim still feared that the group would secretly violate the Toohey Ridge frontier.

"If we could beat CRF to it," Jim reasoned, "they would be forced to adhere to the non-connection policy."

Bill and Pete didn't really buy the immediacy of a CRF threat. They were not dissuaded from their plan.

Pete shut down the discussion. "I think that we can get a lot of survey out there. Come along."

"No, I don't want to," Jim answered. He watched the other two; it was apparent that his pleading was falling on deaf ears. "Shit, it looks like I don't have a choice, does it? You guys are going with or without me. And I'm not in the mood to cave solo."

"Well," Pete said, "I guess you're right. Come with us anyway."

An hour later, the trio was squeezing through the canyons above Coalition Chasm, the beginning of the long trip north to the limits of exploration in Elysian Way. Jim was in a foul mood, grumbling as he tried to keep pace with the legendary speedsters. The traditional rest stops along the route between the Hobbit Trail and Arlie Way were passed in a blur. This arduous traverse was becoming a trade route for the experienced caravans of Roppel Cave explorers. The cavers memorized the obstacles and how to flow effortlessly over them. What was once at least a two-and-a-half-hour journey was now trimmed to just ninety minutes. On one memorable trip, Bill and Pete had thundered into Arlie Way fifty minutes after starting the rappel down the Roppel Entrance pit!

This record of increasingly efficient travel made it possible to push exploration continuously outward. Speed was the best way to melt this increased distance. For Jim, this was insult piled on injury. Not only was he prevented from going to his primary objective, he was also being dragged at breakneck speed miles into the cave to its most distant reaches. His aversion to such long, punishing trips with others' objectives was well known. He simply was not interested in other people's northern cave.

Sweat poured off his nose as he ran, bent over, up the desert-like North Crouchway. His mood further darkened as he continued to sweat his way through the foot-high Kris Krawl and into the scary traverses of Death Canyon.

Everyone rested at the bottom of Death Canyon on the banks of the River Lethe at Station P7. Their plan was to survey upstream in the promising canyon for a couple of hours. Bill and Pete hoped that this route might connect to the downstream end of Black River. Such a linkup would provide a significant shortcut to this area of the cave. A bypass route would end the rigors of the North Crouchway, which was systematically destroying exhausted cavers during the long, bleary-eyed trips back to the surface in the small hours of the morning.

The three set sixteen N-series stations before quitting at a deep pool of water. Jim recalled from the map that it was still over a thousand feet to the surveyed end of Black River, probably too far to reach today. The passage was still big and a strong breeze led upstream, maybe to Arlie Way, but checking it out would have to wait.

They reversed their march and moved downstream through the River Lethe's complex canyon. In less than an hour, they came out of a crouchway tube into the impressive junction with Elysian Way.

"Wow!" Pete said.

Two enormous passages plus the one they had entered through led off the junction room. To the right, a stream issued from a walking canyon in the wall and flowed into the northern, downstream branch of Elysian Way.

They christened this room the Grand Junction. And, indeed, it was grand.

Elysian Way lived up to its billing. It was huge! The threesome climbed up and down over breakdown mountains, picking their way along the continuation of the P Survey. The black rocks, walls, and ceiling swallowed the light from their carbide lamps, increasing the apparent size to more than it actually measured. The telltale white marks where the black coating had been scraped away by cavers' feet kept them on track in the baffling maze of passages. Big leads beckoned in every direction at several levels. Bill followed the more obvious ones a few feet, unable to curb his voracious appetite for discovery. A minute or two later he would return, wide-eyed at the promise of these leads.

Jim remained somber and quiet, unmoved by what he was seeing around him. Everything was so black, so depressing. This was not the friendly, gypsum-encrusted elliptical tubes of Toohey Ridge but some of the most hostile lowest-level cave in the system. Most of the surfaces were damp, either clean-washed from fast moving water or, where the water stagnated, coated with a fine layer of wet silt. Flooding was total and frequent, a condition they had never before had to consider in their exploration of Roppel Cave.

Before long, the trio reached the last station, marked by the thundering waterfall at Station P88. It was getting late, and after setting just another eleven stations, Jim called it a day; he had had enough of this depressing passage and was ready to return to the surface.

Bill and Pete were not ready to head back, not yet anyway. They still felt fresh, and the vastness of Elysian Way invited them to continue.

"Jim, I think we want to look ahead a bit," Bill said. "Do you want to come with us?"

"No, I'll head out. I'll wait for you at S139 at Hobbit Trail. You guys might want to push upstream in the River Lethe to Black River. That would be the first decent place where the routes would converge."

Traveling alone in a cave is dangerous business, not to be done lightly. However, solo travel was common in Roppel Cave, an accepted part of our approach to exploration. Occasionally, someone would want to leave sooner than the remainder of his or her party. If this individual was competent and not overly tired and if the route out was safe, the experienced caver would be allowed to depart alone. If anything went wrong, the rest of the party would be along soon enough. The risk seemed small. However, if for any reason the individual was not comfortable traveling alone, the whole party would call it a day and head to the surface. In fact, only a handful knew the cave well enough to travel alone. All three on this trip were capable.

Pete considered the situation. What Jim proposed was probably at least three hours of solo caving through a relatively unhazardous route, except for the complexity and airy canyon straddles of Death Canyon between the River Lethe and the Kris Krawl. If Bill and Pete decided to try to complete a connection from upstream River Lethe to Black River, they would not be retracing most of the route Jim would be following. This increased the risk to an unacceptably high level. Pete shuddered as he thought of having to double back through the higher route to look for Jim if their connection attempt was successful and then they failed to find him waiting at S139 in Arlie Way at the beginning of Hobbit Trail.

Pete replied, "Okay, you start heading out, but we'll probably come back out through the North Crouchway." He did not want to give the impression that they were not confident in Jim's ability to travel alone. "Besides, we wouldn't have enough time to try the connection if we press on to the north. Getting as far north as we can is more important than finding a connection to Black River."

"Okay, I'll see you later," Jim said as he turned to the south back toward the dry and friendly upper levels of Arlie Way.

Bill and Pete were now at their best—just the two of them and big cave to look at. They had not come this far to turn heels and head out without even a peek. They hurried north following the two sets of footprints that had been laid down ahead of them just a week ago.

After fifteen minutes, they stood at the pool where Don Coons and I had stopped, considering whether they really wanted to get their feet wet. Bill was finicky about dry feet. Wet feet made him cold. On trips where he knew that his feet would get wet, he sometimes encased them in orange Roman Meal bread bags held tight with rubber bands. He had told us before, "Roman Meal is the best because they make the thickest bags. Believe me, I know. I've tried them all."

We would nod in agreement and chuckle as we imagined the Walter family wolfing down loaf after loaf of Roman Meal bread to keep Bill equipped with an adequate supply of these special bags.

Bill wasn't wearing bread bags today.

Pete volunteered to go ahead to see if the water diminished or continued, as suspected, for a considerable distance. He stepped in up to his knees. The water looked deep in places, but he stayed on the bar side of the bends where water would typically be shallow.

A couple hundred feet after first stepping in the pool, Pete reached dry land. The water had not come up past mid-thigh level, and a brief look ahead revealed no additional pools, so Pete called back for Bill to come on through.

They resumed their rapid progress north. The passage continued as large as ever, a high canyon fifteen feet wide. Banks of mud along the walls formed a narrow gully ponded with water down the center. Footing was slick as they tried to stay out of the water. They passed an area of large leads that they ignored as they marched north.

After half a mile of virgin cave, the pools gradually became deeper and longer. They avoided one particularly nasty looking pool by stepping up into a higher and dryer cutaround. The two were finally halted at a ledge above what looked like a waist-deep pool, which may have been an extension of the one they had just avoided. It extended at least a hundred feet to the next corner. Bill smoked a sign on the wall marking their limit of exploration.

"This son of a bitch just doesn't end, does it?" Bill marveled.

Pete let out a long breath, watching the steam as it was caught by the breeze and swiftly blown back the way they had come. "Boy, where does this thing go?"

The passage was more impressive now than it had been where Don and I had quit a week before.

Without pause, they walked rapidly back to Grand Junction and turned upstream, following the O Survey up the huge borehole. After just a few hundred feet, they lost the survey line. After another few minutes, they were pressing the first footprints in the soft mud. They restrained themselves this time, exploring only about fifteen hundred feet of virgin cave. Once again, they turned their backs on big, going cave with a strong breeze following the passage. They had overstayed their agreed time, and it was now urgent to join Jim Currens.

They found him sleeping comfortably at the entrance to Hobbit Trail, wrapped in a plastic garbage bag, the heat from his carbide lamp warming him. He had been waiting for them about an hour.

They left the cave.

Jim Currens still wanted to bind the CKKC to a policy of pushing the cave south. But after what many of us had seen to the north, there was no way we were going to quit the exploration of Elysian Way. Jim's pleading and reasoning had not worked on us, but he was steadfast, adamantly determined not to return to Elysian Way.

Furthermore, he argued that any efforts in that direction would be a departure from the real goals of the CKKC. One way or another, he vowed, he would be doing the right thing: pushing cave that might lead south into the broadest section of Toohey Ridge. We all supported Jim in his desire to find cave to the south, but, so far as I was concerned, our efforts to follow the best leads would continue unabated. The hot leads, for now, were to the north.

As a result of Jim's firm position, his influence on the direction of exploration of Roppel Cave by the Central Kentucky Karst Coalition began to wane. And I could see the beginnings of a rift due to our diverging priorities for exploration. This conflict was exacerbated by the growing number of active CKKC cavers in Roppel Cave, particularly by the existence of the D.C. contingent. Jim's private caving club was changing into a diverse group of independent cavers who could not be controlled by edict alone.

Over the next several months, we relentlessly pushed toward the end of downstream Elysian Way, Lower Elysian Way. Members of every party that traveled the long route to the northwest marveled at what they found, for it was always grander than before.

Ron Gariepy and Bill Eidson had little trouble being convinced of the promise of Lower Elysian Way. "It's endless," I told them.

On 26 January 1980, on what proved to be one of my most difficult caving trips ever, Bill Eidson, Chet McInski, Ron Gariepy, and I made the long march through the cave to continue the survey beyond P143 in Lower Elysian Way. On a trip a few weeks before, Pete Crecelius had surveyed to the big pool where he and Bill Walter had stopped on 28 December. I would continue the main line to the northwest. I do not think any of us were prepared for what we found. At P143, Bill Eidson complained mightily when he saw how deep the pool really was. I was sure I had told him about it, but he called me a liar anyway. At his insistence, we found some higher passages that circumvented the long, deep pool. However, that was not the end of our problems. Beyond, enormous rimstone dams impounded a series of six-foot-deep pools extending for hundreds of feet. We managed to cling to the wall and crawl along ledges to avoid the worst of it. Often, when the ledges ran out, we made long leaps across the azure pools to the opposite side, looking for more ledges.

Although it was a large passage, we progressed at a snail's pace. After several hundred feet of ledge traversing, we reached the other side of the watery obstacle. Ahead, we walked easily in wet gravel along a thirty-foot-high and eight-foot-wide passage. At P187, we called it quits. Bill and Chet were trashed; thoughts of the long trip to the surface drained away their enthusiasm. The passage continued ahead as big as ever. Ron and I could not resist. While the others began the long trip out, Ron and I slogged farther ahead into the unknown.

We walked thirty minutes, and I began to believe that Lower Elysian Way would continue forever. The wind was blowing, the passage was big, and it was heading directly toward the Green River, still many miles away. It seemed nothing could stop this passage. Ron and I stood contemplating the large, black passage in water to our knees, our boots ankle-deep in the underlying mud. It was a long way home, we were tired, and it was obvious we would not decipher the mystery of Lower Elysian Way today. I shouted into the unknown and listened to the echo. More big cave.

It was almost two miles back to Grand Junction, and Ron was fading fast. Because he did not know the way, I would sit and wait for him to catch up at each confusing spot. I usually was sleeping by the time he arrived. On and on we continued in this way, each stop becoming longer as the route became more complicated and we became more exhausted.

Hours and a dozen rest stops later, we finally reached the friendlier Arlie Way. I waited at the junction of Arlie Way, falling into a deep sleep before Ron arrived. An hour later, I woke up cold. Ron lay prone where he had collapsed in a stupor some time earlier.

After four more hours, we reached the base of the entrance pit. Sheets of ice covered the walls and coated the rope while snow blew from above, carried in by the cold blast. Our hands numb, we struggled out into the frigid January morning, our wet clothes frozen stiff from the bitter climb up the entrance pit.

Those trips to the end of Lower Elysian Way were tough! As I shivered in the long walk back to the fieldhouse, I was not certain I would ever find the strength to repeat the trip. I did not know if I even wanted to repeat it. But where did that passage go?

Don Coons and Sheri Engler could not resist the lure of Lower Elysian Way. On 16 February 1980, on a thirty-hour trip, the two of them led a survey party that finally reached an obstacle that marked the end of Lower Elysian Way. Two and a half miles beyond Grand Junction, the main route plunged through the Lost River Chert, one of the more insoluble layers of rock that retards downward cave development in the Mammoth Cave region. This plunging, in itself, would not be a problem except for the flowstone and rimstone that have modified the passages by enclosing pools of water. Chris Welsh describes this in his trip report:

> We found the stream again. It plunged over a series of waterfalls! We were sure that we were going to cut through the chert layer and scoop a romping, stomping (swimming?), reintegrated trunk. We pushed ahead, downclimbing two waterfalls (tricky) and came to a seven-foot-high rimstone dam in an eight-foot-high passage with the water draining through an unseen hole. Upon climbing the rimstone dam, we peered over yet another with a perfectly circular pool of undetermined depth with a rimstone sliding board plunging straight down for ten feet to water level. Completely unclimbable. On the far side, another rimstone dam, this one with its top at water level, completed the landscape with, presumably, another pool beyond. Not wishing to swim for it, we backed off.

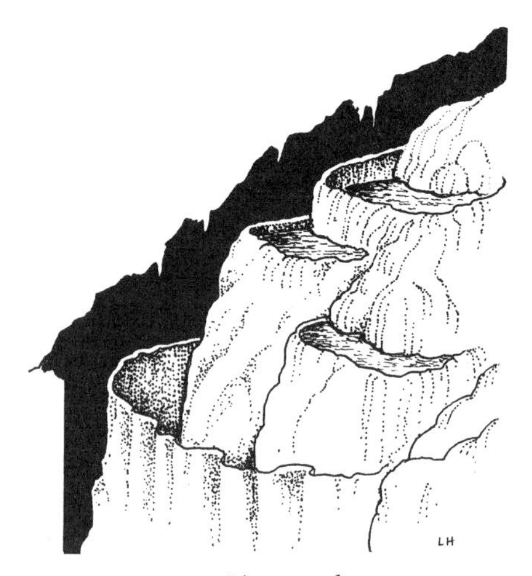

Rimstone dams

The obstacle, although intriguing, appeared too formidable for this trip. Chris Welsh named the obstacle the Watergate because of its deep pools and its apparent impenetrability. Above the beginning of the Watergate, Welsh saw a low passage continuing to the west. While crawling in, he was greeted by a cool breeze suggesting that the passage might provide a bypass to the watery obstacle below. He slid along fifty feet before turning around in the low passage.

The last major push trip to Lower Elysian Way was in May of 1980. Tom Brucker (on his first trip to Roppel Cave), Pete Crecelius, and I continued the survey beyond the Watergate in the upper-level crawl found by Chris Welsh. Enough time had passed so that I was ready again to solve the riddle of the passage. On the way to the Watergate, we traveled through a baffling tangle of crawls and cutarounds that made us feel lost. At times, we would find ourselves completely turned around and heading back toward the entrance. Our survey was in a long crawlway that we followed for nearly a thousand feet. We passed through gooey mud and one six-hundred-foot-long pool of water.

Our crawling ended at a small window thirty feet above the floor of an enormous vertical shaft. With our assistance and the aid of a daisy chain of tied-together pack straps, Pete Crecelius managed to climb to the bottom. We could see high, unreachable passages leading off. A large waterfall fell from the south side of the dome and drained underneath a ledge. Pete climbed down a pair of short waterfalls to find a deep blue pool at the bottom. He could see a submerged canyon leading off ten feet below the surface—the drain. This was base level, and everything was flooded. We set the last survey station of the P Survey, smoking P294 on the wall above the deep pool.

This trip was long but not nearly so long as Chris Welsh's thirty-hour epic, thanks to a bypass we had discovered on our way in. In the crouching-height Hobbit Trail, heading for Elysian Way, Pete and I had discussed how terrible it was going to be traveling once again through the North Crouchway. The prospect of all that hot crawling and stooping made us gloomy.

On an impulse, we decided that we should attempt to find the supposed connection from the downstream end of Black River to the River Lethe. If we failed, our original objective would be out of reach. We would have to scrub our plans to penetrate beyond the Watergate. Instead, we would have to retrace our steps.

But if our search for a connection was successful, we would save many valuable hours of travel time. It was a gamble, and Tom Brucker certainly did not object. We walked out through the intricate canyons of the Black River Complex and waded through the waist-deep, inky water of Black River. Black River flowed through a tall, black-walled canyon with the pooled water for the last five hundred feet of explored cave. I recalled my negative assessment of the lead nearly two years ago. Today, I kept my reservations to myself. Why discourage my companions with my own fear of encountering a sump?

Soon, we were huddled on a mudbank looking at the deep water before us and eating candy bars as we discussed the prospects. It looked even wetter than I remembered. Pete and I argued about who should go first. I was taller, Pete said, and would get less wet; therefore I should lead. Maybe, but I still did not want to go first. I just was not in the mood to get drenched. We continued to argue. Tom Brucker listened without comment.

Suddenly our debate was interrupted by the *whoomph . . . whoomph . . . whoomph* of someone walking through deep water. We looked up, surprised to see Tom, hunched over, moving away from us into the low, wet passage. We watched his progress. The water was now above his waist as he leaned over, hands on the opposite wall, struggling to keep his balance on some unseen mudbank. His helmet scraped blobs of wet mud off the goo-encrusted ceiling.

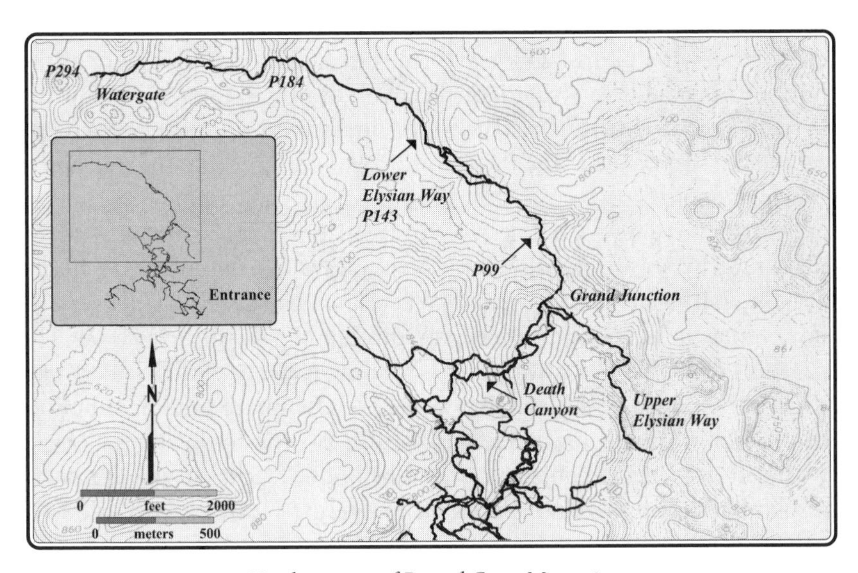

Northern arm of Roppel Cave, May 1980

He rounded the corner and his sloshing increased its pace. A few seconds later he bellowed, "Come on through! The water gets shallow!"

Chagrined by Tom's direct approach, we waded in and were soon walking in a fine stream passage. The mud yielded to coarse gravel as the stream merrily flowed freely to the north. After fifteen minutes, we reached the strip flagging tape tied to a projection marking the end of the survey in upstream River Lethe. Success! We had found the connection shortcut.

The name River Lethe was retired in favor of Lower Black River, since the two streams were now found to be one. We named the low spot through which Tom Brucker had waded the Brucker Connection.

The discovery refreshed us. We could now bypass the horrors of the North Crouchway, Kris Krawl, and Death Canyon and continue with renewed vigor toward our objective in Lower Elysian Way.

The North Crouchway once again lay fallow, unused by cavers.

Jim Currens was true to his principles. In the ensuing months, he continued his relentless crusade to push south, all but ignoring everything else going on in Roppel Cave, particularly the efforts to the north. He was not going to let those CRF cavers beat him to the plum that was south Toohey Ridge. He systematically worked leads to the south in the main section near Arlie Way and Downy Avenue.

We regulars were still not interested; we had big cave to the north. So Jim led trip after trip of new and gullible cavers to those leads. He had already attacked the easier and more open passages such as West Lexington Avenue and Western Kentucky Parkway. All these were either blocked or else reconnected to known sections of cave with no progress made to the south. Jim was finding new cave, but not what he was searching for.

Later, he selected an obvious lead at the southern end of Arlie Way at the point where the passage withered away into a low, muddy crawlway. This passage blew a lot of air and had impressed us since the first breakthrough into Arlie Way. In view of the relative proximity of the presumed upstream continuation of Logsdon River beyond the sump in Proctor Cave, we speculated that it was possible that this lead at Station S163 might be related in some way to the river. Jim led a winter trip to try to push this passage, but his party was thwarted by deep water after just fifty feet. The passage was probably sumped. A wetsuit and lower water levels would be needed to push this slimy prospect any farther south. Jim had no desire to go caving in a wetsuit, so he wrote off the lead at S163. On most of these trips, his new CKKC cavers would emerge totally spent and thoroughly unimpressed by the effort. Often they did not return.

Jim's focus eventually returned to the southern end of Downy Avenue where a large and dry passage was filled with shattered limestone. Cool air streamed through the gaps in the rocks, suggesting more cave beyond. This was the objective that Jim had wanted to pursue when he had been dragged into Elysian Way by Bill Walter and Pete Crecelius a few months back.

"This lead is the key!" Jim repeatedly preached to anyone who would listen.

We agreed that it probably was a good lead. However, I had more important things to push. Besides, I was not interested in digging in any more piles of rocks. Bill Walter, on the other hand, did not like to get wet. He had not returned to Lower Elysian Way since the push on the trip with Jim; there was too much water. But Bill liked to dig, and he liked dry cave.

Jim arranged another trip to the pile of rocks. This time, there was no big discovery to subvert his plan, so with Bill Walter and Hal Bridges, they attacked the offending pile of rocks.

Bill was the best digger and pusher among them, so he was unanimously nominated to lead. Swinging a hammer, he was able to squeeze forward. The breakdown was yielding, but unfortunately, it was also hundreds of feet long, not the few feet that had been hoped for. Jim and Hal followed along behind, they too hammering to make more room for their larger frames.

They passed one particularly difficult spot and emerged into open space. Breakthrough! They raced out of the collapse, surveying as they explored.

Unfortunately, the breakthrough was short. It was large passage, but the breakdown was so plentiful that it was more like being in a chain of large rooms separated by breakdown chokes. After five or six of these rooms, the new passage that they had decided to call Currens Corridor ran into a jumbled pile of wet sandstone. This was the end. A few leads did remain, assuring at least one return trip.

Breaking through into south Toohey Ridge would not be so easy. However, the CRF had not made any further progress in penetrating east into Toohey Ridge, either. Jim reconsidered his plan of attack, keeping an ever-watchful eye on the CRF presence he so much abhorred.

The rest of us continued caving, unfettered by any burden of politics. By early 1981, Roppel Cave quietly passed twenty-five miles in surveyed length.

To celebrate, Bill Walter formally announced the opening of a new fieldhouse, providing this advertising flier:

Roppel Hotel

The daily rates are now one dollar per day. The money is to be used for improvements of accommodations(?) and maintenance problems. [Free! That would be an improvement!]

Think of the Good Points!!

- Fresh air daily—(Except near that box across the road)
- Running water—(When it rains)
- Heat—(In the summer) Extra heat (winter) is available if you can get Borden and Currens to discuss . . .
- Scenic View—Be sure to bring your camera and get a picture of Lord Borden and his neatly arrayed caving gear (if he remembered to bring it)
- Beds — Count Currens will let you sleep in his canopied bed (when it is empty, of course)
- Bathroom—Wimpy Walter will pose free on camera in the outhouse with door shut (one dollar charge for open door shots)
- Recreation—Pistol Pete says there will be no charge for throwing cow chips if you forgot your Frisbee
- Sport—Dangerous Dave will let you dig in his sand box in the new entrance. No cats please
- Entertainment—Happy Hal has a free night course on how to fly over tables in the dark (he is the only one to make a successful landing on his face and still smile)

Dying crayfish littering the riverbank after a gasoline spill

16

Power Play

A Gasoline Spill Changes the Rules

Cave City is one of three small rural towns along the main CSX line of the former Louisville & Nashville Railroad. They grew largely in response to the growing tourist industry of Mammoth Cave National Park. Along with Horse Cave to the north and Park City to the south, these communities still compete for the tourist dollars.

Cave City is the southbound gateway for tourists on their way to visit Mammoth Cave National Park. Traffic swarms over and ensnarls the poorly designed secondary roads. Acres of asphalt funnel the runoff water that ultimately flows into and through the passages of the Mammoth Cave System. Cave City's storm water and its flotsam immediately find the way underground through the thousands of sinkhole depressions that dot the landscape. Unfortunately, a cavernous limestone aquifer such as that which underlies the Mammoth Cave region conducts water speedily, without the natural filtration that occurs in gravel aquifers. Underground streams are thus littered with trash, high fecal coliform populations, and agricultural chemicals.

Two spectacular events demonstrated the vulnerability of the ground water to hazardous contamination. On a slick and rainy July day in 1979, a gasoline tanker truck lost control and jackknifed on Interstate 65 in front of the busy Cave City interchange. The truck skidded, coming to rest on its side along the shoulder. Several people were killed in the violent collision. The tank ruptured and its contents then drained into an adjoining sinkhole. The spill was contained, the local papers said; no evacuation was necessary. A potential major disaster was narrowly averted. In just a few short hours, the excitement had passed and the truck was hauled away. Life in the town of fifteen hundred people resumed its normal pace almost as quickly as it had been interrupted. Accidents along this busy stretch of road were not uncommon; this one was soon forgotten.

At the same time, an underground gasoline storage tank at a Texaco gas station, also in Cave City, was mysteriously losing its contents. It had a leak.

As a result of the accident and the leak, thousands of gallons of petroleum products flowed into the underground streams and began a journey to the Green River. Left in the wake was an ecological catastrophe of unimaginable proportions. What was once a biologically rich underground stream became uninhabitable; large populations of blindfish, crayfish, and other aquatic cave fauna were snuffed out in a few hours.

The hydrocarbons flowed freely and quickly from Cave City, beneath Toohey Ridge and Mammoth Cave Ridge via the underground Logsdon River, and then southwest toward Proctor Cave in Joppa Ridge. The plume of hydrocarbons resurfaced into the Green River at Turnhole Spring, ten miles west of Cave City. The spill petered out, but the damage remained.

A strong whiff of petroleum greeted Jim Quinlan's researchers as they made the long climb from the lower levels of Proctor Cave into the upper-level trunk passages. Quinlan, running a water tracing program to map the flow of water through the region, had placed charcoal dye traps in large streams to intercept dye put in sinkholes miles away. In some cases, such as in Proctor Cave, retrieving the dye traps involved a lot of work. As his field workers moved toward the pits in the upper level, the pungent odor told them that something was wrong. When they reached the bottom of the rope in P17 Pit and stood in Hawkins River, they immediately saw what had happened. Dozens of dying crayfish were crawling up the steep mudbanks, trying to escape the contaminated water. A thin oil slick moved slowly past them.

It was heartbreaking. Scores of dead and dying blindfish and crayfish littered the river and its banks. Each footstep in the polluted passage churned the settled goo, further increasing the stench. The river was dead.

In the following weeks, many people made the long crawl to Hawkins River. Photographers took pictures; hydrologists took water samples; biologists surveyed the mortality rate of the cave life. All this evidence strengthened the case for initiatives to protect the groundwater of the Mammoth Cave region.

Within a week of the tanker truck spill, hydrocarbon fumes from the polluted water permeated the breezes that moved through Arlie Way. When Bill Walter and I first encountered the smell after stepping out of Hobbit Trail into Arlie Way, we were overwhelmed nearly to the point of nausea by its potency.

"What the hell stinks?" Bill had a frown on his face.

I puzzled over the unexpected odor. The air was blowing from south to north. We knew that the unexplored cave beyond S163 was the major conduit for the wind that blew through Arlie Way. The smell had to be coming from those unknown wet passages south of S163.

"Do you know what that smell is?" I asked Bill.

He said nothing. He apparently did not know about the fuel spills.

"You're smelling gasoline from a tanker truck spill," I said. "The smell's from the main drain of Turnhole Spring!"

He still said nothing.

"We always thought S163 might connect to a base-level river; we were just too lazy to check it. It was too wet! Not only does it lead to a base-level river, it leads to *the* base-level river—the upstream continuation of Logsdon River in Proctor Cave!"

Bill was more concerned about the gasoline odor. "This smell makes me feel sick. It's not safe to stay here. We might blow up."

I wasn't finished. "There could be miles of cave down there!"

Bill was gathering his stuff.

I relented. "Yeah, I guess it's pretty bad. Let's pack out of here."

The odor was nauseating, and I feared that the impact to the cave life had to be severe. Yet I smiled as we walked north in Arlie Way. The cards had now fallen in our favor. A calamity had struck the caves, but it had proven beyond any doubt that Roppel Cave and Mammoth Cave shared the great underground water course named Logsdon River. It would not be easy to connect, though; we knew that from the CRF's discovery of the sump beneath Toohey Ridge. But we might be able to show that the caves were close, and if they were, the CRF would then be bound by their own non-connection policy not to explore cave in the vicinity of the sump. This would further ensure that the cave of south Toohey Ridge would be ours for the taking. The pressure would be off.

However, I also knew that if this scenario were played out fully, the CKKC would have to face the inevitable. From that moment on, the question was not whether the caves would connect but when the connection would be found.

The possibilities of discovery beyond S163 were intriguing, especially in view of the possibility of finding an unknown segment of a large underground river. Unfortunately, the winter rains came early, and the water levels rose. There was no chance now to push the base-level passages, either for us or for the CRF. Through the winter, the lead gnawed at me.

In the spring of 1980, I began planning my trip to S163. The water levels would be dropping soon, and I was eager to see what lay beyond to the south. Wearing or carrying a wetsuit through the hot S Survey would be horrible. Memories of the hell of hauling the camping gear into the cave the previous Thanksgiving were still fresh in everyone's minds. The effort had been exhausting, and nobody yearned to repeat that struggle.

I spent my spare time at home in Bethesda, Maryland, selling the idea of a push beyond S163 to the strongest of the Roppel cavers. Chris Welsh and Linda Baker were veterans of the cave camp, and their enthusiasm for hard trips had not been diminished by the experience. They were always up for a challenge and quickly accepted my proposal. My plan was to carry as little additional gear as possible into the cave. We would wear enough to fight off the cold from the water in S163 but not enough to allow for the slow pace of a survey. Instead of a full wetsuit, each of us would take only a wetsuit top. We would then push into the lead without surveying and see what we could find.

This plan departed from our normal policy of surveying as we explored; but in my opinion, this lean approach was justified to enable us to discover where it went. It was a gamble. If our results were inconclusive, we would accomplish nothing and might not return to explore the lead for a long time. If the lead ended quickly, it again would likely never be surveyed. Nobody would want to drag a wetsuit into the cave just to survey a grim and already explored dead end. But I was sure that a determined-enough push would find the big cave. If that happened, there would be no problem about later completing the survey.

On Memorial Day weekend, Chris Welsh and I were sweating our way through the narrow passages of the S Survey. Each of us carried an additional pack containing the wetsuit tops for the push south beyond Station S163. Linda Baker and Ron Gariepy were a few feet behind, the pair of them easily gliding through the tight canyons with their relatively small loads. Linda had cleverly taken advantage of Chris's woeful financial state: she hired him to carry her wetsuit load through the cave. They had set an appropriate hourly wage, and now Chris carried not only his own wetsuit but also hers. I shook my head in amazement after hearing of this arrangement, but both parties seemed satisfied.

My original plan not to survey had continued to disturb me as the date for the trip approached. I arrived at a suitable compromise with myself. We would go to Arlie Way with the wetsuits, park them, and then survey a dry gypsum passageway that overlay Currens Corridor. We would work a full day, then return to the wetsuits and resume with the plan to push the wet lead. No matter what the outcome of the push, we would have a productive survey effort to show for our time in the cave. Once we completed the survey, Ron Gariepy would go back to the surface alone, leaving us to our wet mission. He had no desire to push these wet passages.

The first phase of the plan was a good one. The four of us surveyed twelve hundred feet in a stunningly beautiful elliptical tube. Massive white crystals of gypsum and calcite lined the walls and ceiling of the sandy-floored crawlway. We found a small hole in the floor that reconnected to the main passage through a previously unseen ceiling cavity, and this connection avoided the need to retrace our path back through the long crawl.

We quickly made our way back to Arlie Way and our waiting packs. We had already been underground for fifteen hours, and the packs did not appeal to me now. I had lost my enthusiasm for phase two of the plan. I cringed at the thought of putting on a cold wetsuit and getting soaked at this point in the trip.

"Do you guys really want to do this?" I asked.

"Sure," Chris Welsh said. "I didn't drag this crap in for nothing."

"Well, it is a bit late," I said. "Perhaps doing the survey was too much to bite off."

"What?" Chris railed, appalled to hear me wimping-out on my own plan. He could not tell if I was being serious.

"Maybe we should just head out," I said.

"Hey! You are going to push that lead!" he demanded.

Chris walked over to the packs, grabbed mine, and flung it at me. It bounced off my chest and fell to the floor.

"Let's go!" he said. Chris grabbed the pack again, this time thrusting it into my arms.

Ron watched, silent and amused. I winked at him. He knew how it worked.

After animated haranguing from Chris, we began the walk down Arlie Way. Chris marched behind me like a prison guard.

When he glanced away, I quickly tossed the pack aside, hiding it behind a rock. I continued several steps before Chris noticed my empty arms, the wetsuit missing.

He had my number. "Hey! What the hell are you doing?"

Chris went back, found the discarded pack, and dramatically slammed it back into my chest.

"Now, keep walking!" he ordered.

We repeated the scene several times. The energy of our mock conflict lifted all of our spirits. Finally, we stood on the last dry sandbank in Arlie Way, 150 feet north of S163.

The game was over.

While Ron looked on, we donned our wetsuit tops. We shuddered as we pulled the cold, clammy rubber over our warm bodies. Sand lodged under the rubber, grating our skin. Ron smiled.

Chris, Linda, and I walked down to the muddy stream that flowed into the wet passages beyond S163. Ron stopped. "Good luck!" was the last we heard from him as he turned and began his solo trek back to the surface.

At Station S163, the ceiling dropped two feet and the trickling stream flowed into a pool that stretched out of sight. A piece of flagging tape tied around a projection marked this last point. I crouched and looked into the low and gloomy lead. Mudbanks rose steeply out of the water to meet the mud-covered ceiling just eighteen inches above. An ugly scum floated on the water.

"Yech."

We eased ourselves into the dreadful pool, each of us howling in turn as the fifty-four-degree water reached our chests. The passage was a crawlway, three feet tall and half filled with water. We sunk into the mud as we crawled, the cool breeze further chilling our already shivering bodies. After just fifty feet, we were forced even deeper into the pool by a large chert projection that jutted from the ceiling.

"This is really discouraging," I said.

I paused to peer underneath another ledge. The ceiling was lower, now just ten inches above the water.

"Well, it still goes." I reported back to my silent followers.

I lay down in the water and scooted ahead in the wet fourteen-inch-high crawl. The waves from my movements made foreboding lapping sounds against the ceiling, indicating lower airspace ahead. I knew it would not sump because the strong breeze ruled out that possibility.

Around the next corner, another ominous black chert lump hung close to the water's surface.

"Shit! It's getting low!" I said.

"Does it look like it still goes?" Linda asked.

I cocked my head awkwardly above the water just inches below my nose. "Well, it does blow air. Looks like it gets a little higher in just a few feet.

"Keep moving!" Chris bellowed from the back of the line. "This is no time to stop!"

I squeezed ahead, the side of my face now in the water. The muddy floor gave way to sharp rocks. As I forced my way beneath the obstruction, the sharp corners of the rocks ground into my chest and water sloshed into my eyes.

Past the chert obstacle, the passage was more spacious and comfortable. The airspace increased to ten inches in the eighteen-inch-high passage. After several hundred feet of this, I saw a shadowy slot in the left wall.

"I think we might have something here!"

The opening was almost too low to squeeze through. The mudbank pinned my chest against the corner of the slot as I sought purchase with my feet in the water behind me. A couple of good pushes allowed me to pop through. I scrambled up a steep bank and looked to see what I had found. The lower passage had fortuitously connected with a higher-level parallel passage. In two directions, a crouch-high tube led off into blackness.

"Come on through! It's big!" The echoes of my voice were confirmation enough of a breakthrough.

Linda and Chris plowed through the last of the obstacles below and soon stood by my side.

South was the way to go. The three of us walked onward in the enlarging passage, wind in our faces. We lumbered through an easy walking passage floored with hard mud. Three hundred feet ahead, the floor dropped away into a large room with a small stream of water falling from the ceiling. On the far side of the fifty-foot-long chamber, a mudbank led up to a continuation of the passage we were following. To the right at floor level, a stream issued from a low passage, crossed the room, and disappeared into another similar crawl on the opposite side.

Beyond the room, we continued in the virgin corridor.

At a mud ramp leading down to a lower level, I suddenly froze in my tracks.

"Quiet!" I said. "Can you hear that?"

We stood, listening. Sounds of a large stream greeted us.

We stepped onto the steep bank to investigate. Simultaneously, we all lost our footing and careened on our rears down the bank into the unknown passage. With resounding splashes, we landed on our feet, knee-deep in a large pool of water.

"Look at this river!" cried Linda.

We all whooped and hollered. Echoes from our shouting reverberated.

To our left, a round tunnel fifteen feet in diameter led off. A large stream gurgled downstream among rocks the size of watermelons. Behind us, the water flowed from a deep blue pool. The ceiling dipped beneath the level of the water—a sump. Exploration upstream was blocked.

"This is the river we thought was down here." I beamed as I confirmed the prediction. We had probably found what I had promised (and hoped for)—the upstream Roppel Cave extension of Logsdon River.

"Let's go!" Chris led the way downstream in the big passage.

The stream was alternately knee-deep and waist-deep and was moving fast. After three hundred feet, the passage made a sharp turn from due south to

Jim Borden

the northwest. The fast-moving water had cut a deep trench in the floor, and we had to swim to continue in the passage.

We soon came to a large breakdown where the stream disappeared below the rocks. Plastic milk jugs, shreds of plastic bags, and old rusty cans were wedged among the large blocks.

"This trash comes all the way from Cave City," I said. Cave City was three miles away, but it was the most likely source.

Above us, the blocks were larger. We climbed upward between them easily. Twenty-five feet above the stream, we emerged into blackness to stand on the floor of an enormous passage. The ceiling was sixty feet above us and the walls were vertical and thirty feet apart. It was huge!

We scurried among the boulders on the floor to see where the giant canyon would lead. As we proceeded, the piles of rocks rose higher, eventually reaching the ceiling and completely sealing the end of the passage. We stood at this eastern end of the passage looking down the long, sweeping hill sixty feet to the level of the river. The rocks we had climbed up through were faintly visible in the distant gloom. We perched on a large rock that jutted out like a pulpit overlooking the giant passage.

We moved down the breakdown mountain, searching the opposite wall for a way down to the river. We threaded our way between the boulders to a steep mud slope back into the murky river water.

Linda and I were done. Exhausted. Satisfaction with our success was not a substitute for energy. Chris, with his boundless energy, disappeared downstream, reappearing after fifteen minutes. He had been stopped where the ceiling lowered to five inches above the water. The water was over his head and the current strong. To move on would be too dangerous to attempt alone.

We began the long trip out of the cave, pausing only in Arlie Way to pack our wetsuit tops. Chris was still on retainer with Linda, so he crammed two tops into his bulging pack. I started into the S Survey first, Linda and the heavily-laden Chris following a few minutes later. Insensitive to any hardships Chris and his oversized pack might encounter, I moved out alone and never looked back. I did not see the two of them again until I opened my eyes the following morning after a fitful sleep in my bunk.

When Chris and Linda finally struggled through the Entrance Series to the surface, the sun was high in the sky and the weather was warm. It was 1:00 P.M., three hours behind my exit: they had been underground for twenty-six hours.

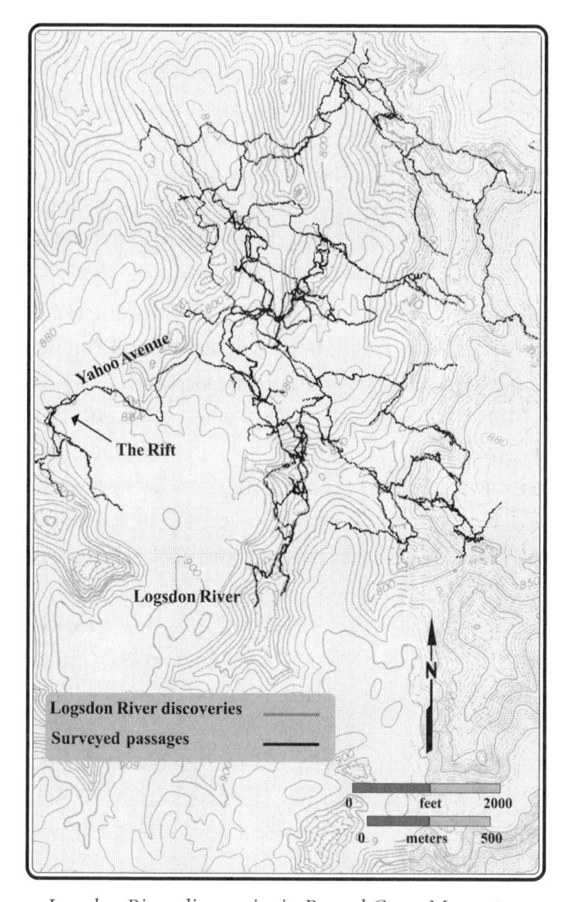

Logsdon River discoveries in Roppel Cave, May 1980

It wasn't until fall of that same year, 1980, that the memories of the pain of the S163 push trip had dulled sufficiently to be outweighed by thoughts of the wondrous river. By October, we had put together a party of five—Chris Welsh, Linda Baker, Pete Crecelius, Bob Anderson, and myself—to return to the new discovery. Unlike that first trip, surveying now had to be on the agenda. Wetsuit tops would not do; this time, each of us would have to haul in a full wetsuit. It was imperative to see where the river lay and how close we could penetrate downstream toward Mammoth Cave. This was a dangerous game; the results could be devastating if misplayed. We could not tip off the CRF to our plan before we completed it.

Linda Baker still did not want to carry a wetsuit through the S Survey but was determined to survey the discovery. Unfortunately for her, Chris Welsh had lost interest in being her porter. She had to solve this problem on her own. To the astonishment of all, her solution was to don the rubber suit on the surface and wear the bulky thing into the cave. Madness!

While the rest of us painfully struggled with our large and unruly wetsuit packs, Linda sweltered her way to Arlie Way, nearly succumbing to heat prostration. Salt tablets and copious amounts of water saved her. On the sandy floor of Arlie Way, she collapsed, drenched in sweat. To cool off, she poured bottles of water down her top.

After a rest, we climbed into our constrictive suits, the cold, clammy rubber even less tolerable than it had been the last time. We split into two survey teams. Given the trouble we were going through, we wanted to survey as much as possible. The river had to be pushed to its end. We gathered our packs and proceeded along to S163.

While Pete, Linda, and I surveyed through the water crawl south of S163, the lead party of Chris and Anderson followed the river downstream, quickly passing the limits of Chris's previous exploration. This month, the conditions were dry and water levels were low, much lower than in May. What had seemed an impressive river before had dwindled to just a flowing stream. The long swim where Chris had quit was no obstacle this time. They moved rapidly through several thousand feet of wide river passage. The water was pooled and often deep; the tunnel featureless and stark.

The reason for the ponding was evident when they reached a breakdown that ended their rapid progress. Unlike the breakdown explored on the previous trip, the rocks here were not massive. The water gurgled away to be lost underneath small pieces of shattered rock. Wide, low crawlways continued through the jumbled blocks, but the breakdown looked extensive, and if there was a way through, it was not obvious. The once strong airflow now drifted lazily past them, dissipated in the vastness of the collapse zone.

Chris and Anderson set a permanent poker chip survey marker—which would last in flooded passage—and began their survey back upstream. Two thousand feet upstream from the collapse, they linked with my survey line coming downstream toward them from S163. The survey tied, we added up the numbers. It had been a big day; between the two groups, we had surveyed more than six thousand feet. We had pushed the river to an end.

However, it was not the end we had expected. We had anticipated finding the water gurgling into the sump that ended upstream Logsdon River in Proctor–Morrison–Mammoth Cave to the east. We thought that the airflow would lead us to south Toohey Ridge, completing the stratagem of outflanking the CRF cavers. Unfortunately, the airflow came from the breakdown, still from the direction of Mammoth Cave. A breakdown? What was going on here? Pete and I were disappointed. While the river had extended Roppel Cave far to the south, the gateway to the vast cave we knew had to exist still eluded us. Over a mile of cave surveyed and hardly any leads. How could this be?

We struggled back to the surface to complete the twenty-four-hour trip.

When the survey data was plotted and added to the map, we confirmed it was all but certain that the new river was an extension of Logsdon River in Mammoth Cave. Although we had yet to see a map of the river passages beyond the park boundary to the east, we knew that the sump that terminated Logsdon River in the upstream direction was beneath the western flanks of Toohey Ridge. Our maps showed that the downstream breakdown in the river in Roppel—also named Logsdon River—was probably about fifteen hundred feet short of the now-famous sump.

The hand had been played and won. We had successfully outmaneuvered the political windbags of the CRF, beating them at their own game. Their organization was effectively shut down to the east. In just one day of cave exploration, the principal gap between the two caves had closed from well over a mile to a span of barely over a thousand feet. No map was available to us of Hawkins River and Logsdon River; it was a CRF secret. How-

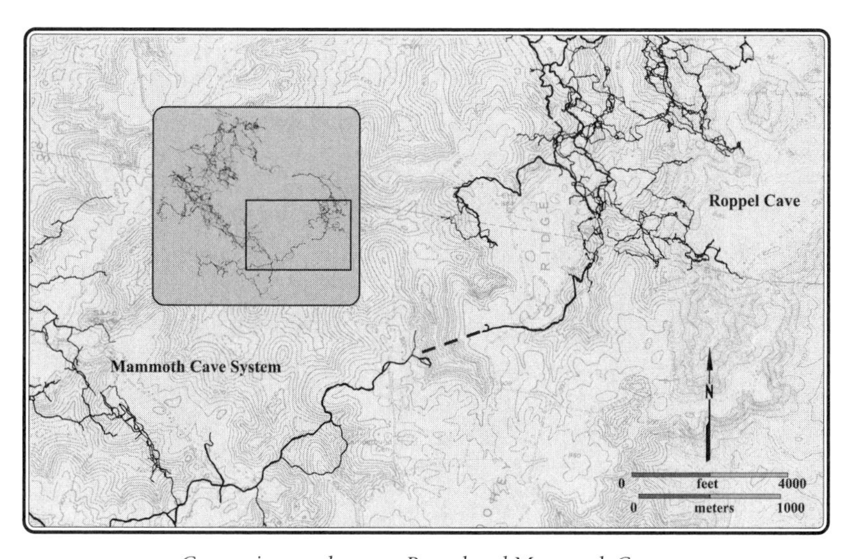

Connection gap between Roppel and Mammoth Caves

ever, I believed that if our map were joined to it, it would show over five miles of river with only one small gap. This was certainly the main river system of the vast underground Turnhole Spring Basin that drained about fifty square miles. The relationship between the two caves was unmistakable: they were the same.

The political situation was volatile, much more so than we had thought. Two large caves looked to be on the brink of connection. Our map was not a secret, and knowledge of the fifteen-hundred-foot gap between the two caves spread through the caving community like wildfire. We were continually questioned about when we would connect the caves.

The CRF was intrigued by the connection possibility, making it very clear that we were quite welcome to connect as far as they were concerned. However, after such a daring maneuver of outflanking the CRF, the last thing we wanted to do was to connect the caves. Roppel Cave now had an identity. We liked our big cave, and we wanted to make it bigger. We had just snatched south Toohey Ridge from the CRF; we did not intend to connect it to Mammoth Cave.

Jim Currens watched me carefully to make sure my conviction not to connect was firm. It was—for now.

Nevertheless, the connection area was a temptation to any caver greedy for glory. Such cavers abound. Controlling them would be difficult, and the idea of legislating exploration was abhorrent. Had we not built our organization on the bedrock of our hatred of the CRF's regimented style of caving?

Roberta Swicegood

It galled us to reverse our policy, but the alternatives seemed even uglier. In October 1980, we openly declared a moratorium on exploration beyond S223 in downstream Logsdon River in Roppel Cave. Cavers from all over the country howled. The CKKC was accused of playing the same dastardly political games that the CRF played. To those not involved, a moratorium was unthinkable. The accusations stung, but we ignored them.

We kept our fingers crossed, though. For now, the intervening sump seemed a sufficient enough obstacle to allow us to set the issue of connection on the back burner for awhile. But questions remained and festered: Where did the airflow come from at the end of the river in Roppel Cave? Where would the upper levels above the sump in Mammoth Cave lead? We hoped no one would learn the answers for a long time.

Roger Brucker peering into the drain where he is convinced the connection between Roppel Cave and Mammoth Cave will be found

17

Secret Trip to Morrison Cave

Diana Daunt and the Bruckers Infuriate Everyone

"Dad, we want you to come along on a special trip to Morrison Cave . . . to the river," Tom Brucker said to me—Roger Brucker—over the telephone. This was unprecedented behavior by my son, who often patronized me but seldom tried to con me.

"Who is 'we'?" I asked, hoping not to betray suspicion.

"Diana Daunt and I have been discussing a trip that requires your seasoned experience. We want you to enjoy a once-in-a-lifetime trip. After all, you can't continue caving indefinitely," Tom said. There it was, the harpoon of patronage cloaked in velvet flattery. Worse, the appeal had a "before-you-die" overtone that pissed me off.

In the past months, I had trained myself to the absolute peak of my caving abilities and endurance. I was able to do the toughest trips because I ran several miles most mornings. I had participated fully in extensive wetsuit survey trips into Morrison Cave, both up and down Logsdon River. I had led the survey to the upstream sump of Logsdon River and had noted several small leads that might go on from that remote termination.

"Sure, I'll go," I said. It was April 1981, and the rains that had drenched Kentucky had abated. I met Tom and Diana at Diana's house near Fort Knox, Kentucky, on Friday night, 10 April.

The trip was to be on the quiet, Tom said. Nobody must know about it. Tom had arranged with Mrs. John Logsdon to expect us Sunday morning. We would park out of sight and, being fully prepared before setting out, would alight from the car, sweep up our gear and rope in one graceful movement, and disappear over the hill with nobody the wiser.

Our trip was to be a comprehensive reconnaissance of the possibilities of pushing upstream beyond the sump in Logsdon River. Unknown to Tom, Diana had confided to me that Tom's unspoken purpose was to try to find the connection to Roppel Cave. He might—or might not—stop just short of it so the official discovery could be made at a time of our own choosing. I told Diana this was overly optimistic. Connections are not made that easily.

Unknown to Diana, Tom had told me in a low voice that Diana's purpose was to find the connection between Morrison Cave and Roppel Cave. He wanted me along to restrain her if and when the time came. He apparently believed that I could exert some sort of moral leadership or ethical appeal if we found the connection. I told Tom that in my judgment, we would not find the connection. I guessed it must be two thousand feet away.

What was my motive in joining the trip? I knew that there were some leads at the sump and wanted to know where they went. I had negotiated with the CKKC representatives on behalf of the CRF, and our organization had unilaterally agreed to refrain from searching for a connection between the caves. On the other hand, the CRF had asserted its intention to explore the cave passages wherever they went. Of course, these policies were in conflict, but it set up the CRF party leader as the sole arbiter of where exploration left off and connection began. In a superior way, I knew I could make such a delicate distinction, but could any other leader be relied upon? Certainly John Wilcox and Pete Lindsley could, and maybe one or two others—but Tom?

Doubtful. Diana? Impossible. So my motives were as pure as driven snow. As the party leader of experience, maturity, and political sensitivity, I was the obvious and only pick.

I had other motives. We had stayed away from the river too long due to the rainy weather. I relished caving with Tom and Diana, both capable and great companions in tough caving. Also, this reconnaissance allowed us to discover the possibilities of going on without the fanfare of a large, heavily organized party. So if our motives were so lofty, why the conspiracy and secrecy?

First, it wouldn't do to let the CKKC know we were going up to the end of the river for *any* reason. Just the mention of such a trip would produce an uproar. Second, we didn't want Don Coons to know about the trip. He had a mixed reputation—he kept his extensive explorations secret when it suited him but also told secrets to almost anyone, including the CKKC, when it suited him. Third, I didn't want some in the CRF to know about the outing, particularly Richard Zopf, who seldom approved of other people's secret trips. He would conclude that we were trying to connect the caves if he got wind of the trip. I didn't tell Scooter Hildebolt, the CRF's operations director and our representative to Mammoth Cave National Park, to provide him deniability in case of a screw-up.

There was still another person we wished to keep out of the know: the Cleveland Grotto spy Bob Nadich, who had "just happened" upon our engineering efforts when we installed a gate on the entrance to Morrison Cave the previous year. His breathless account of discovering a CRF crew putting the finishing touches on a formidable steel gate to Mammoth Cave had appeared quickly in the *Cleve-O-Grotto News.* Never mind that we had fed him such eye-opening information; the fact is we had been under surveillance and word could travel at light speed.

Finally, we did not want John Logsdon inquiring about what we had found. He already had an exaggerated idea of what the federal government might pay for another entrance to Mammoth Cave, and we wanted to avoid feeding his notion of riches with yet another story of fabulous discovery. We timed our entrance and exit to miss him.

I did tell Pete Lindsley, the CRF's president, of the impending trip and the reasons for it. He warned me to be careful and to make a full report of what we found.

We rigged the Morrison Entrance drop and locked the gate behind us. After reaching the river, we detoured upward into a large breakdown room to check our emergency cache of food and first-aid supplies. The CRF had put emergency provision dumps in likely entrapment areas since the mid-fifties. Here, the threat was entrapment by fast-rising water.

We came close to the intersection of the route to Mammoth Cave, the now-famous French Connection. I showed Tom and Diana the red poker chip I had wired to the wall in prediction of where the Mammoth connection would be made. My prediction was off by two hundred feet; I had thought we would connect through a dome in the ceiling. Around the corner we saw the sandbank with footprints made by John Wilcox and Tom Gracanin and the subsequent survey party I had led.

We waded and walked and floated for ninety minutes. Diana said, "When we connect with Roppel, let's build a big cairn. We can write 'SUCKER' on it in big letters."

"Diana, we're not trying to connect," I said. "We're following the cave wherever it goes, as we said we would. In the name of science."

"That's a crock of shit!"

We were having a great time. Tom, who had not seen the upper reaches of Logsdon River, was impressed by the length and breadth of it.

We reached the sump, which neither of the others had seen. On my last visit, at the termination of the survey, I had examined the sump carefully. Previously, the water had formed a pool five or six inches deep, backing against the rock wall that extended a few inches underwater to a lip. Then, as now, there was no air over the water. But now the pool was a foot deep! The underwater lip of the wall and continuing passage were more deeply submerged than before. If the pool level could fluctuate, it meant the sump might open or flood, depending on rainfall. Or maybe with a little engineering, *we* could drop the pool level.

I examined the downstream bank of the pool, which was made up of rocks cemented with flowstone. We kicked at the rocks, then collected some loose ones farther downstream to pound with. In a few minutes of vigorous pounding, we failed to breach the bank in any significant way. We discussed the possibility of blasting, but dynamite had never been the CRF's style, due to the abhorrence of blasting cave passages in a wilderness-protected national park. The CKKC seemed in love with explosives, a blasting immorality we disdained.

No matter; there were leads to be examined. We retreated a few feet downstream to a hole in the right wall. We climbed into a small muddy canyon, a little above the level of the stream passage. After 120 feet, the canyon branched to the left and right. The floor sloped up to passage two feet high by four feet wide going two hundred feet to a higher canyon that branched left and right.

Roger Brucker

We went to the right eighty feet to a set of vertical shafts. We climbed around over rocks and down under boulders, looking for abandoned or active drains we might enter. There was several hundred feet of cave passage here, and we searched aggressively for a way onward.

Tom and Diana pushed a tiny crawlway, while I returned to the first shaft, which was about ten feet in diameter and fifty feet high, the first of a chain of five or six such shafts. They were aligned parallel to the flooded river passage, about fifteen feet below me. I was trying to climb up into what might be an abandoned drain, which was no more than a dark shadow into which my light beam disappeared, but I couldn't reach it.

I returned to the intermediate level of the shafts and searched for a shaft drain. The largest stream of water drained through a four-inch hole. Other drains were too small or blocked. I found a tight, twelve-inch-high squeeze heading promisingly downward for ten feet. It led to a small room about four feet in diameter. Several drains joined in this room and combined to flow out the other side through a tube two feet high by eighteen inches wide, with a nine-inch depth of wall-to-wall water. The slope of the drain leveled out. I figured that this was the master drain. It would connect with the river in a few feet, and the airspace indicated the ceiling of the river passage had risen above the present level of the sump.

What was the significance of these clues? Just this: I had found the way to bypass the sump! Nearly thirty years of looking at cave passages convinced me of the certainty of my conclusion. My mind was racing: If you're so certain, why not just slip through and confirm it? Go ahead, you're dressed for it.

My conscience responded: If the river does continue open to the air and you can bypass the sump without diving, haven't you as good as connected to Roppel Cave?

My son, Tom, had been in this position when he had found Pete Hanson and Leo Hunt's smoked initials and arrow in Hansons Lost River in 1972. He knew he had found the connection between the Flint Ridge Cave System and Mammoth Cave. He was so certain that when John Wilcox had offered him a place on the connection trip, Tom declined. He'd done it, and besides, he had to study that weekend.

My mental war continued: You're too pooped and too old to do it. Not wanting to connect is a cop-out. Face it, you're not *that* sure, and you can't stand being wrong.

No, responded my conscience, that's not true. Turning back now *is* an ethical decision.

Ethics? Don't kid me, you're so crooked you bring up ethics only when it's convenient.

True, but how can you have any self-respect when this certain connection prize is sacrificed on your altar of self-gratification?

My mind-war escalated. What will your friends think?

You won't *have* any friends if you find the connection. Even Tom and Diana will be pissed off. And how will *that* look when you write the book?

That's another cop-out. Which are you, a real caver or a self-important writer?

No, it's not a cop-out! Back in 1954 on the C-3 expedition to Floyd Collins' Crystal Cave, I did, however, write the final entry in the Lost Passage camp log with an eye to how that would look in the book when I wrote it.

So, you admit it! That's self-gratification! Ego drives you, my conscience said.

No, it's a sense of the dramatic. I'm a storyteller at heart. Any book I write, they'll have to read. Finding the connection now is climaxing too soon, bad form in writing and everything else.

That settled it for me. It would make a better book not knowing and not connecting. I would not press on but would act on the certainty that this new drain would bypass the sump when the time was right. Besides that, lying prone in the drain passage peering into the water, I was getting cold. And, alas, older.

I told Tom and Diana about the drain passage, and Tom went back to look at it. He concluded it was not nearly so promising as I knew it was.

The time had come to head out if we were to miss John Logsdon. We sloshed downstream in the river and up the tributary stream that led to the Morrison Entrance.

The fire-storm reaction to our secret trip began ten minutes after we parked our car near John Logsdon's machinery shed. Don Coons spotted it. He was investigating and would broadcast the news with relish. The following day, he called Diana, who confirmed that there had been a trip, primarily to introduce her to the cave—a tourist trip to show off the new stuff.

Don then called Tom, without mention of his conversation with Diana. Tom flatly denied there had been a trip!

Next, Don called me. Without reference to his calls to Diana and Tom, he asked me if there had been a trip. I confirmed that there had been and told him the substance of what we had found but omitted the master-drain discovery. I overemphasized our search for the source of a considerable quantity of water that the river picked up somewhere along its course.

Tom, feeling guilt and panic, called Jim Borden, telling him everything, including his intention to connect with Roppel Cave if possible.

What was going on, anyway? The CKKC cavers were thunderstruck: here was confirming proof that the CRF hypocrites were trying to make the connection. Why else would they go secretly? Of course their sneak trip was to connect, and it wouldn't matter what cover story they feebly offered.

Richard Zopf also believed that this had been a trip to connect and was angry about the breach of policy and about my dissembling remarks to him to cover the real purpose of the trip.

I apologized to him and gave him a lengthy account of the trip and its circumstances.

"Your apologies are a dime a dozen," he wrote.

It was a long time before Richard would forgive me, my apology notwithstanding. I had shattered a trust, and my guilty feelings led me to conclude that I had gone beyond the limits of usefulness as far as representing the CRF was concerned. The new CRF administration might do better without me.

There was a further complication. Diana told me that Tom Gracanin, Cal Welbourn, and she were planning another trip, a trip to look at the biota in the river. The real purpose of their trip, Diana said, was to "end this bullshit" and connect the caves. Now I was horrified. I immediately confronted Gracanin for ninety minutes. My ultimatum to him was to subject this trip to my control, including veto power, or expose it. He relented and agreed.

After a night of no sleep, I wrote Gracanin and Diana that the trip was canceled. I phoned Pete Lindsley to declare that the whole situation stank and to urge him to tell Tom Brucker that his trip was canceled by official order of the CRF's president.

Tom reported later that Gracanin and Diana were still planning the trip. Pete then ordered Tom to change the lock on the cave gate. He did, and apparently, that was the last of it.

Cal Welbourn, a CRF director, was angry about my participation in this secret stuff—or was it because I had scuttled his own secret stuff? Tom Gracanin was mad. Diana was furious. Don Coons was seething, and he enlisted John Branstetter's sympathy. Richard Zopf was angry. When Scooter Hildebolt heard the scathing condemnation of my role, he expressed disappointment—in no small part because he had not been invited—but also wrote a letter saying the harm was not permanent.

Don Coons's letter warned me about the risk of alienating the CKKC. He stressed the importance of maintaining good relations in the future.

The final outcome: when my term of office as a CRF director expired, I withdrew from consideration for another term. In hindsight, I had made a bad decision and felt the responsibility deeply for the anguish I caused. I was happy that Pete Lindsley placed a moratorium on all future secret trips.

My last policy recommendation to the CRF's operations people was that we ask the CKKC to furnish a party leader for any future trips to the upstream end of the river. If they accepted, then any connection—accidental or otherwise—would be the CKKC's responsibility.

Excavation of a new entrance to Roppel Cave

18

Further Separations

Worn-out Cavers and Outside Pressure

Sometime in late 1980, I—Jim Borden—realized that Roppel Cave had achieved the remarkable distinction of being the longest cave on earth with just a single entrance.

Every cave that was longer had at least two entrances. Mammoth Cave had seventeen! The event would have passed unnoticed except that the trampled route between the Roppel Entrance and Arlie Way—the S Survey—was finally beginning to wear out the patience of the cavers who had come to know it so intimately. Many of us were just plain sick of the S Survey.

Roppel Cave was hard; it would have been harder were it not for the elaborate set of linkups and shortcut routes we continually discovered. Yet, round trips to the frontiers of Roppel Cave involved nearly ten miles of caving, and there was no bypass to the S Survey. Ten miles was longer than the entire length of all but a handful of the world's caves.

As a result of the distance and tough travel, cavers returned in April 1982 to the long-abandoned part of Roppel Cave known prior to September 1978—the Old Cave. For some time, I had been lamenting the many good (but difficult) leads that we had left prior to the breakthrough into Arlie Way. I gushed about the beckoning Fishhook Canyon and Grim Trail, blowing passages on which we had turned our backs. Except for Jim Currens and me, none of the currently active cavers at Roppel had ever experienced the Old Cave. They were too new to know!

I thought these newcomers could be lulled into overlooking the evil reputation earned by this seldom-visited section of Roppel. There was cave to be found in the Old Cave!

Both Jim and I led trips into the old passages. The trips were hard, they were hot, and they were longer and more difficult than I remembered.

God, I thought, how could we have suffered this trip after trip? Was I getting old?

It was impossible to walk in the Old Cave; cavers either had to crawl or relentlessly squeeze their bodies through tight canyons or over sharp corners of breakdown. Nevertheless, cave discoveries were made almost immediately. The new cavers scoffed at the old-timers—Jim and me—when they found an overlooked crawlway in the floor of Vivian Way. This lead was in such an obvious place. How could we have missed it? After a few hundred feet of hands-and-knees crawl (the Rape of the Sabines), it broke into big cave. The green cavers whooped as they ran down the big and dry elliptical tube. They called their new discovery Abracadabra, since finding it surely had to be by magic. There was even a hint that this could be a portal to the south if the elusive air could be followed.

Dumb luck, I thought.

The Old Cave, unfortunately, was true to form. Abracadabra was short-lived, extending only five hundred feet before ending in a pile of sand. However, strong air blew from a tall canyon heading to the southeast, hinting of big cave somewhere. Cavers combed the canyon, but they could not find the source of the air.

Leads did remain. Pete Crecelius, Bill Walter, and Jim Currens followed narrow canyons below the Rape of the Sabines to climb down through tight slots to lower levels. After one last climb, they joined a small stream. Downstream led to footprints and a connection to the wet passages beyond Tinkle

Shaft in the C Crawl first explored by Don Coons, Bill Eidson, and me. I recalled that years ago, I had considered many of the leads promising. When my best lead from that time ended with the stream falling through a hopelessly small crack, Bill Walter named it Borden's Folly. Jim had laughed, while I had winced at the humiliation.

Later, it felt strange to crawl easily through this newly discovered route to passages that, previously, were nearly beyond my abilities. It was as if a legend had been shattered, some of the mystery unshrouded. However, old

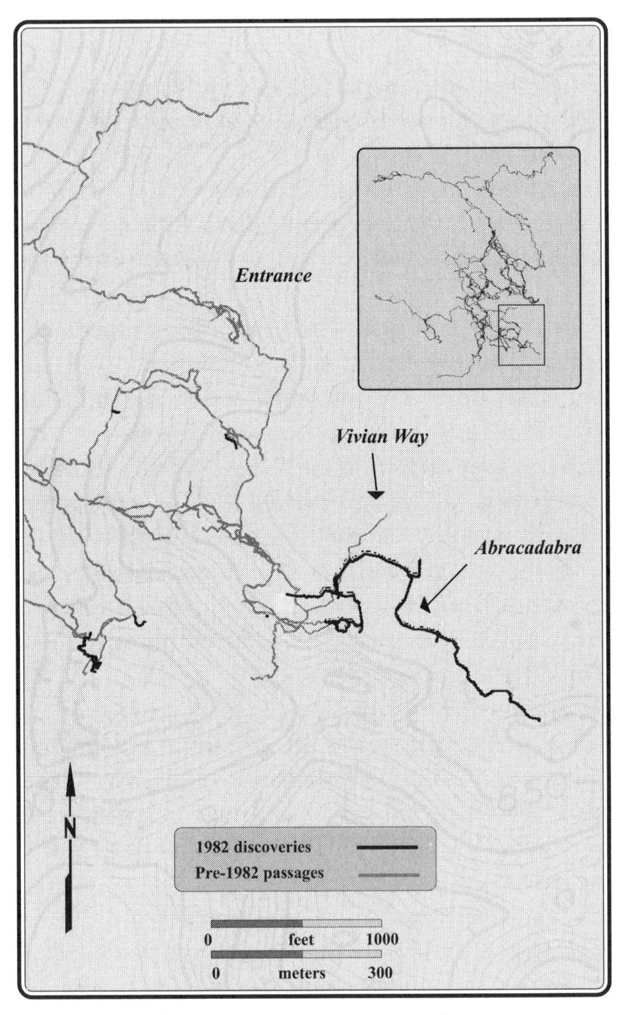

Old Cave extensions in Roppel Cave

mysteries were replaced with new when Bill Walter dug through at the end of a long crawlway to emerge in walking cave with one set of footprints. Their origin is still unresolved, the lead never pursued.

The trips continued, but like before, the cavers eventually wore out. The cave was small, difficult, and unforgiving. Once again, the Old Cave section of Roppel went to sleep, still stubbornly clinging to its secrets.

The tedium of the S Survey influenced different people in different ways. Some cavers just quit coming altogether. The difficulties of this route coupled with the extremely long trips were simply too much to endure. Others avoided the long trips; they reserved their strength for the strenuous exit and seemed satisfied to leave the frontier leads to others. Most of us just put up with the reality of tight, terrible cave and gritted our teeth at the prospects of returning each time back through the S Survey. No doubt about it: we were suffering heroes (or was it bumbling idiots?).

Jim Currens was fed up with the S Survey also. But if his toleration had worn thin, he had no intention of quitting. The cave would not beat him. He would solve the problem as he had many times before. If an obstacle was too difficult or a situation too formidable, he would either overcome it or circumvent it altogether, finding creative solutions to these challenges. His performance in the breakthrough to Arlie Way was a result of his can-do attitude.

The prospects of finding a suitable bypass to the S Survey by some undiscovered entrance or another route were not promising. So far, we had found just one alternative to the S Survey, technically a bypass, but by that route it took eight hours of hard caving to reach Arlie Way. We had also spent countless hours walking the surface above Roppel Cave looking for any hole that could possibly connect down into the main cave. Our luck was no better now than it had been during the period prior to the discovery of Roppel Cave. We found lots of holes, but almost nothing we examined went anywhere, and none went into Roppel.

The best of the lot of these holes was one I had found while walking outdoors along the base of Dry Valley the day of the discovery of Roppel Cave, 3 April 1976. At that time, a small spring flowed into a low entrance blowing cold air. I shooed away cows cooling themselves in the breeze and excitedly squeezed over sharp rocks into the opening, elbows digging deep into liquefied cow manure. After a few feet I was able to sit up in a room, but the stream flowed into a four-inch-high, gravel-floored crawlway. No way. I called it Dry Valley Cave. As a rule, natural entrances to large caves in the Mammoth Cave area are associated with perennial springs. These springs usually emerge a few feet from the cave entrance and flow into the cave, keeping its entrance clean of silt and debris. Dry Valley Cave was true to type.

One day, I noticed on the most recent Roppel Cave map that upper levels recently explored and surveyed above Pirates Pot, a beautiful vertical shaft near Black River, wandered to within a couple hundred feet of a sinkhole. The cave was high and the sinkhole deep; they might be related. After further studying the map, I realized that this was not just any sinkhole—it was Dry Valley Cave.

But Dry Valley Cave was a dig, and my cave-digging days were ancient history. Knowing that Jim Currens was manic these days in his quest to open another route into the main part of the cave, I was happy to pass on to him what I knew about the spring-fed entrance.

Jim wasted little time. He was soon crawling through cow manure into the low entrance, flashlight in hand. As the airflow teased him, Jim started digging out the four-inch-high crawl; he soon pulled fist-sized rocks out of the wet gravel. In just one hour's work, he extended the cave four feet. However, as he lay on his belly, peering through the opening ahead of him, he could see that the low passage continued at least twenty feet. This was not going to be easy digging. Dry Valley Cave was put back on the shelf, at least for the time being.

The next month, Pete Crecelius, Bill Walter, and I headed into Roppel Cave to survey in the upper levels above Pirates Pot. Several interesting leads had been found that we all wanted to check. We followed the now-familiar interleaved canyons of the Black River complex to the lip of Pirates Pot. Across from our perch and at the same level, a glistening stream of water arced like a silver ribbon from a clean-washed canyon, crashing twenty feet into a sea of spray directly below us. Wow! We chimneyed directly up the canyon to a point where we could traverse across the pit on a series of wide ledges above the floor forty feet below. Once across, we slid back down to the canyon floor and walked sideways upstream in the narrow passage. We squeezed around the tight bends, feet sloshing through the clear, six-inch-deep water. After six hundred feet of water-scalloped canyon, we chimneyed up toward one of several holes leading into blackness. Up between the wide walls, we followed the ledges. Thirty feet above the level of the stream, we hauled ourselves into a sandy elliptical tube. Its floor was incised by a deep, meandering canyon carrying the stream between the walls of the twenty-five-foot-wide passage.

This elliptical tube had been named Kangaroo Trail because of the required leaps across the wide canyon in the floor. For long-legged and non-acrophobic cavers, the leaps were exhilarating. Bill, Pete, and I hooted with delight each time we hurdled across one of the many chasms.

Fifteen hundred feet east along Kangaroo Trail, the broad tube forked into two slightly smaller tubes. We took the left branch and rapidly moved to Station C21. From there we continued the C Survey to Station C59 at the edge of a pile of wet limestone that blocked the passage. Returning back toward the junction with Kangaroo Trail, we checked the leads we had turned up. Most were small and went only a few feet; however, Pete checked a body-sized tube at Station C3 and after forty feet emerged into the base of a large dome with dry, fluffy sand on the floor. The waterflow that had formed it was long gone, and now the walls sparkled with fine gypsum crystals. It was a breath-taking sight. Pete named it Sand Dome.

A short climb up the east wall took him to a high canyon going three directions. He turned to the right, following the stiff breeze. After three hundred feet, the canyon forked. The air whipped up the narrowing left branch. Pete forced himself forty feet farther, but because he was alone, he was reluctant to continue jamming his body through the narrow crack. Rusty tin cans and sticks lay scattered on the floor, washed in from the surface during heavy rains by the small stream in which he was now crawling. He shouted into the passage, listening for an echo that would indicate larger space ahead. There was no such echo, but an opening to the surface had to be there.

Pete backed out and returned to his waiting party.

After hearing about the rusty tin cans and sticks, Jim Currens took a party in the next month to map the passage. They could squeeze no farther along the narrow canyon than Pete had, but they completed the survey.

At home in Lexington the following day, Jim held the enlarged topographic map against his living room window and overlaid it with an updated map of Roppel Cave at the same scale. The bright sunshine backlighted the contour lines of the topographic map, making it easy to see the interrelationship between the cave and the surface features. As Jim studied the map, a broad smile spread across his face. Pete's rusty can lead was heading directly for Dry Valley Cave, only 150 feet away. The tin cans and sticks must have been washed in by the stream that flowed into the entrance of Dry Valley Cave. With all that trash—it had to connect!

Dave Weller, a caver from nearby Louisville, Kentucky, came to Roppel Cave on an invitation from Bill Walter. Dave's first trip into Roppel had been with Bill and me in March 1979, six months after the discovery of Arlie

Way. Dave was six feet tall and two hundred pounds and carried a heavy steel ammo can as his cave pack. I had cringed when I saw that army-green metal box.

"Why?" I cried, pleading with Dave not to take the damned thing into the cave. I held up a spare canvas pack, offering it to him.

"My stuff stays dry and doesn't get smashed," he answered, rejecting my offer.

"But how can you stand to carry that heavy thing?"

"Oh, it's not too bad."

"But . . . you've never been in Roppel Cave before." I imagined how awful it would be carrying that bulky, unyielding box in small passages. He would be sorry.

Dave took his armored box.

He derived one advantage from his ammo can: it was stuffed so full that there was no room for community gear. Bill and I were stuck carrying all the survey equipment.

Two hours into the trip, the interminable *clunk! . . . clunk! . . . clunk!* wore on my nerves like fingernails scraping a blackboard. I bit my lip, resisting the urge to scream, "I told you so!" Dave was having a bad time moving that rectangular, heavy metal monstrosity through the non-rectangular S Survey passage. Sweat poured off him in rivers. No, I did not need to humiliate him further.

He doggedly continued without complaint. At Arlie Way, we took an extra-long break to allow him to cool down and regain his strength.

Dave was a slow caver, but the three of us had a fine trip surveying through spectacular canyons and domes, closing a large loop of Arlie Way.

Then came that mind-numbing *clunk! . . . clunk! . . . clunk!* over and over again on our agonizingly slow trip back to the surface. The return through the S Survey took over four hours. Dave was as tired as anyone I have ever seen emerge from the Roppel Entrance pit. He unclipped from the rope and collapsed, sound asleep.

I would have bet a million dollars that we had seen the last of Dave Weller.

I was wrong.

Dave did return for another go. I was horrified, though, when he again produced his ammo can. Why hadn't he jettisoned it? No matter, he was not going with me on this particular trip, so I would not have to endure it. The box sported a smashed-in corner that hadn't been there the first time I had seen it—a Roppel scar, I assumed.

Dave again had a desperate trip. He was even more exhausted than he had been his first time. Bill Walter and Pete Crecelius dragged him three miles to the distant end of Walter Way, Arlie Way's most significant side passage. But they had a splendid time, coming out with a book full of survey data.

Dave had fallen in love with the cave, even though he found it to be very difficult. He hated the S Survey. He had already begun to suggest seriously that we should begin to dig a second entrance. Anywhere.

Jim Currens recruited Dave to help on the Dry Valley Dig Project, and Dave turned out to be the paragon of a cave digger. He was persistent and methodical, and his every action was carefully planned. He was a self-taught engineer, masterfully using any tool he could find to its maximal advantage. He had already earned his reputation as a cave digger by helping to excavate a back entrance to a large cave system southwest of Louisville. Moreover, he was sufficiently motivated to seek a solution to his problems in the S Survey (although, in my opinion, just buying a new cave pack would have made all the difference in the world).

On his first trip to Dry Valley Cave with Jim, Dave agreed that it would be a long dig, a major effort.

"We'll have to start digging here," he said, pointing to a spot six feet outside the drip line of the entrance, "and dig our way into the cave all the way to the low spot."

"Why do we have to start digging out here?" Jim asked incredulously.

Dave explained matter-of-factly. "For a horizontal cave dig, your work area at the headwall of the dig has to have plenty of room. Otherwise, our progress will stall when it gets too difficult to drag out our tailings. I want to have at least four feet of headroom the entire way. It has to be high enough for a wheelbarrow. We'll need to start trenching back here to do that."

"Okay." But Jim was not convinced. A wheelbarrow! He had envisioned beginning immediately at the end of his original dig. If the advance slowed, the diggers could fade. That was the common wisdom of us rainy-day diggers. Dave's way would take several days of digging before they even made an inch of forward progress. Impractical!

For the remainder of the day, they hurled shovelfuls of dirt, rock, and cow manure out of a growing trench leading toward the low entrance. At some invisible point, the dig changed from a Jim Currens project to one managed and owned by Dave Weller. Jim may not have noticed this transfer at first, but as the dig progressed, it became apparent that Dave—and only

Dave—had the determination to carry the tunnel through to its successful completion. Jim deferred. The final result was what counted. Jim became Dave's assistant.

As the dig progressed, I remained skeptical. Sure, Pete Crecelius's rusty can canyon was close to Dry Valley Cave, but it was still over 150 feet and at least one level below; it might as well have been a mile! To link the two would take a lot of luck.

"No problem," Dave assured me. "We'll just work harder and use more explosives. You may have to help," he added, aware of my hatred of digging.

Dave had already acquired dynamite, primer cord, spools of fuse, and many varieties of blasting caps.

"This dig may take awhile," I said.

"It'll take a long time," he agreed.

Thereafter, I supported Dave. We soon became close friends. His contributions to the CKKC went far beyond digging his way into Roppel Cave. His counsel, enthusiasm, and hard work became indispensable to the functioning of the coalition.

The Roppel cavers steadfastly continued caving through the S Survey while Dave Weller tunneled through the dirt and rock that filled Dry Valley Cave. The exploration of Lower Elysian Way was in full swing while Dave wheeled out load after load of debris to dump on the growing pile. After a few months, we could walk in a four-foot-high passage to the back of the first room—a distance of about thirty feet. At this point, the ceiling dipped and the tunnel floor was ten feet below the entrance. Here, Dave began trenching a two-foot-high tunnel through the nearly filled passage. Digging was difficult. The fill was ancient and well consolidated and the passage narrow. Crowbars and hammers were necessary for prying up the resilient surface. Complicating the affair were rocks distributed through the silt like lumps in gravy. Many of these rocks would not yield, even to the crushing blows of a ten-pound sledgehammer. Stymied, the digging crew would be forced to return to the surface.

Dave's solution was to walk up the hill to his van jammed full with equipment. He took dynamite from a case, a blasting cap from the glove box, and a spool of fuse from under the back seat. At forty-seconds-per-foot burn, three feet of fuse was usual. He slid the fuse into the open end of the cylindrical blasting cap, then crimped the end tight around the fuse with his teeth. After jamming a large nail into the end of the dynamite stick to make a hole about six inches deep, he inserted the blasting cap and fuse. He pulled the free end of the fuse down along the stick of dynamite, securing it with black electrician's tape so the fuse could not be yanked out accidentally. The explosive charge was then placed over the obstruction. He packed the charge with rocks and mud, sculpting the mound to maximize explosive force in the proper direction. The fuse was lighted with either a match or cigarette lighter. Cavers love to light the fuse on explosives, and Dave used this attraction to recruit helpers. It lured me! Nothing was more satisfying than the loud *whump!* that sounded after several anxious minutes. We would grimace, bracing ourselves for the inevitable concussion, then, after the billowing plumes of white smoke billowed from the cave, we would rush excitedly into the dig to assess the damage.

For removing the excavated dirt in this lower section, Dave fashioned a sled that looked like a portable mortar box, like the ones used in mixing small amounts of concrete in building construction. Long pieces of rope served as a harness to drag the trough both in and out of the dig. The full sled was dumped into buckets in the entrance room that were then hauled to the surface. As the dig progressed, the debris was moved out of the cave in stages, which required one person at each station.

One digger scooped at the headwall; a second behind the first would load the sled; a third would haul the sled to the entrance room and transfer the load; and a fourth would haul the buckets to the surface. Ideally, an additional one or two people would be available for relief, making an effective digging complement of six. Usually, though, Dave would have only one helper, and occasionally he worked alone.

Progress was painfully slow. Many of us helped from time to time, but Dave returned most weekends. I called him nearly every Sunday evening to see how things had gone. He would report that they had made six inches, or that they had spent their effort enlarging the entrance room. I was always disappointed, hoping for a breakthrough or at least a sudden burst of six feet with the expectation of easier going ahead. But it was not to be. At least I maintained an upbeat cheeriness on the phone and helped whenever I had the chance on a caving weekend.

Dave's large dynamite supply amused many of us during our off-caving days. In exchange for our unconditional help on his dig, Dave tolerated this blatant waste of explosives. We experimented with turning logs into missiles. Similar to putting a small can over a firecracker, we placed a log vertically over a shallow hole containing a stick of dynamite. The task was not as easy as it would seem. In most cases, the log would detonate and blast wood shrapnel from the launch area. Only once were we successful in our efforts. The sight was spectacular as the slender log shot off into the sky, gently turning end-over-end as it rapidly gained altitude. In three seconds, the log reached apogee and began its downward descent. Two cavers, awakened by the loud report, sat bolt upright in the back of their truck, puzzled by the upward-gazing faces. The spectators suddenly scattered; apparently something dangerous was falling from the sky. The two shrank into the corner of their truck. The ripping sound of shredding leaves and snapping branches announced the imminent impact.

Out in the field, we watched with horror as the missile concluded the last few feet of its descent.

Thud!

The five-pound log had struck the ground just fifteen feet from their truck.

The hooligans never confessed.

Another time, Pete Crecelius slid a short-fused blasting cap beneath my bunk. A party was going on, and I was struggling to sleep. The sound of a sputtering fuse sent me running out into the main room with no clothing on, only my sleeping bag wrapped around me. Howls of laughter from the pranksters increased my indignation as the blasting cap exploded in the next room.

Our practical joking was frequent and was a diversion from the rigors of the long cave trips. Most of us never took the pranks too seriously, yet no one would let another get away with anything without retaliation. I had vowed that Pete's mischievous deed would be paid back with interest.

Often, when we had a big crowd, Pete slept in his tent instead of in the packed fieldhouse. On one occasion, while he was in the cave, three of us carried his tent up to the gently pitched roof of the building. We put a lamp inside to make the nylon dome glow an eerie blue and decorated the outside with blinking Christmas lights. It was a sight to behold—you could see the "blue moon" from all over Toohey Ridge! Several of our neighbors came by to investigate the spectacle.

The next morning, Pete was not to be found. After a short search, we found him sound asleep in his tent on the roof.

When he climbed down from his perch, he commented, "Well, the tent was easy to find; too bad my keys weren't. They were back on the ground beneath where I pitched my tent!"

I guess we were even.

Meanwhile, Dave Weller methodically continued toward a goal still over one hundred feet away.

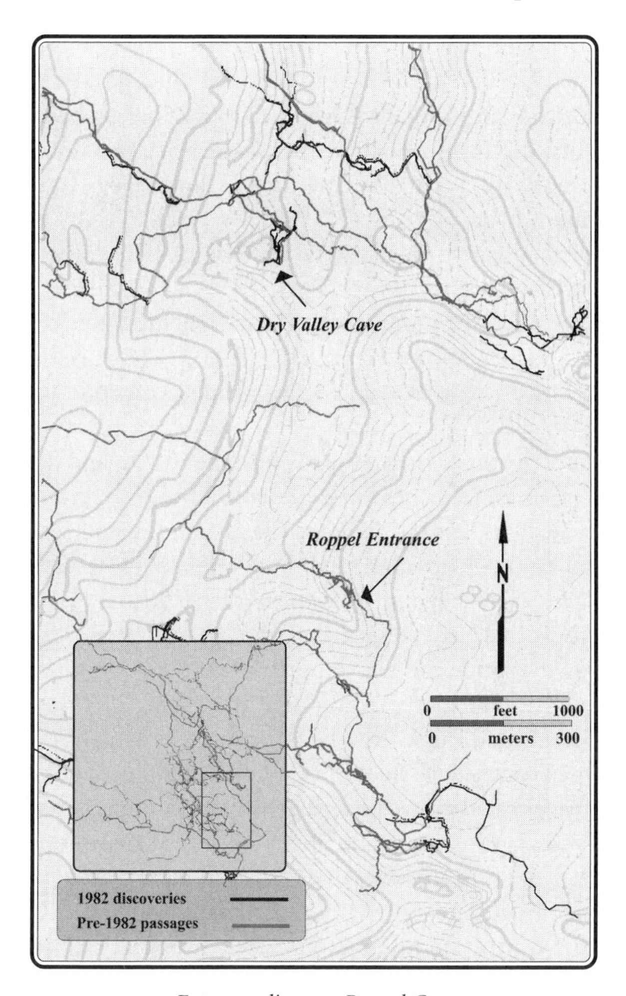

Entrance dig area, Roppel Cave

During the coming months, discoveries were dramatic. Survey parties poured through the fast and easy Brucker Connection in Roppel Cave, and the horrors of the North Crouchway became a dim memory. The abrupt end of Lower Elysian Way at the Watergate was certainly an enigma, but for now, we did not care. We were exploring the many leads along the vast length of the passage. The Watergate could wait.

The first and most significant of these discoveries was made by Pete Crecelius on a trip into a promising passage heading to the east from Station P152 in Lower Elysian Way. Pete had discovered and briefly explored this lead on one of his early survey trips in the area. The lead was not huge, but it bellowed a large volume of air. It had the looks of having been at one time a major tributary to the now inactive main passage. Deep pools covered its floor, and its walls were coated with dramatic flowstone cascades. The lead was spectacular.

Pete named the passage the Black-White-Orange-Blue Passage (BWOB)—"gorgeous side lead with potholed black ledges, white formations, orange rimstone dams, and blue pools," he had written in his trip report. His party surveyed over three thousand feet in the BWOB. They discovered splendid, clean-washed canyons, dazzling waterfalls, and good leads at every turn.

On trip after trip, they sloshed their way to the BWOB and into the ankle-deep axle grease mud of Muckwater Canyon. They traversed over the glistening and menacing Turbine Blades of Clearwater Canyon, walked past the three waterfalls of the Elephant Dome, and ran through the long and broad Muddy Tube to points beyond. This chain of spectacular passages went on for over ten thousand feet, the strong breeze undiminished, beckoning them ever further to the southeast. The opportunity for discovery in this new area seemed immense. Our maps showed that the passages of the BWOB area crossed under a broad valley into the northwestern flank of Eudora Ridge. Unlike the small passages of Old Roppel, a mile distant in the southeast corner of the same ridge, leads here seemed ripe for a significant breakthrough. They would surely take us into the large upper levels that must underlie Eudora Ridge and lead us far to the east.

The discoveries did not stop with the breakthrough in the BWOB, however. Upstream from Grand Junction, where Lower Black River flowed into Elysian Way, Elysian Way headed south, big and walking. Large gravel banks and wide pools led nearly two miles to emerge into a complex area of tall canyons and domes that lay less than a thousand feet in a straight line from the Roppel Entrance, yet five arduous miles underground.

Also, just a few hundred feet south of Grand Junction, an elliptical tube with a narrow stream canyon in the floor took off from the south wall. The lead had been noted but not explored. While heading out after a long survey trip, Bill Walter crawled into the small passage. After several hundred feet of muddy hands-and-knees crawl, he came to a spot where the ceiling broke up into a higher passage.

Bill climbed up into a large canyon, big and walking. In one direction, he walked a couple of hundred feet; in the other, the ceiling broke upwards again.

He climbed into the inviting higher level where he stood in another canyon, like the first, although a little smaller.

He ran down the flat-floored, six-foot-high canyon. In one direction was the same scenario as before—a hole upwards to more walking canyons. Cave everywhere. Satisfied, Bill retraced his steps to the waiting party.

"Well, what did you find?" Hal Bridges asked between mouthfuls of a candy bar.

"Oh . . . not too much." Bill shrugged his shoulders. "Just a bunch of canyons—nothing too spectacular."

Hal did not notice that Bill was downplaying the lead. Roppel Cave was full of canyons. Hal finished his candy bar, and they closed their packs. They rose and began the now standard high-speed sprint to the entrance.

Bill decided to keep his discovery to himself, for the time being. But on 24 January 1981, he proudly led two parties into his small side passage. He told each party where to begin surveying. This was well planned.

"How come you didn't tell anybody about this?" I asked.

"I didn't want to get everyone too excited. It may not go anywhere."

I looked around the half dozen canyons that led off from where we were sitting. Big cave. The simple fact was that Bill Walter just didn't want to get scooped.

"You're kidding, right?"

He smiled and said nothing.

The new canyons, named Freedom Trail in commemoration of the release of the Iranian hostages, led north and south. On that first day, we surveyed over three thousand feet, ignoring many walking-sized leads. Over the next several months, Freedom Trail was extended to the south to connect with the shafts in upstream Elysian Way and north to connect with a maze of upper levels near the big waterfall at P88 in Lower Elysian Way. Lots of cave.

The initial breakthrough into Elysian Way was pay dirt for the CKKC. There was cave everywhere, and we were immediately intoxicated by the immensity of it all. We surveyed mile after mile, expanding the cave system in multiple directions. It looked as though it would never end.

Throughout all this fervor of exploration, Jim Currens was true to his word. He had vowed never to return to the cave to the north, and he did not—no matter that miles of passage were being discovered and surveyed. After a while, we stopped asking him to come.

Jim continued to follow his plan of beating the CRF into south Toohey Ridge. They could not be trusted, he assured us. Who could argue with him now? The CRF's blatant disregard of the connection moratorium was vindication of his long-held position, a reinforcement for his distrust of them. It had not taken long for us to find out about the secret connection trip taken by Diana Daunt, Roger Brucker, and Tom Brucker to the sump in upstream Logsdon River. Tom Brucker knew that the cat was out of the bag and had called me before I could hear about it from other sources. I was aghast as Tom told his tale.

"What the hell did you guys think you were doing in there?"

"Well," Tom said sheepishly, "I wanted my dad to finally have the opportunity to be on a connection trip, I . . ."

"What! How could you!"

"I regret it now. I know we shouldn't have done it."

"It's a little late for that, don't you think? What would you have done if you had connected?"

"I don't know . . . sorry."

I thought about it. Roger Brucker was sneaky. He had set us up with the connection moratorium agreement. Follow the cave wherever it goes, I said to myself. Such bullshit. He knew better; he had to. He was just playing on his son's emotions.

"Shit, Tommy, I can't believe you did that. We trusted you."

I could tell that stung, but he deserved it. Tom was silent. "But, I am glad you called to tell me," I said.

I forgave him, probably too easily. I knew that the likes of Diana and his father together would be hard for anybody to resist. Tom had been seduced.

I was furious! I had trusted the CRF, and now I had been screwed by them. For two years, I had been lambasting Jim Currens for his CRF paranoia. Much as I hated to admit it, he had been right. Not only had the CRF broken the rules, but they were broken by the person who had written the rules. I had been the group's strongest advocate among the cavers of the CKKC; now, my credibility concerning CRF politics was destroyed. I now knew that the CRF could not be trusted. We had to do everything we could to stay one step in front of them.

I shot off a handful of letters to Roger Brucker and the CRF president expressing my rage at this blatant attempt to connect the caves. I received no satisfactory answer except to hear that the political repercussions on the CRF side were considerable also.

Given the fallout of the ill-fated endeavor, I felt some comfort that there was little likelihood that the connection attempt would be repeated, for awhile anyway. Again, little could be done about it except protest. Meanwhile, there was cave to explore. I looked to the future.

Jim Currens's resolve hardened. He concentrated his efforts in a series of passages overlaying Logsdon River on the upper level. At the end of a long crawlway appropriately named Tylenol Trail, Jim explored a complicated network of small canyons. Tylenol Trail headed south off a small domepit at the southern end of Currens Corridor. After a few feet of pleasant passage, the lead withered to a low, dry, tedious crawlway. The small canyons beyond the six-hundred-foot belly crawl went south, but vertical shafts, breakdown, and dangerous traverses over deep pits made the going slow and difficult. These were tough trips, and the big discovery never came.

The Rift was the last remaining major area with leads that pointed to the south. Ron Gariepy and Bill Eidson, pioneers in the exploration of this area, had left an enormous borehole unsurveyed. There was no trip report and no information to build on. Bill and Ron, like many others, had not documented their efforts, so many things they had learned were inevitably lost. Some of their work had to be redone by later parties. Neither caver could recall substantive details about the borehole other than that it ended at a dark, menacing pool. It looked like a good lead on the map.

Jim Currens's party stood in front of the deep pool in May 1981. It was painfully apparent why Ron and Bill had stopped; the location was miserable. A strong breeze roared through the twelve-inch gap between the ceiling and the surface of the slimy pool. Undaunted by the water, Greg McNamara splashed into the pool and eased through the low, wet passage. A few moments later, his party heard his echoing whoops.

"Come on through! It's huge!"

This was too good to be true. The rest of the crew sloshed through after Greg and descended a steep mud bank into a large river passage. The river emerged from a high canyon and drained away from them to the south.

Jim's face beamed with excitement. Here was a big passage, thirty feet wide and high. They were heading due south and a strong breeze blew. What more could you ask for?

"Let's survey!" Jim ordered.

They pulled out the survey gear and began the march south. But after only eight hundred feet, the ceiling lowered and the stream disappeared into a gravel sump. They were stunned.

Defeated, they worked their way back to the north, looking for any missed side leads. There were none except for a few high on the wall they could not reach and the canyon from which the stream flowed. The air blew from this canyon, but it did not head south. Jim was not interested. They surveyed dutifully through the pool of water and linked the survey to the main line at the Rift.

The last of the known leads that might have led into south Toohey Ridge had died. I did not see Jim Currens for many months. It was time for a break, he said.

Over the summer of 1981, Dave Weller extended his dig in Dry Valley Cave to over one hundred feet. The tunnel he had dug was long and low for the entire distance. As is usually the case in most digs, the floor gradually sloped upwards to the headwall to where the clearance reduced to sixteen inches. If anything, the passage was getting narrower, and the pace of advance had slowed considerably. Not only was moving the tailings to the surface more difficult as the distance increased, but the floor was also more resistant to our digging efforts. Blasting was now necessary nearly every foot of the way.

Compounding the problems, every time it rained, a stream flowed into the entrance, pooling in the enlarged space of the excavated passage. It then was weeks before the goo dried out enough that we could resume work.

Dave Weller

One day, Dave proved that the rusty can lead had at least an air connection with Dry Valley Cave. He used a homemade smoke generator to create thick plumes of white smoke that were carried in by the breeze blowing through the entrance. Pete Crecelius volunteered to sniff for the smoke in the cave, like a bloodhound, and follow it to its source at the rusty can lead.

But this proof was not enough. The rusty can lead was much deeper down in the cave than the level of the tunnel Dave was excavating in Dry Valley Cave. The vertical difference was a problem. Air could move through lots of places that would be too difficult to dig through.

In early October, Bill Walter and Hal Bridges forced their way into the rusty can lead and, after seventy-five feet of narrow canyon squeezing, emerged into a small dome. Fractured rock walls from frost action showed that they were near the surface. At the far end of the small dome, they chimneyed between the narrow walls to reach a level twenty feet higher. A wide passage extended thirty feet to the south before becoming totally choked with rocks washed in from the surface. It was not obvious what they should do next. As they considered their options, they heard the distinctive *thump . . . thump . . . thump* of someone digging.

Dave!

The digging noise came from all around them. Low frequency sound carried by the surrounding rock was nondirectional. However, in just a few minutes, at the prearranged time, Dave would begin shouting into the cave from the headwall of the dig.

They waited. The noise stopped.

Then, the distinctive tone of Dave's voice greeted their ears. This time, the sound did not come from everywhere. Bill and Hal looked up; Dave's voice emanated from a previously unseen hole in the ceiling. Bill and Hal shouted, but because voices do not necessarily carry equally well in both directions in caves, Dave could not hear them. The hole above them was small and out of reach; deep red flowstone draped over its edges. They could hear Dave continuing to shout as they stacked rocks into a tower so they could reach the hole.

With flailing legs, Bill reached an arm over the lip and pulled himself up and through. The passage above was parallel to the choked passage. Bill crawled toward the sound of Dave's voice, now quite close. After twenty feet, he reached a round room with a high, narrow canyon snaking off to the left. Ahead, the passage continued but was too low. Lying on his belly, Bill squeezed in a couple of feet to look through the low passage, where he saw the glow from Dave's lamp. Dave was around a corner and very close—perhaps within ten feet. They were able to talk together in normal tones.

"Almost there," Bill said. Determined to worm through, he scraped at the dirt with a flat rock. But after fifteen minutes and two feet of progress, he decided he would not get out that way today. Bill and Hal went back the same long way they had come in.

Now it was just a matter of time before Dry Valley Cave would be connected to Roppel Cave. During the next two weeks, enthusiastic diggers narrowed the gap to three feet. However, the difficulties of continuing the dig for even that short distance were formidable. I proposed that digging could be done more effectively now from within Roppel Cave where there was considerably more elbow room to work. Any tools required could be passed through the tiny opening to the cavers inside the cave, avoiding the necessity of lugging tools the long distance from the Roppel Entrance. I guessed that only a few hours would be needed to finish the project. Roberta Swicegood and I volunteered for the trip.

I thought this was a hero trip, and I was surprised that no one else expressed interest in being a member of the in-cave party. Roberta and I began making plans.

With Cady Soukup, Roberta and I were the main CKKC force in the Washington D.C. area. The three of us were together frequently, meeting no less than once a week at my parents' house on Tuesday evenings. Although I had recently moved to Frederick, Maryland, forty miles to the west at the base of Catoctin Mountain, we were still using my folks' house as our base of operations. We had taken over the entire basement. Wall-to-wall tables were littered with cave maps, stacked boxes of survey notes, file folders of trip reports, and other documents having to do with the cave. In one corner, we had set up a typewriter with stacks of handwritten trip reports next to it and layouts of future issues of the *CKKC Newsletter.* On those Tuesday evenings we would get together, talk, and even do some work. Every aspect of the CKKC day-to-day operations was getting done. As Jim Currens had steadily slacked off administrative duties, we had taken them over with zeal.

Our initiatives infuriated Jim. Some of his duties we stole—the production of the newsletter was one—because he was not publishing it. Jim continued to carp at us, accusing us of deliberately cutting him out of the CKKC. We pooh-poohed his allegations as silly. Our discounting of him made Jim even angrier. He shut up after seeing that the *CKKC Newsletter* as edited and published by Roberta was of such high quality that no one could complain about it.

On the morning of the planned trip, Halloween 1981, Jim Currens showed up at the cave. After more than four months of not seeing him around, I had written him off as a contributing caver. He was angry at me for telling everyone that he had quit.

"Well, hadn't you?" I asked.

"No, I was just taking a break."

"A break? You've got to be kidding. A break from what?"

"From you."

His bluntness stung. I had become his loudest detractor recently. I was dismayed by his actions in mishandling the adversarial situation with the CRF and had let him and everybody else know of my disapproval. Perhaps I was unfair to have bad-mouthed him so roughly. Maybe I was unnecessarily harsh because he had no defense.

There was not much else to say, so I offered, "I'm sorry you feel that way." It wasn't much of an apology. I felt sorry I had made him angry, but I still disagreed with his withdrawal as a way to avoid conflict.

As Roberta and I started walking down to the cave entrance, Jim said, "You know, you're only taking this trip to steal the glory from Dave. You don't deserve it."

I stopped in my tracks. I was seething. "Okay, asshole, you want to go on the trip? I didn't see you volunteer." I was hot at his cheap shot.

"No, I don't." Jim turned and walked back to the others.

"I thought not."

Roberta and I continued toward the entrance.

Jim's remark had the ring of truth and made me uncomfortable. I did want the self-gratification of being on the trip that finished off the dig from the inside. But he was absolutely wrong about me wanting to steal Dave Weller's glory. Dave was the hero. It was Dave who had steadfastly worked at the dig for over a year, riding through the ebb and flow of our enthusiasm that sustained and threatened the work.

We had wanted quick success. When success had not come quickly, our enthusiasm and our participation too frequently disappeared. Often, Dave would continue his obsessive dig alone. Then his small victories would fuel our enthusiasm, and we would come to help, upsetting his methodical plans. But he would just smile and let us do what we wanted. Without Dave Weller, there would have been no second entrance to Roppel Cave.

Why was I going into the cave? I wanted to do whatever it took to bring the dig project to a useful conclusion. Dave deserved that, and I wanted him to enjoy the limelight of success. I also knew that if someone had offered to take my place, I would have happily stayed topside, because I hate digging. I was so concerned with the risk of failure that I would relinquish my place on this trip to anyone.

Hours later in the cave, I was lying on my side along the floor of a narrow and twisting canyon that lead to the inside of the entrance dig. My knees were wedged tight against the right wall, unyielding to my efforts to make them bend backwards. As I craned my neck, trying to look ahead, my denim coveralls soaked up water from wet streambed gravel. My muscles straining, I saw that the passage bent back again to the left and looked to be narrower still.

"Shit!" My back cramped as I tried to force my body around the next corner, left shoulder plowing uselessly through the wet gravel. I ground to a halt. My body just wasn't meant to bend this way.

Bill Walter and Hal Bridges had reported that this canyon was not too bad. Not bad? This canyon was pure hell. Resting again, I realized what the problem was. Hal and Bill were both shorter than me. My length was the problem. My six-foot, one-inch frame would not bend around the sharp turns of the canyon. I listened to Roberta behind me. She was struggling but was getting through with far less trouble than I was. She was overweight, but short. I was the opposite. So I knew what the problem was, but that knowledge was no help in this case.

"Damn, I don't know if I can make it!"

Roberta tried to be sympathetic. "I know it's tight, but just keep trying. You can do it!" She had to be cold too, stuck behind me in the narrow, wet passage.

Roberta knew I hated that kind of patronizing encouragement. I imagined her chuckling to herself behind me as I struggled.

In the narrow canyon, I gave another push and was around the corner. I could continue more easily now. There was black space above me, so I grunted and groaned my way into an upright position to where I could climb into the dome overhead. I was through!

Roberta was quick, climbing out of the wet slot just a minute or two behind me.

"I sure don't want to have to go back through there!" I told her.

Roberta agreed the passage was terrible. My coveralls were shredded, and we both were wet and miserable.

We followed the trail Bill and Hal had blazed up to the small room at the dig. I had to boost Roberta up through the tight hole in the ceiling. We must have looked like Abbott and Costello as I tried to push on whatever I could to force her up through that hole while I tried to dodge her wildly swinging legs.

While we were traveling through the cave, a large crew had been working to enlarge the passage leading up to the headwall, just three feet from our noses. Bill Walter handed us a sledgehammer, a digging trowel, and a crowbar through the tiny opening that connected the new entrance with the small room we were in. With a glint in his eye, he asked me how I liked the canyon drain.

"Pretty bad. I didn't think I would make it."

"I thought you'd like it," he said.

Roberta and I took turns working at the dig while Bill watched from the other side. Digging was slow and exhausting. When we were not digging, we were freezing from the cold wind blowing into the cave. Bill soon retreated to the entrance. After several hours, we were nearly in a hypothermic state, and the last twelve inches of the dig resisted all our efforts. It was time for desperate measures. We yelled our orders to the surface.

In fifteen minutes, four sticks of dynamite were placed into my outstretched hand. I shakily placed them at the last obstacle. The explosives had already been set up with a blasting cap and were taped into a tight bundle; a long wire stretched to the surface. Instead of a powder fuse, Dave Weller had begun using more reliable electric blasting wire. Too many times during the project, the dig had to be abandoned for the day when a powder fuse failed to detonate the charge. We never knew if a powder fuse had fizzled out or continued to smolder, so we would just leave it to burn out. Sometimes it would go off in fifteen minutes; sometimes it would not. Better to be safe. With electric detonation, if the charge did not ignite, we could immediately diagnose and correct the problem.

I could not properly tamp the charge in the confined space, but four sticks ought to be enough, I figured.

After we placed the explosives, we agreed on a time that Dave would touch the exposed wire leads to the battery poles to set off the dynamite. Unfortunately, there were not many places for shelter. Roberta and I were not willing to climb down the hole back to the lower level—that was too much trouble. Instead, we selected a twisting canyon leading off the room at the beginning of our dig. The multiple turns would surely deflect the concussion of the blast. The two of us squeezed into the narrow lead, forcing our bodies as far as we could. Three corners and twenty feet—not much distance between us and four sticks of dynamite. We hoped for the best, covered our ears, and waited for the anticipated blast.

Ker-blam!

The explosion was deafening. The immense concussion wave stunned us and forced us down hard to the floor.

Neither of us moved for almost a minute. I tried to clear my head.

Where was I? I slowly opened my eyes. Nothing: it was pitch black. My ears rang. I shook my head. I dreamed I was at home in bed in a deep sleep. I felt warm.

Someone was shaking me.

"James, are you okay?"

I had taken the bulk of the shock wave impact, my body cushioning Roberta from the blast.

The fog began to clear. I was in Roppel Cave. But what was that horrible smell?

"James! *Are you okay?*"

I shook my head to clear the last of the cobwebs. The smell—dynamite. It all came back to me.

"I'm okay, I think."

My back stung painfully, probably from broken rock ejected from ground zero.

"That was the third time I asked. I was beginning to worry."

"Well, I'm back. That was one hell of a shock. Maybe we were too close."

"You took the brunt of the concussion. To me it was just goddamn loud."

My ears were still ringing. I shook my head. The ringing continued.

"Yeah, it was loud all right. I hope we don't go deaf."

We felt around for our carbide lamps, which we had held in our hands but then dropped at the time of the explosion. The blast had extinguished them.

The smoke had cleared by the time we got our lamps lit. We crawled back to the dig. The smell of explosive-shattered rock was intense, and chunks of broken rock littered the passage. As we started pulling rocks away, someone came up to the other side of the blast pile.

"Who's there?" I asked.

"It's me!"

What an answer. "Who's 'me'?"

"Oh . . . sorry. It's Davey."

Dave Weller had a son, also known as Dave Weller. We all called him Davey to avoid confusion. Davey was fifteen years old and often helped his father on the entrance project.

Davey worked on the entrance side of the pile, passing the larger rocks over to where I could reach them and pass them back to Roberta, who then tossed

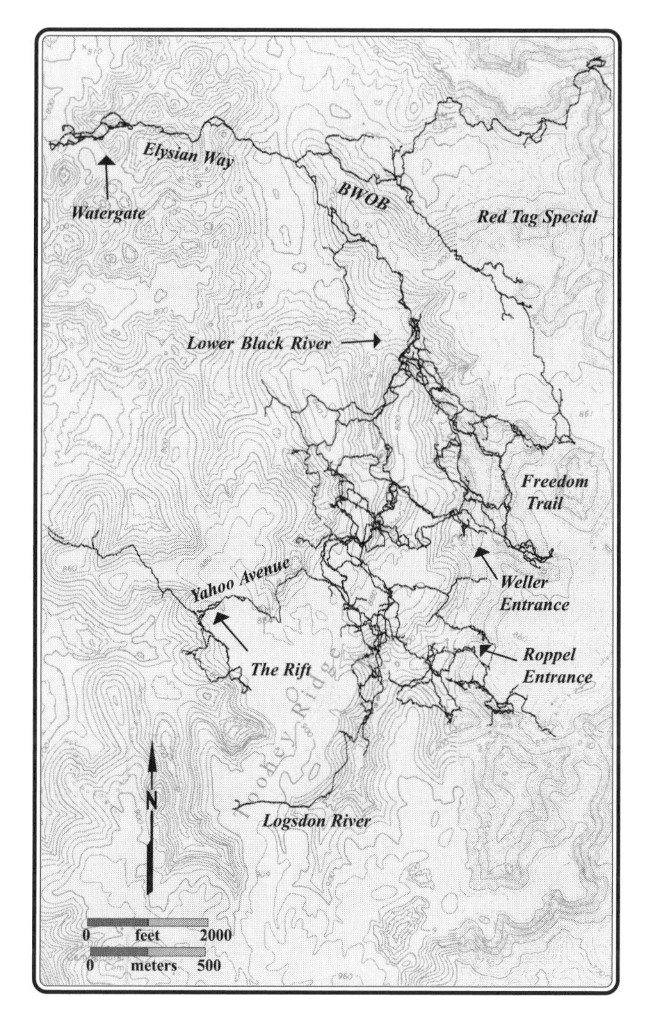

Roppel Cave, 1982

them out of the way back in the cave. In short order, we had a passable gap. Since Davey was all the way in, it made sense for him to come on through so he could turn around. A hard worker, he deserved the honor of being the first person to make the connection between Dry Valley Cave and Roppel Cave. Davey crawled through to outstretched arms and warm congratulations.

Roberta and I were shivering and ready to head out. With Davey in the lead, me in the middle, and Roberta bringing up the rear, we crawled toward the surface. The gap was certainly tight! It took us fifteen minutes to get around the narrow turns at the rear of the dig—a distance of ten feet. The sharp, irregular floor dug into our bodies. Wind whistled through the small space and extinguished our lamp flames. But we finally crawled out to brilliant camera flashes and a crowd of cavers who had gathered for the anticipated connection. Dave Weller smiled and handed us each a plastic goblet of champagne.

Breakdown

19

House of Cards

A Breakdown Pile and Political Chaos

Dave Weller's monumental dig was named the Weller Entrance. However, this new, second entrance was far from practical. A series of formidable and time-consuming obstacles delayed quick entry into the main part of the cave. The excavation was 120 feet long, almost half of which was low belly crawl, and took fifteen minutes to traverse. Once into the cave proper, we were faced with the difficult down-climb through the hole in the ceiling. We could not see where to put our feet. Some cavers would just let go, dropping the remaining three feet to the floor. The result was often a jumbled mass of a caver and toppled stones from the teetering tower of loose rocks built to shorten the free drop. Next came the short, tricky chimney down to the bottom of the dome above the rusty can lead. We called this dome Connection Pit since it had been the key discovery that had allowed the entrance to be completed. Roberta Swicegood named the rusty can lead out of it the Dred Drain. "This thing is dreadful," Roberta had said in a deadpan voice.

Cavers were spent before their trips had barely begun.

Dave Weller saw two principle problems. One, the Dred Drain had to be enlarged. Nobody wanted to crawl through that terrible canyon to begin (or finish) a long trip in Roppel Cave. It was too wet and too miserable. Two, the flooding had to be stopped. During rainy periods, the entrance became a muddy sleazeway.

Dave wanted to start with the Dred Drain, but first he had to enlarge the excavated entrance tunnel to allow his large girth to pass. This took several weeks. For the Dred Drain, he planned to use sledgehammers, crowbars, and dynamite—whatever it took to widen the canyon to an easy crawl for its worst thirty-five feet. I didn't think Dave's plan could be successful. It was difficult to blast solid, smooth-walled rock, because there were no cracks in which to jam the explosives.

On D-day—dynamite day—five of us gathered at the foot of Connection Pit to help Dave. He had over a dozen sticks of dynamite to set off, so on this day he was quite popular. He studied the problem at hand. The Dred Drain consisted of a narrow canyon (where one could fit) overlaid by a low and wide tube (where one could not fit). The cross section looked like an exaggerated mushroom. This tube followed the canyon for about ten feet before veering off and disappearing. Dave decided to work along the top of the canyon at the base of the tube (the top of the mushroom stem). There would be less rock to move and more work space. He would open this wider level for as far as he could, then drop to the floor of the Dred Drain and continue from there.

Dave pried big chunks of rocks from the wall and floor and set up the explosive charges. Connection Pit offered good shelter from the blasts. Dave did not have any electric caps, so we lit the fuses, rushed to crouch in a corner, and cleared out the debris after the blast.

We spent anxious moments when one charge did not go off, wishing we had the more dependable electric caps. Several minutes after the expected detonation time, we crept to the corner and carefully peered around. The fuse had burned all the way into the explosive charge but had failed to ignite the blasting cap. I shook like a green recruit in a bomb squad. Had the fuse fizzled, or was the blasting cap flawed?

We waited some more. What if the fuse was still ignited? A firecracker sometimes plays dud until some nerd picks it up. We could not afford that mistake with two sticks of dynamite. I crawled up to the unexploded charge and yanked the burned fuse; the blasting cap came out with it. It was a stupid thing to do, but I was lucky. Now we could continue with the project.

I was horrified when I saw Dave scraping the broken rock from a blast into the underlying Dred Drain.

"What are you doing?" I howled.

Dave looked confused. "Throwing the rocks into the canyon."

"You're blocking the route into the cave." The only way down to the bottom of the Dred Drain was at the edge of Connection Pit. He had packed it with rubble and made it impassable.

Dave looked at the rock in his hand, peered down the canyon, then blinked at me. With Olympic grace, Dave ceremonially heaved the rock in a sweeping arc over his right shoulder and into the Dred Drain. He laughed. "Don't worry. You'll never use that route again."

"Hope you know what you're doing."

"I'll have this fixed in no time." He continued clearing the rocks.

After more blasts, the drain was brimful with shattered rocks. However, we had progressed almost fifteen feet, nearly halfway through the obstacle. Maybe there was hope.

It took two more weekends for Dave to bypass the Dred Drain. He tunneled along the top level to a corner where he was able to drop straight down to the floor. After a few more shots at key corners, we had an easy way into the cave.

The project was not completed, however. Dave's mania for route building possessed him. As the months passed, the route into the cave changed continually. Roberta and I had discovered a lower level below the entrance passage. When I showed the map to Dave, he smiled broadly. He quickly blasted a new pit into the rubble floor just inside the entrance. We were astonished when we saw the once menacing stream now falling harmlessly into the lower level Roberta and I had mapped.

The flooding problem now solved, Dave worked on the floor of a shallow pit just beyond the Dred Drain. The route from Sand Dome into Kangaroo Trail was a body-sized tube called the Gunbarrel, which was too tight for Dave. This shallow pit was close to another section of Kangaroo Trail, and Dave was confident that he could punch through its floor and make an effective shortcut.

A few blasts proved him right again.

Next, Dave recruited us to drag sections of fabricated steel ladders into the cave. As the ladder we were dragging jammed hard in one especially tight curve of the Dred Drain, Roberta growled, "This is going to be a goddamn commercial cave before he gets done."

We pleaded with Dave to use restraint, but the ladders kept coming. We tried to impose limits on his project by restricting his work to the entrance side of the new pit connection to Kangaroo Trail. We yearned to dictate a halt to the cave modification—even we had our limits of how much we could stand! The conflict between cave conservation, volunteer enthusiasm, and engineering challenge raged. Dave was an engineering zealot who now resented our attempts to bridle him. After all, he knew he was helping the project.

He eventually relented to the pressure and promised to exercise restraint. However, he visited one last insult upon the cave: a twenty-five-foot ladder fixed top to bottom in Pirates Pot. We nearly cried as we viewed the red-painted indignity in the middle of the glistening-walled Pirates Pot. With that eyesore installed, Dave promised that the project was completed.

The Weller Entrance was like a mine portal. Explorers could reach Arlie Way in ninety minutes, about an hour faster and with far less energy expended than if they had taken the old route through the S Survey. The Weller Entrance shaved three hours, one way, for access to the cave to the north. The route to Pirates Pot, by way of Dave's ladders, was so fast that explorers resumed their northward push with vigor. The old Roppel Entrance fell into disuse.

I was sad: we had reached the end of an era. It was as if we had put down a favorite old dog and were left with only fond memories. The old Roppel Entrance was now as good as dead. Would the memories die also? I never wanted to lose them.

Several of our hardest-charging cavers scolded us severely for opening the Weller Entrance. How dare we reduce the cave to our own level? It was our obligation to rise to the challenge, not blast it into oblivion. The engineering triumph of the Weller Entrance meant the loss of one of the most challenging caves in the country. Now, they said, Roppel Cave was for wimps!

Some were so serious in their protest of this flagrant violation of the ethical spirit of adventure that they never returned to the cave. Their determined withdrawal saddened us, but we were practical, were we not? For the project to advance, we had to have that entrance. Were we hypocrites? Well, maybe all caving is hypocrisy. On one hand, doctrinaire cavers said we should subject our bodies to the challenge, but they thought nothing of smashing a rock that blocked their way into fragments with a hammer. Call us hypocrites? Conservation-minded cavers—including many in the CRF—railed at us for destroying cave features by blasting. I am interested in protecting the cave, but where does one draw the line? Cavers all over the world cry for strict cave conservation; the caves must be protected at all costs from the vandals, they say. Vandals? Who are the vandals? True conservation is pure preservation—no use of or access to the cave whatsoever. Effective conservation is measured and controlled change. Many cavers were quick to condemn us but immediately justified their own efforts elsewhere as necessary exploration.

Maybe the CKKC was a little more hypocritical than others. Well, people could complain. We had a vast cave to explore, and we were doers, not debaters. Nevertheless, old Roppel, that killer cave, was gone forever.

We had not foreseen the problems that such an easy entrance would bring. When the cave was so difficult, the cavers who arrived were strong and responsible; now, nearly anyone could cave in Roppel. The number of aborted trips increased, the quality of survey dropped, and the cave showed more wear and tear. The green cavers came by the carload, alarming us; we had a real management problem on our hands. We established rules about who could come. We formulated membership policies, the kind of bureaucratic nonsense for which we had criticized the CRF. This disconcerted all of us. And despite it all, we lost control.

We no longer knew where people were going or what they were doing. I was dismayed by the number of tourist trips to Yahoo Avenue that were taken under the guise of special "science" trips. The cave was wearing out. Maybe the cavers who called this a wimp cave were right.

Our amateurish, soft policy became embarrassing and ineffective soon after the opening of the Weller Entrance. Our lack of control tested our resolve to maintain the Mammoth Cave Connection Moratorium. We had condemned those deceitful CRF cavers after their bungled secret connection trip in April, loudly proclaiming that we were superior to that group. And now?

Just one month after the opening of the Weller Entrance, we learned how like the CRF we really were. On 28 November 1981, Bob Anderson and Linda Baker entered the Weller Entrance wearing full wetsuits, en route to Logsdon River. The two had asked me to come along, but at the time I was uninterested in a long trip or one that required wetsuits. They said they did not feel comfortable finding their way to anywhere but Logsdon River, so I approved their proposed trip. Anderson and Linda planned to probe downstream in the river to determine if there were any possibilities of a route into south Toohey Ridge. Strong airflow in Logsdon River was still an enigma, especially since we knew that the two caves were separated by a sump. Find where the air goes, and you'll find south Toohey Ridge, I proclaimed.

"But, don't push the breakdown too far!"

Anderson nodded as they ducked into the low entrance and disappeared into the cave.

Linda and Anderson had never used the Weller Entrance before and were unprepared for what they found. Fully expecting an easy route into the cave, they were dismayed at the unending series of obstacles that made the Weller Entrance difficult during the first weeks after its opening. They grunted through the excavated entrance tunnel, sweating profusely in their rubber suits. The small stream on the floor of the Dred Drain helped cool them off.

It took them three hot and difficult hours to reach the pool of water at S163 at the south end of Arlie Way. They eased into the murky water and crouched and sloshed south to Logsdon River, then turned downstream. The stretch of river beyond the Pulpit Room was worse than Anderson remembered. They had to swim through water over their heads, the waves from their strokes lapping at the walls. Beyond, the passage was barely four feet high, two feet of it water. Submerged sharp rocks cut their shins and continually threatened to trip them. Their backs ached as they bent over in the low, wide passage. Linda discovered that it was easiest to travel by lying in the water and pulling herself along. Better still, she could keep her legs above the underwater rocks. After two thousand feet, a dark breakdown loomed ahead.

They sat comfortably on the rock pile. It had been over a year since Anderson had sat, dripping wet, at this spot, but the memories of it were still fresh. This breakdown was unusual. Instead of being located beneath an overhead valley, the headwall where they were now sitting was well beneath the caprock of Toohey Ridge. This suggested that an undiscovered upper-level passage might have caused the collapse zone. The collapse consisted of enormous black slabs stacked like a tall layer cake. The stacked layers extended high out of sight into the ceiling.

The twosome first explored in the lower levels of the room where the gaps between the slabs were the greatest. Surveying as they went, they penetrated a few hundred feet along each side of the collapse. To the right, the openings lowered to tight belly crawls that looked as though they would soon become too low to follow. Linda remarked at how clean the breakdown was; there was no mud at all. They had been crawling on broad, flat rocks with a jet black coating, leaving behind a trail of contrasting white scrape marks where their feet had removed the coating from the rock. It would be decades before signs of their passing disappeared.

On the left side of the collapse, the slabs were broken and jumbled, fully blocking any way to squeeze through. No go.

Anderson then tried to push a route at stream level. The Logsdon River flowed to the base of the pile and forked to flow not only through but also around the pile. A main route was not evident, and other options closed down as well. Nothing.

They headed back toward the entrance, ignoring the other inviting openings along the river between the collapse and the Pulpit. Fourteen hours after going underground, they reached the surface and climbed the hill to the fieldhouse.

Anderson described the trip to me the next morning.

"We went straight to the breakdown . . ."

"What! You didn't push it far, did you?"

"Nah, just a few feet. It doesn't look like it'll go."

"How far did you go?" I was beginning to worry about the ramifications of what they had done.

"Not far. Maybe a hundred feet or so. It doesn't go."

"Are you sure?"

Anderson looked annoyed. "No, it doesn't go . . . anywhere!"

"Well, did you check anything else along the way?" They were supposed to check any leads that might lead south, and there were leads to check.

"No. We didn't feel like it."

"Why not!" I cried. "You were there! You had a wetsuit! Surely, you had time!"

"We just . . ." Anderson paused. "We just didn't."

I was puzzled. Why hadn't Anderson checked the other leads? What was up?

Although Anderson's capabilities as a caver were renowned, he had a reputation for controversy—not that he was ever at the center of controversy; he was too smart for that. However, Anderson was famous for fanning the flames of contention to bonfire level. He would do this while avoiding slipping into the middle of the fire himself.

After his and Linda's trip, I knew something was wrong. They had pushed into the breakdown in violation of the intent of our agreement with the CRF. Now, it could be that Anderson was not aware of the specifics of the agreement; but he had acted strangely when I spoke to him after the trip. What was it that made me uneasy?

In the coming weeks, I knew a storm was brewing when I heard the specifics of a trip report that Anderson was writing for the *Potomac Caver,* a monthly publication of the Potomac Speleological Club. I was sure that the statements in the trip report would be a catalyst for controversy. I called the president of the caving club to try to have the story pulled, but I was ridiculed by him for attempting blatant censorship.

When Anderson's story appeared in the *Potomac Caver,* I was unprepared for its provocative slant: "Who knows, we might even stumble upon the footprints of the 'lost' CRF cavers!"

What made the impact of his story so significant was that it immediately followed an unsigned short editorial, probably also written by Anderson. "An Attempt to Connect Roppel Cave to the Mammoth Cave System" gave a detailed account of the flubbed connection effort by Roger Brucker, Tom Brucker, and Diana Daunt. The editorial was laced with hearsay accounts the editor had gleaned concerning the motivations of the participants, what they had found, and the political fallout resulting from it. Reading the two accounts together, one would be bound to conclude that a retaliatory connection attempt to counter CRF ignorance and stupidity was justified! There was no mistaking his advocacy.

I was appalled. What would the CRF think? What would the rest of the CKKC think? I knew one thing: the story would be seized upon and interpreted as a defiant attempt by the CKKC to connect with Mammoth Cave. Something had to be done quickly, and the CKKC leaders scrambled to minimize the damage. Our greatest fears were that the CRF would interpret Bob Anderson's trip as an aggressive act by the CKKC meant to junk the Connection Moratorium. The CRF might then think that the way was now open for connection attempts to proceed, unimpeded by past agreements.

I recommended that we disassociate the CKKC from Anderson's publicized connection attempt. Pete Crecelius, the coalition's president, publicly ostracized Bob Anderson and Linda Baker for their flagrant violation of the Connection Moratorium and asked them not to return to Roppel Cave and Toohey Ridge.

A political tempest raged for weeks, unlike anything we had ever experienced in vituperation, accusation, and malice—and I was in the middle of it. Cavers from all over the Washington, D.C., area spoke vociferously and with disgust about what they viewed as injustice to Anderson and Linda. I was branded as the attacker and victimizer. Waves of verbal broadsides crashed in my ears. I was unprepared for and outclassed at moral warfare, bungling my attempt at response to their stinging assaults. Had I given permission for the trip, or not? I said I had. But I certainly had not given permission for a connection attempt or even a push of the breakdown, for that matter. The crux of the issue was that Anderson's flagrant public revelation could be construed as indicating that the CKKC sanctioned the connection attempt.

It did not matter what the twosome intended; what mattered was how the event was perceived by those who counted. The CRF would see the written record and draw their own conclusions. We might not swing the sword of moral indignation skillfully, but we could sever the offenders from the CKKC. Our cutoff and banishment of Bob Anderson and Linda Baker was our attempt to solve the problem.

The controversy roared on. I hunkered down, sticking to my story. I claimed that although I gave Anderson and Linda permission to go downstream, they had taken advantage of my vague instructions. They knew they were not supposed to try to connect, so they should be held totally accountable for their deed.

The strength of my position was weak, softened by my fuzzy memory. I further damaged my position by responding to an editor's request to comment on a story written by Linda for the *D.C. Speleograph*, a publication of the D.C. Grotto of the National Speleological Society. I objected to what I regarded as inflammatory parts of the story and urged the editor to cut the offensive sections. The editor took my suggestions without question. However, I had not consulted Linda, nor had I sought her permission. The edited story appeared, and Linda was outraged at the changes.

After months of letters between the CKKC and Bob Anderson, Linda Baker, and the Potomac Speleological Club, as well as numerous others who had been drawn into the fracas, the issue finally headed toward a climax.

On Memorial Day weekend of 1982, at the annual meeting of the CKKC, Linda personally pleaded with the CKKC board of directors for due process and justice.

I shrank in my seat as the tears began welling in Linda's eyes as she laid out her case. Her sobbing and sincere conviction that her position was right seemed to convince everyone—even me. I felt ashamed of my part in what had happened. What had begun as inadequate communication—I should have told them not to push the breakdown—had led to events and interpretations that had spun out of control. Through a series of errors, miscalculations, and the sensitivity of the various issues, it was apparent that Linda's personal feelings had been mowed over, and she was now paying a high price indeed. As I looked around the room, the long faces of my friends revealed that they felt the same sympathy that I was feeling. Linda was telling the truth. However, despite her pain, most of us remained suspicious of Bob Anderson. Was he really innocent? Was he counting on Linda's distress and protestation of innocence to mask the truth?

We stuck to our guns and did not rescind Pete Crecelius's banishment order. Anderson's story in the *Potomac Caver* was too much for that. Linda, regardless of her innocence, was bound to Anderson. Differentiating between the two of them would dilute the strong effect we hoped our public actions would have in explaining ourselves to the CRF.

To attempt to mitigate the extent of the personal damage, the CKKC board of directors voted to permit Anderson to document his story to the CKKC for presentation to the CRF. If he availed himself of this option, it might convince us of his goodwill that Linda so vigorously defended.

Weeks passed, and no statement came from Bob Anderson. With his silence, the issue was quickly forgotten. In a way, I felt vindicated. Conclusions drawn from missing evidence are weak, but Anderson's silence suggested to me that his motives were not what Linda portrayed them to be.

I then realized that, in fact, apart from the offending newsletter stories, Anderson had never made any public statements concerning the affair. Linda undertook to speak fully on his behalf. I reflected that regardless of where the blame lay, the damage resulting from the affair was substantial. I wanted to forget how much ill will the controversy created.

Bob Anderson and Linda Baker did not return to Roppel Cave. The CRF was openly amused by the series of events, remarking that this proved that we were no less sneaky than they. Their teasing tone and morally superior attitude rankled us. The CKKC's support of the Connection Moratorium had not changed; had we not been victims of an individual who could not be controlled?

But was not this same lack of control the root cause of the CRF's secret trip?

Cavers and circumstances cannot be controlled. Both the CRF and the CKKC had learned this the hard way. The more I thought about it, the more I began to realize that our two organizations were in many ways the same.

This was a vexing and frightening thought.

Jim Borden secretly photographing John Wilcox's map

20

Comfortable Ignorance

The Roppel Cavers Try to Ignore Painful Problems

Bob Anderson's report on the breakdown at the downstream end of Logsdon River was simple and to the point—it did not go. I was suspicious. Was he saving something for his next trip? It just did not make sense. Anderson's story had too many inconsistencies. No matter, I decided. Anderson was gone, and the breakdown didn't go. We would return to the pile of rocks at the end of Roppel Cave—sometime—but there was no urgency about it.

What were the possibilities of a connection to Mammoth Cave now? Were they small? Maybe. Nonexistent? Definitely not. What about the upper levels that Roger and Tom Brucker and Diana Daunt had found above the sump on their secret trip? Everyone knew that almost all the caves in the Mammoth Cave area would be found to connect—eventually. This wasn't just hope; passageways migrate downward. If they are closed on one level, they may be open on another. They had to connect because the cave went everywhere! There were just too many possibilities for Mammoth and Roppel. A connection would certainly be found sooner rather than later.

Explorers had known for years that Mammoth Cave sprawled beyond the geographic boundaries of Mammoth Cave National Park. Over the past ten years in Flint Ridge, CRF explorers led by John Wilcox had quietly surveyed the lower levels of Salts Cave to the southeast. The passages were large and exploration rapid as Salts Cave was extended over one mile toward Toohey Ridge. At the time, any discussion of Mammoth Cave extending beyond park boundaries was considered taboo. CRF cavers said they told the National Park Service about passages extending outside but did not furnish maps and, indeed, were silent about any specific extensions. However, the CRF's expressed policy was to follow wherever the cave would lead.

Wilcox was good at keeping a secret, and he planned for every contingency. His CRF trips were taken only on non-expedition weekends and were limited to a small circle of cavers who could keep their mouths shut. There would be no leaks. Wilcox kept all the survey books and took them home with him to Columbus, Ohio. He added the new cave to his map and stored the survey books separately from the CRF survey-book archives. But these maps were available to no one. Although cavers always suspected that Salts Cave extended to the east (anyone who looked at the map would wonder why all the passages stopped just a few hundred feet short of the park boundary), there was never any evidence. Why such secrecy? CRF cavers said it was so there would not be a repeat of George Morrison finding the New Entrance to Mammoth Cave off Mammoth Cave lands. Personally, I thought the CRF just liked to keep a secret; it felt good to be in the know when others were not.

During my work with the CRF during the mid-1970s, John Wilcox and I became close friends. I had undying respect for him, and he instructed me. His success and capabilities were legendary. He was precise, methodical, even-minded, and a hell of a caver—all traits to admire. He took as a professional challenge the task of shaping me according to the CRF mold. I was a tough study. Through his training, I discovered a passage that ended very close to Great Onyx Cave. Cavers had been walking by that lead for decades. The discovery was a vindication of my belief that the CRF needed fresh new blood, like me.

During my time with the CRF, Wilcox took me into his confidence and told me about Salts Cave's eastern reaches outside the park. Since Salts Cave lies relatively close to Toohey Ridge, he thought I should know. The exploration of Roppel Cave was still in its infancy, hemmed into that small corner of Eudora Ridge, but Wilcox knew that our scope was the whole of Toohey Ridge. He said he was impressed with my effort and dedication and offered these morsels of information as helpful instructive data.

A couple of years later, I visited Wilcox at his new home in Coolspring in the rolling hills of northwestern Pennsylvania. I had worked with his wife, Pat, entering Mammoth Cave survey data into the computer in Boston, Massachusetts. I had been a student then with time to kill, and Pat had invited me to stop over for a couple days. The two of them shared the CRF chief cartographer duties. Pat thought I would be interested in seeing their new cartographic operation, and indeed I was.

One sunny morning, we were sitting at the table sipping cups of hot coffee. I was still thinking about those enigmatic passages in Salts Cave, and I asked Wilcox about them again.

I listened carefully to his description, then asked the next obvious question.

"Is there a map?" I knew he had to have a map.

Wilcox looked thoughtful, then slid his chair out from the table. He stood up.

"Come on."

He walked through the living room and up the stairs. He went into a bedroom and reached beneath the bed to pull out a stack of maps neatly pressed between two large sheets of cardboard. He slid one sheet out.

"Here," he said.

I studied the map carefully. One main passage snaked across the entire length of the four-foot piece of Mylar drafting film. Along this main passage were a number of short side passages and cutarounds. There were gaps shown in the walls indicating unexplored or unsurveyed cave. At a scale of a hundred feet to the inch, there was a lot of cave here.

Wilcox was an excellent cartographer whose drawing brought the cave passages to life. I imagined myself traveling through the twisting passages as I followed them along on the map. I looked up enormous vertical shafts and squeezed into the many side leads. I longed to visit this place. Wilcox had carefully added the overlying topography on the map. You could see everything!

"What's this?" I pointed to one particularly enormous room, more than one hundred feet in diameter, in the middle of the map. Many passages radiated from the big landmark in the remote cave.

"Flint Dome."

I spanned my thumb and index finger over the gap between the two walls. It was over an inch. "Is it as big as it looks here?"

Wilcox smiled. "I'm sure it is the biggest vertical shaft in the system—over 120 feet in diameter and higher than you can see. I named it Flint Dome since it is the largest in Flint Ridge."

"Leads?"

"All over the place. Lots to survey."

I continued studying the map. I noted carefully where the eastern terminus of the passages were relative to the surface. I counted sinkholes northeast from a prominent finger of ridge. I wanted to be able to reproduce the endpoint on my map at home.

I asked about the end of Salts Cave.

"It takes about six hours of tough caving to get out there," he said. "Lots of water and mud, and the wind makes you cold—sucks the heat right out of you. We mapped for a long time and still didn't finish it. We were too cold. After we quit, we explored awhile, trying to warm up. We saw another five hundred feet of passage, all in four feet of water. The passage was only six feet high! I would say it sumped, but the wind was blowing. I don't know how far it goes like that."

Wilcox was laconic and serious as he told his tale. He never dramatized things. It had to be something wonderful to evoke this much feeling from him.

"It took us a long time to get out of the cave. Eight hours I think—I don't really remember, I was too tired. I think that I was as exhausted as I have ever been on a caving trip. I was so pooped I could hardly climb up all those canyons to the bottom of Dismal Valley. I thought I would never get out. The trip was twenty-six hours long—a real marathon."

"But it still goes?"

"Yeah, it still goes."

I felt a chill. Wilcox was a strong caver. If he had become that tired, it had to be a tough trip. The lead was obviously safe. I promised myself that one day I would see where it went.

"Can I have a copy of the map?"

"No."

He laid the map neatly on top of the stack and slid the whole pile back under the bed.

"How come?"

"Nobody has a copy of this map—too risky. We don't want to start any trouble with the park or other landowners."

I nodded, unconvinced.

I was at the Wilcox home for several more days. The possibilities shown on the map lying beneath the bed upstairs gnawed at me. One afternoon, I could no longer resist the temptation. I crept upstairs to take just one more look.

As I studied the map again, I realized that knowing where the end of Salts Cave was not enough. I wanted a copy of that map!

I considered borrowing it and driving to neighboring Punxatawny twenty miles to the north to have a copy made. However, there were several risks to that scheme. Punxatawny was a small town, and the blueprint shop owner undoubtedly knew John Wilcox, who visited frequently. There were not many cavers in this part of Pennsylvania, and not many people drew cave maps. Worse, what if the blueprint machine ate the map? It was not much of a chance, but the results would be devastating if I were forced to return to Coolspring with a crumpled and ripped sheet of Mylar.

I decided to photograph the map secretly. It was perfect—low risk and guaranteed success. I got my camera and flash from the car and arranged the map on the floor. I estimated the optimal distance to hold the camera above the map, the angle of the flash, and how many exposures it would require to cover the entire reach of passages.

My hands were sweating. I looked at the map and thought about what I was doing. It pained me to break a trust with Wilcox, but I had to have the map and thought the chances of getting caught were nearly nil.

I took a mosaic of snapshots that I could later stitch together to make my own map. With each click of the shutter, a bright burst of light from the flash flooded the room. My heart was pounding.

Six snapshots and the crime was complete.

I could not keep a secret. Within a week after returning home, I called Don Coons and told him what I had done. My feeling of elation was incomplete until I had bragged to someone. Don asked me a little about the map and seemed pleased with what I had done. I knew he would understand.

Unfortunately, his wife at that time, Diana Daunt, overheard enough fragments of the conversation to know something devious was up. In short order, Diana browbeat the information out of her husband, who crumpled immediately under her relentless pressure. With just one phone call, I had blown it. *Tick, tick, tick . . .*

The next weekend, I attended the Labor Day CRF expedition. In the intervening days, Diana had telephoned just about everyone in a position of responsibility in the CRF about my crime. On Sunday morning, the phone in the Austin House rang. The call was for Roger Brucker.

I was sitting out on the picnic tables in the warm morning sun talking with the others about the previous day's cave adventures. Ten parties had been underground, and there were many lies to swap. I knew nothing of Diana's activities since my conversation with Don. As far as I knew, my secret was still secure. *Tick, tick, tick . . .*

My conversation was interrupted by a loud bellow coming from the open kitchen window.

"Borden! Get in here!" It was Roger Brucker. "Now!"

"What the hell does he want?" I mumbled.

I had no idea what this was about as I walked into the house. Roger had his arm extended toward me, phone receiver in hand.

"Here." He thrust his arm at my face. "Talk!"

I took the phone from Roger, slowly bringing it to my ear.

"Hello?"

It was John Wilcox. My heart began to pound and my spirits sank as I listened. This was not a conversation; this was a tirade. Wilcox had found out about my covert picture-taking operation. His words cut like knives. He was very angry, and my violation of his trust injured him deeply. I breathed in short gasps, speechless. I stammered, offering a few guilty excuses, and said how sorry I was.

I handed the phone back to Roger and slumped down in the chair.

Shit.

I was sorry, but the damage had been done. For the next twenty minutes, Roger and I engaged in a shouting match, the details of which escape my memory. He was pissed and exploded with indignation. I became defensive. The entire camp heard the exchange. Probably Floyd Collins was turning over in his coffin below us in Crystal Cave.

The exchange slid into insults and threats. It was ugly.

I was humiliated. I had been caught red-handed, and Roger had twisted the knife in my festering wound. I had had enough.

I stomped out of the house and threw my gear into the back of the car. Gravel flew as my wheels spun.

In the heat of the moment, I had capitulated, agreeing to return both the prints and negatives of the map. I was ashamed. I had failed as a spy, destroyed my credibility, and, worse, had hurt John and Pat Wilcox in the process.

Diana Daunt later told me that she had been the one who sold me out. She gloated as she explained that it had been her duty to report me. She was a member of the CRF; I was just a Joint Venturer in the CRF, like everyone else. I glowered at her as she snickered about how clever she was.

Regardless of the outcome, I still had all the information I needed. I had paid a high price for my deceit, had hurt a dear friend, and was universally unwelcome at the CRF as a result. Unintentionally, almost befittingly, I was now devoted full-time to the work in Roppel Cave.

I leaned back in my chair, rereading Bob Anderson's trip report. Something smelled fishy, but nothing clicked. Disgusted by the recent turn of events, I reinserted the document into the file folder labeled "1981 Trip Reports" and threw it onto the cluttered desk.

With the outlook of pushing the breakdown in Logsdon River apparently grim, we discounted this threat of connection, for now. Instead, we concentrated on widening the frontiers of Roppel Cave, aided now by the easy access that the Weller Entrance provided.

Even with this new way into the cave, the far northwest end of Elysian Way was still remote by most standards, and trips were still infrequent. Besides, with the going cave in the BWOB as a temptation, who could blame us? I kept my counsel about Wilcox's Salts Cave passage he had mapped heading straight for the Watergate. No use doing more damage.

Nevertheless, I kept reminding Roberta Swicegood of the Watergate's allure, appealing to her sense of challenge. There had to be large cave out there to the northwest. A passage as large as Elysian Way does not just end; a continuation had to exist. After months of cajoling, Roberta and Don Coons could no longer resist the blowing passages at this far corner of the cave. They armed themselves with lengths of rope and artificial anchors and began a long trip to conquer the Watergate.

After eight hours, they finally stood on the first of the rimstone dams. They looked down a shimmering orange ramp that dropped ten feet into a dark, blue pool. A small hole in the wall looped around the upper five feet of the dam to where Don could hang onto some holds and lower himself into the deep pool. Roberta fixed a line around a pillar and payed it out as Don continued down to the next dam. The second dam was easier, something like a child's playground slide. He slid down into the next pool and swam across to the top of the next dam. Beyond was only blackness. One more climb down led him into a large room fifty feet wide. An enormous rimstone dam ten feet high extended from wall to wall across the room, like a barrier placed specifically to guard the cave beyond. The pool had long since drained away. They had to climb over the wall by pulling themselves up, throwing a leg over, then lowering themselves on the other side of the dam. Beyond, the cave was large and complex. They were now below the Lost River Chert. A thousand feet ahead, they intersected a large underground river of nearly the same flow as Logsdon River. Although this river immediately sumped both upstream and downstream, there were many low-level leads that were wet and blew air, teasing with the possibility of continuing to the west. The gap between the secret end of Salts Cave and the Watergate was now less than two thousand feet.

Around the same time, Dave Black was pulling rocks from under a ledge just a few feet from the base of the rope in the Rift. He and Danny Dible had felt a cool breeze and were wildly digging, trying to find its source. After a few minutes of removing round sandstone cobbles, they opened a down-sloping hole beneath the low, muddy ledge. The strong breeze cooling their sweating faces hinted of extensive cave ahead.

They continued the survey line below the Rift into the new lead and on that first trip added nearly three thousand feet to the map. Just past the dig, they followed a low, wide, cobblestone-floored crouchway that ended in a magnificent white flowstone mound. They called the passage Rocky Road.

A chert-floored chute under the south wall a few feet before the termination showed promise of bypassing the flowstone mound. They crawled . . . and crawled . . . and crawled. On that first trip, over a thousand feet of crawlway was pushed. It led on endlessly, the black chert walls, floor, and ceiling soaking up the light from their carbide lamps.

Over the next several months, Dave Black and Danny Dible led a series of trips to push the crawlway, now known as the Lunatic Fringe. They gave up trying to reconnect to Rocky Road and concentrated instead on going west. The passage continued relentlessly, nearly always crawling. The map showed that they were crossing under a broad valley and were approaching the eastern flanks of Flint Ridge. The possibilities spurred them on. This section of Flint Ridge east of Colossal Cave had for years resisted CRF attempts to discover cave in it. We snickered, thinking of a preemptive strike against the CRF, beating them at their own game as we explored miles of new cave beneath Flint Ridge.

The trips were difficult and long, requiring nearly a mile of crawling over chert to reach the limits of exploration. Each long trip back toward the surface was punctuated by the ninety-foot climb up the rope to the top of the Rift. The climb was made in three stages by way of broad ledges; the last stage was forty feet. On one trip, as Dave Black was reaching the top, the prussik knots he was using to climb began to slip down the muddy rope. Instinctively, he grabbed the knots, trying to arrest this unplanned descent. This only served to make the situation worse, and he began to glide uncontrollably down the rope. As he gained speed, the friction of the knots sliding against the standing rope caused them to melt and separate from the rope. He was now in free-fall. The rapid descent ended in a crash as he hit the bottom. He landed on his feet and fell backwards to a sitting position in a shallow pool of water. Blackness. The air stank from burning nylon.

As was customary on these trips, Dave had brought up the rear. The rest of the party had continued from the top of the Rift into Yahoo Avenue, where the waiting was drier and warmer. He knew that there was no use yelling; there would be no one to hear. He was sore but thought nothing was broken. He groped around for his lamp, finding it in two pieces beside him. He remembered that he had a small flashlight on a lanyard around his neck.

Dave squinted at the fuzzy scene now bathed in light from his flashlight. His glasses! Where were his glasses? He groped around and eventually found them a few feet farther away in a pool of water. They were still in one piece although smeared with mud; one of the lenses was scratched badly. He put them on, then he reassembled his carbide lamp and reamed the tip. *Pop!* The bright glow from the flame was comforting.

He stood up.

"Ouch!" He fell back to the floor. Bolts of pain shot through the heels of both his feet as put weight on them.

Dave breathed slowly and deeply. "Shit, I hope I didn't break anything."

As he was contemplating cutting the rope to make new prussik knots, a glow appeared at the top of the drop.

"Dave! What's going on?" It was Danny Dible.

"I fell. My knots slipped."

"Are you okay?"

"Yeah, I think so. My ass and feet are pretty sore—just bruised, I guess."

"Anything we can do from up here?"

"No, just send down some gear so I can climb out of here."

One hour later, Dave Black was crawling down Yahoo Avenue, unable to put any weight on his badly bruised heels. "The longest stretch of walking passage I'd ever crawled through," he later remarked. "I couldn't wait to get to the S Survey where everyone crawled and I could keep up."

Fortunately, Dave had not broken anything, and after a few weeks he was ready to go at it again. He returned to the lunacy below the Rift.

The breakthrough into Flint Ridge appeared to be imminent at last. But when the passage finally snaked beneath the ridge's broad sandstone cap, it necked down to impassable dimensions. The breeze had diminished, lost somewhere along the crawl's great length. The explorers had bypassed many leads that hinted of lower levels. There was more to find in the Lunatic Fringe, but it thwarted us for now. Exploration of the lower levels still awaits the hardy caver.

The cave system still seemed to have no limits. Everywhere we looked, we widened the boundaries of Roppel Cave. These penetrations were not without their price, however. After we extended one lead from the BWOB over two miles to the foot of Fisher Ridge northeast of Eudora Ridge, we found ourselves embroiled in a new wave of caving politics.

Since 1981, cavers from Michigan had been enthusiastically exploring a cave beneath Fisher Ridge. Optimistic from the beginning, they had

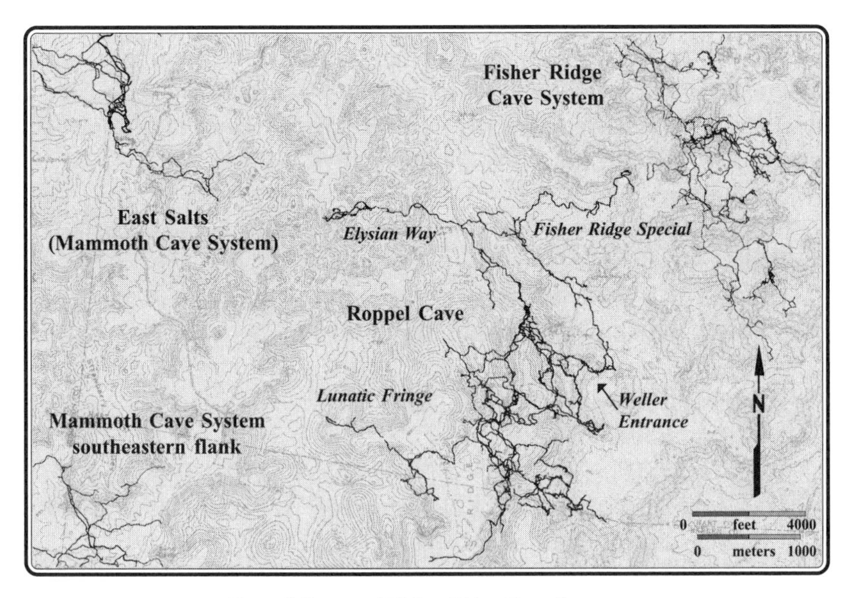

Roppel Cave and Fisher Ridge Cave System

named the network of shafts and canyons they discovered the Fisher Ridge Cave System. Their boldness did not jinx their discovery; the Fisher Ridge Cave System quickly grew into a fine cave with many miles of passage.

The Fisher Ridge cavers cried foul when we penetrated beneath their ridge. Worse, though, we had explored to within twelve hundred feet of a connection to that cave. Now, the shoe was on the other foot.

Insults were hurled and threats exchanged, a moratorium was proposed, deals were made. With little to lose, the Fisher Ridgers fought dirty. The landowner of their cave was dragged into the political quagmire and was requested to enforce his rights to protect his property to keep us at bay. We were appalled at their treachery! The section of Roppel Cave beneath Fisher Ridge was far away, and that fact alone led us to agree to yet another connection moratorium. We shook our heads with disgust at the turn of events. Now we had a moratorium on two fronts: one we had demanded and the other demanded of us. We felt stuck, jammed between the risk of yet more hypocrisy and the risk of turnabout if we should seek a connection to Fisher Ridge. What a joke. Cavers all over the country laughed at the CKKC, the CRF, and now the Fisher Ridge cavers for engaging in exploration-bridling politics. Limiting connections by agreement began to suck, in our estimation. Was the posturing worth it? If the caves connected, then so be it!

Politically, things looked largely under control on the connection front, but the CRF had a few more cards to play. After our discovery of the Logsdon River breakdown in Roppel Cave, the Mammoth Cave Connection Moratorium seemed finally enforceable, if only because of the breakdown and a sump. However, the CRF explorers were not yet ready to give up their desire to know more about that vicinity of the cave. They had learned their lesson after the Brucker connection attempt. Now, they courted Pete Crecelius to lead CRF trips up the river. They were clever. If Pete, the president of the CKKC, led the trips, then the spirit of the moratorium would be preserved. Any efforts now would be considered joint efforts—at least, that was how the CRF viewed it. The self-serving ploy was obvious to me, but Pete took the bait anyhow, "in the spirit of cooperation." I could not dissuade him—he was playing right into their hands. All I could do was grit my teeth.

Pete led a number of trips to the upper levels above the sump first explored on the Bruckers' connection reconnaissance. There was a lot of cave, but breakdown hemmed progress farther to the east toward Roppel Cave.

Pete did not find Roger Brucker's obscure drain to bypass the sump. However, on one trip in February 1983, he made the discovery that abruptly terminated our complacency about connection. On this particular day, he noticed that the level of sump in upstream Logsdon River in Mammoth Cave had lowered to the point where one inch of air space now existed between the water's surface and the ceiling. As Pete crouched to peer through the low gap, he felt a cool breeze on his face and heard sounds of a small waterfall.

The sump was draining and was now open—barely. Once again, the situation had changed and we were forced to rethink our strategies.

Ferguson Entrance

21

A New Entrance to Secret Passages

Growing Complexity and Changed Relationships

I—Roger Brucker—learned in 1979 from Scooter Hildebolt that John Wilcox had been running low-profile cave trips since 1976 south of the Frozen Niagara section of Mammoth Cave. From Scooter's hushed, excited accounts, I knew that Wilcox had penetrated beyond the boundary of the park.

"We haven't been surveying. John has made careful compass and pace estimates, so he knows where this cave lies in general. But there aren't any surveys or maps!" Scooter said. "I saw a footprint of a boot in the bottom of a pool of water in a passage way out there. Not a Vibram sole mark, but a smooth print, like a farmer's work shoe. Couldn't have come in the way we did. Think there could be an entrance out there?" he wanted to know.

More secret exploration, I thought. I asked Wilcox about it. He explained that there were no surveys or maps of these passages that led outside Mammoth Cave National Park; this preserved deniability. Whose deniability? The National Park Service, for one. Any Freedom of Information Act request for details about south-going passages could be truthfully answered; the park possessed no information. Furthermore, the CRF could deny the existence of surveys and maps. Nobody could disclose what they didn't know. Ever since the days of George Morrison's secret surveying in Mammoth Cave that led to his own New Entrance, there was the possibility of enterprising landowners gaining a marketable access through a new entrance to Mammoth Cave. John Wilcox wanted to avoid that. But he was forthcoming with me when I asked.

Wilcox had run trip after trip out East Bransford Avenue and Cocklebur Avenue in Mammoth Cave. Lynn Weller and Scooter Hildebolt had participated. They had found an endless succession of canyons and crawlways heading mostly southeast. These passages certainly extended outside the park under the upland of the plateau. There was no natural obstacle between Mammoth Cave and newly connected Logsdon River in Proctor Cave. The French Connection was an accomplished fact, so as the months of 1979 turned to the fall season, CRF teams revisited and carefully surveyed the secret passageways. Our surveying on the occasion of the August 1979 connection trip was an extension of the Third A Survey with excursions into promising side leads. There was going, hot cave all around the Bransford-Cocklebur area of Mammoth Cave.

After Mammoth Cave and Morrison Cave were joined, the CRF had to survey the connection route to tie the caves together and to map the rest of what we were finding. I led a trip into Morrison Cave to survey the connection route just two weeks after Wilcox and Tom Gracanin had made their nude swim.

Scooter found sticks and organic matter plastered against the ceiling of Logsdon River. "Do you think it floods much?" John Branstetter and I said we didn't know. It set Scooter to thinking.

"How fast do you think the water can rise?" he asked. Again, nobody knew. We waded upstream toward Station Z7.

"But if the water comes up, we could get trapped in here!" He was beginning to cause us to look at the river flotsam more carefully.

"Shut up, Scooter. You're scaring these folks," John said.

"Yeah, but ten years from now my daughter, Nea, may ask her mom, 'What was Daddy like?'"

"Shut up, Scooter!" we yelled.

Scooter and John had not seen this part of the cave, and Scooter marveled that anyone could have squeezed through the pool with four inches of air space.

Scooter forced his wet way through the squeeze on his belly and complained loudly. I went through the low spot on my back, but the water seemed dangerously high. Scooter's concerns were getting to me. I spent one anxious moment slipping down the sloping pool bottom toward certain total submersion. My fingers caught a ceiling nubbin and halted my slide.

"Scooter, how did you get through this?" My voice quavered a little.

"I had to submerge one ear and one nostril to pass through," he replied. John said he'd watch what I did—and do something else.

The main work was to assimilate the ten or more miles of unsurveyed cave in and adjacent to Logsdon River. Each trip required wetsuits, rope work, and eighteen to twenty-two hours. At age fifty-one, I was, objectively, too old to do this kind of caving. But I felt the so-called male menopause rush that has been blamed for causing otherwise sane men to reevaluate everything important in their lives: their physical condition, families, marriages, jobs—everything! I felt I was at the peak of my caving ability, able to complete the toughest trips carrying rope and wetsuits.

Yet, I feared that at any minute I might become too old. My friend Red Watson had lorded his excellent physical condition over everyone by running miles each day and eating bran. Then he essentially quit caving, although he kept on running. God, I thought, I don't want to fizzle out like that. And Watson was a couple of years younger than me. Old age could sneak up!

My kids were mostly grown up. I caved with my son, Tom, now and then, although I felt he subtly patronized me ever since he had discouraged me from going into Proctor Cave. My daughters Ellen and Emily were finishing college near Seattle; Jane was at home planning to marry. Soon the nest would be empty—and my marriage had been a problem for several years. To be honest, caving—the hardest kind of far-out caving—was a sublimation of my unexpressed anger regarding my marriage.

That summer of 1979, I made five wetsuit trips through the Morrison Cave Entrance, surveying the Logsdon River and tributary passages, including the T Survey. Chris Gerace, Lynn Weller, and Tom Gracanin were key workers on these arduous endurance marathons. The surveys were characterized by one-hundred-foot shots, surprise dunkings in deep spots in the river, body surfing in the rapids, and good humor.

We made a spectacular trip in July to survey up the right-hand fork of Hawkins River in Proctor Cave. We wore wetsuits and bobbed on inner tubes, surveying through clouds of flies rising from the river. After several hours of heading toward Park City, Kentucky, we came to a sump. Inner tube surveying gave me a glimpse of what it must feel like to be a cave diver, night parachutist, or solo astronaut. The circumstances are threatening, the environment alien, and one's position is unstable. The visible world ended at the edge of my cone of light. I was utterly dependent on my bag of life support equipment. The least slip, and I would be history. It was ironic to feel this way when most people would see no distinction between a caver's madness and that of other risk takers. But I hated a continuous dependence on equipment. I was uncomfortable abandoning all to the performance of my gear.

The official announcement regarding the connection between Mammoth Cave and Proctor Cave was made by the Park Service regional public information officer at a press conference at Mammoth Cave National Park in October 1979. Superintendent Robert Deskins then remarked on the historic importance of the expansion of the cave and the vast regional drainage basin whose pollution could threaten Mammoth Cave, and CRF president Cal Welbourn discussed the scientific importance of studying such an extensive cave. Members of the connection party told their individual versions of the events before, during, and after the connection. We passed out packets of photos, background materials, and a map. Why had we waited so long, from 11 August to 1 October, to break the news? The Park Service information officer fielded that one. We had wanted to survey the connection, evaluate the reports, and prepare an announcement, all of which took time.

Later that month, a Cincinnati newspaper reporter wanted to write a first-person follow-up story. He claimed to be able to rappel, mountain climb, dive, and parachute jump, so we took him through the Morrison Cave Entrance to accompany a survey trip. Scooter Hildebolt, Tom Gracanin, Stan Sides, and I—all regulars at river surveying—expected him to be able to handle himself. Wrong! The first tip-off was his boast of how fearless he was; the second, his brand new red coveralls and vertical gear. We taught him to rappel on the spot. He learned the practical application of the theoretical on his sixty-five-foot slide to the bottom of the first pit. We surveyed fifty stations in walking cave. When it was time to leave, we taught him how to use his new

Jumar ascenders. His story, more sensational than factual, carried the title "A Mammoth Cave Scare—I Was Trapped and Knew I Was Going to Die!" It was typical of the breathless style of adventure hacks who write about caves with no experience or discernment.

On New Year's Day of 1980, I received a telephone call late at night from my son, Tom. He had been on a trip into the upper-level passageways above the Third A Survey in the Bransford-Cocklebur area where we had made the connection back in August. His party had felt a strong, cold breeze in the passage they were surveying. When they could survey no longer because of time constraints, he and Richard Zopf had pressed ahead. The canyon was tight and led to the bottom of a shaft about forty feet high. There were many dry leaves there. The wind was so strong in this shaft that it blew out an unguarded lamp flame. The pair briefly checked climbing possibilities, then started up a crack on one side of the shaft. They used opposition holds and layback moves to clear the seamless parts of the wall. Near the top, they traversed around the edge.

"It looked like dirt and leaves being blown in by the wind," Tom told me. Was there a hole, an opening to the surface? He couldn't be sure in the gloom.

The following day, Tom made a rough estimate of where the shaft might be located in relation to the surface topography. It was a bright, mild January day when he, Kathleen Womack, Claire Weedman, and George Wood hiked through the trees in the approximate area. Before long, they found a canyon-like opening in the hillside, twelve feet wide by fifty feet long. The rock walls dropped precipitously fifteen feet to the bottom, except at the down-slope end of the crack where the explorers climbed down a dirt collapse. They moved into the darkness, climbed down the face of a breakdown block, and walked a few feet to a funnel of rocks and leaves next to the cave wall.

In a few minutes, they moved enough leaves and rocks to feel a sudden increase in airflow. Rocks chucked into the newly opened black void crashed on the bottom of a drop.

The discovery presented a dilemma. It might—or might not—lead to the dome Tom had climbed a few hours previously. An entrance to Mammoth Cave off national park land would attract a lot of attention. How could it be protected? They didn't know the landowner, nor had they received permission to tramp though the woods. Tom and Kathleen decided to keep the existence of an entrance confidential until an approach could be worked out. They pulled down rocks and leaves to leave the funnel plugged as they had found it.

A later discreet check through a third party revealed that the land over the new entrance was not for sale.

In 1980, the Mammoth Cave length was 214.5 miles. Planning had started for the Eighth International Congress of Speleology to be held in Bowling Green, Kentucky, in July 1981. The CRF shifted its focus to logistical and political matters, away from breakneck exploring. Sarah Bishop, one of the CRF's brightest, started preparing a nomination of Mammoth Cave as a World Heritage Site, a United Nations honorific designation for the most outstanding natural and historic places of the world.

Exploration continued in Logsdon River through Morrison Cave, and also in John Wilcox's secret E Survey in Salts Cave. The E Survey headed outside the park directly toward Roppel Cave. The New Discovery section of Mammoth Cave continued to grow, but we were still not close to connecting with anything that interested me.

Dr. Nick Crawford, director of the Center for Cave and Karst Studies at Western Kentucky University, asked me if I would teach a field course in speleology the week of 9 June. Crawford was launching a summer educational program in cooperation with the National Park Service and had already recruited such prestigious caver-professors as Dr. Art Palmer, Dr. Derek Ford, Dr. Tom Barr, and Dr. Jim Quinlan. Mr. Brucker would be in good company!

The prospect of an international congress of cavers and a speleology course may have seemed a diversion from my purpose of wanting to connect caves, but my first love through this life of adventure had been teaching. I taught a Sunday school class of teenagers one time when nobody else wanted to and enjoyed it immensely. I'd been a YMCA camp counselor on several occasions and had a ball. I taught marketing as an adjunct professor at Wright State University. What could be better than teaching people how to understand caves and to find more of them? I could teach people how to connect caves! The experience proved to be life-changing for the students and for me. The enthusiastic reaction of those students convinced me that teaching caving was, in many ways, more fun than caving itself. I have taught the speleology class for WKU for twenty years since 1980, influencing about 250 students.

Near the end of June 1980, I packed up everything I owned and left home. It was a precipitous desertion, a coward's way out. My wife, Joan, was devastated. She said later she hoped I never allowed anyone to fall in love with me because I was too much in love with myself to have room for anyone else. At the time, she was absolutely right. My mother blamed my failed marriage on too much caving. I knew in my heart it was my own anger and unwillingness to deal with it as an adult that motivated my decision. Then there was always the fear of age sneaking up. First, it was a hernia, then cancer, then . . . what? From this point forward, I told myself, I would select among all the alternatives life presented and be responsible for the choices I made.

In July 1981, when all eyes were focused on preparations for the Eighth International Congress of Speleology, two cavers from Bowling Green were surface walking some of the valleys around Mammoth Cave National Park, looking for signs of cave. Jim Carter and Billy Matlin found the same canyon entrance that Tom Brucker and Kathleen Womack had found six months earlier. Since Carter and Matlin had dug away rocks and leaves to expose the drop, they naturally assumed the new entrance was their original discovery. They asked Don Coons, who knew nothing of the new entrance.

Carter approached Louise Hanson, who owned the property and several others along the road to Mammoth Cave National Park, and asked to lease the entrance so he could explore the cave. Mrs. Hanson agreed to allow his use of the entrance.

Don Coons and Jim Quinlan quickly learned of Jim Carter's coup in tying up an entrance. After Carter cleared away the rocks and rigged the entrance drop, it was only a matter of minutes before he found survey station marks deeper in the cave! Carter asked Don if he knew anything about that survey.

Don guessed that the survey was probably part of the Bransford-Cocklebur extension of Mammoth Cave, where the connection of Mammoth Cave and Proctor-Morrison Cave had taken place the previous August. If this were true, it was a shortcut to the connection and was indeed another entrance to Mammoth Cave. It required a gate, no matter who controlled it. Don called Tom Brucker at his home in Nashville to ask what he knew about a J Survey in a tight canyon. J30 was the last marked station.

Tom was flabbergasted. His secret entrance had been found before plans could be laid to control it by the CRF, and it was now in the hands of an unknown third party! Don had also alerted Jim Quinlan, who was even now having Dave Weller fabricate a gate for this new entrance to Mammoth Cave. Quinlan had funds as well as an appreciation of the scientific value of an entrance so close to the vast Logsdon River drainage of the cave. Dave, from the CKKC, had appointed himself as the civil engineer of the gate building project, which was going forward at high speed.

With alarms going off and the circumstances spinning the decision making out of the CRF's hands, I contacted Jim Carter by telephone. He seemed surprised by my direct approach but knew that there had been a lot of interest by a lot of cavers in his discovery. I appealed to his noble motives: the new entrance had to be protected by a gate as quickly as possible since cavers wanting to yo-yo in for a look-see could be expected at any minute. I told him about the CRF's successful effort to negotiate a lease for the entrance to Morrison Cave and how we had installed a fabricated gate just in time for Cleveland caver Bob Nadich (on a spy mission) to see us removing the last of the concrete form boards from that gate. A significant entrance *had* to be protected with a gate before the spelunkers and vandals arrived.

Carter volunteered that his intent was to allow Quinlan, the CRF, and the CKKC to use the new gate and that all parties would receive keys. But what about the larger public relations issue? What would happen after the landowner had time to reflect on the implications of owning an entrance to Mammoth Cave? "Is she interested in the cave?" I asked Carter.

"Yes, very much so. I've shown her some photos," Carter said. "She doesn't want to go in it."

Thank God, I thought. Louise Hanson was a remarkable person. Micky Storts had interviewed her as part of the CRF's oral history project in 1959. She was a sister-in-law of Pete Hanson, who together with Leo Hunt had found the New Discovery section of Mammoth Cave. In 1938, those two explorers also had crawled out Hansons Lost River, the eventual connection route between Mammoth Cave and the Flint Ridge Cave System. Mrs. Hanson was interested in caves, but if she saw those pits and canyons, she might close the entrance. Nobody needed an attractive and dangerous nuisance like that entrance to precipitate litigation.

"Next time you see Louise Hanson, tell her we're working on a map to give to her. It shows all of the cave under the new entrance. We should name it the Ferguson Entrance. Wasn't that land in the Ferguson family, and wasn't Louise a Ferguson?" I put out all the lures I could think of.

"We've already named it the Ferguson Entrance," Carter said. "I'll tell her about the map." Carter wondered if we could supply some labor to help install Dave Weller's new gate. I said sure. A secure gate would give us some breathing time and allow the CRF to prepare our pre-Congress field camp and trips for international visitors.

Through the year, I had been talking occasionally with Jim Currens about the desirability of holding a seminar for project caving. We set a date, 12 and 13 September 1980, in Lexington, Kentucky. The NSS was interested in co-sponsoring it as the Third Cave Project Workshop. Two similar meetings had been held in the east with some benefit for the growing cadre of

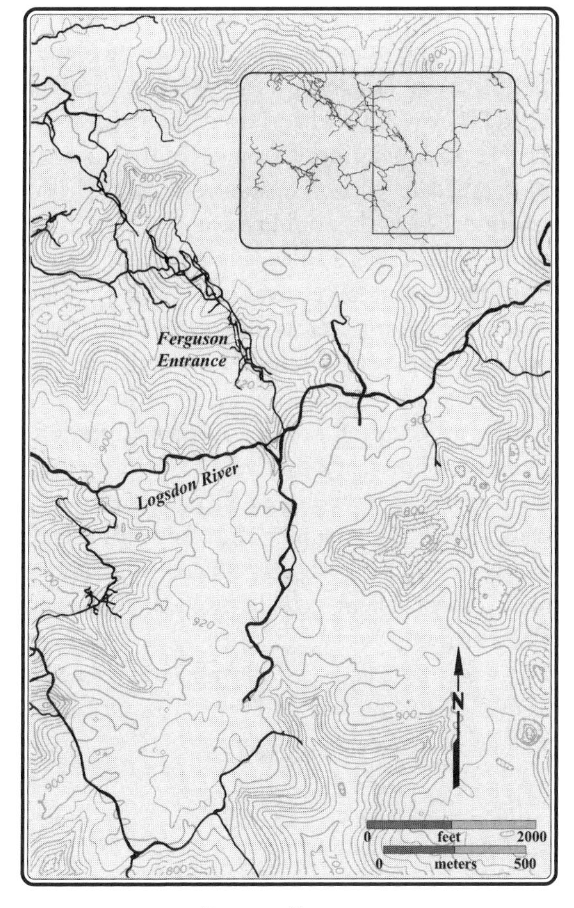

Ferguson Entrance

cavers in the country whose interest extended to a sustained series of trips into significant caves. The CRF had more experience than any other group and had practically invented project caving. When we set up the CRF, it had been designed to avoid the worst characteristics of project caving in this country. The CKKC had been evolving along the CRF model, although Jim Currens and Jim Borden claimed to deplore the CRF's regimentation and elitism. Currens was genuinely interested in how a caving club could be transformed into a professional institution that would endure and sustain excellence. Both of us knew many of the movers and shakers of project caving in the country, so it was fun putting together presenters and lists of cavers to invite.

It was the teacher in me that loved this aspect of caving. Cave projects in the U.S.A. suffered many common problems. One such was how to get rid of the charismatic firebrand leaders who started a project, then became irascible as they aged. They wouldn't hand over "their" organization to young Turks who "haven't paid their dues." More often than not, such leaders became disgruntled, rolled up "their" maps, and disappeared, leaving the troops with no information. The cycle would repeat.

Another example: How could surveying and cartography problems be handled without chaos in a large cave system? Organizing and keeping track of the data is a major undertaking. Microcomputers were becoming popular, and there was wide interest in plotting cave maps using computers.

Finally, how could a monolithic project caving organization accommodate the varying interests, abilities, and motives of a diverse group of cavers? Of course, we sought cave pushers, but the CRF had also deliberately set out to champion diversity in cavers, developing and using each talent in assignments that best suited that individual. We encouraged women to assume leadership roles. The primary challenge was to provide opportunities for cavers, not to order them around.

The workshop was a big success. Jim Currens still seemed to have unspoken suspicions of the CRF, but we had worked together and enjoyed building bridges at all levels between our organizations.

"Hello, hello?" Jim Quinlan's cheery voice greeted me on the telephone. "Don't wish to impose. Could you help us with a radio survey to locate Logsdon River and fix a position for a new entrance?" Jim Quinlan was well known for his comprehensive survey of all the water inputs, outputs, and drainage basin divides in the Mammoth Cave region. It was one of the best scientific efforts ever undertaken in the region, and I would do almost anything for Quinlan.

He explained that Frank Reid, an Indiana caver, CRF Joint Venturer, and cave radio expert, was coming down to the cave. Quinlan wanted us to take the radio antenna and transceiver into the new Ferguson Entrance to Mammoth Cave. Frank would roam on the surface and operate his direction-finding cave radio. The test would fix the exact surface location and vertical distance down to a survey station deep in the cave.

"I may be able to get some money to put in an entrance in Doyle Valley to Logsdon River," Quinlan said. I had been on these radio-locating trips several times before; the most recent had been in Morrison Cave. It was interesting and important work, but there were problems. Often there was a lot of farting around establishing radio contact. The main problem was the bulk and weight of the cave antenna. Frank's early rigs required carrying a hernia-producing twenty-pound coil of wire mounted on a two-foot piece of plywood.

"Sure, we'll help. But it may be tough getting the antenna through the J Survey at the Ferguson Entrance," I said.

"Already thought of that. I'll give you Geary Schindel. He's one of my crew. He'll carry the antenna. You won't have to," Quinlan said. "Just show him the way, because I don't think he could get there by himself."

Geary Schindel, a young, strong graduate student in Nick Crawford's cave and karst studies program, wore his wetsuit into the cave. He flipped the heavy antenna around with apparent ease. We pushed through the U-Tube into Logsdon River, then ate a meal and rested. Lynn Weller asked what Geary thought about the Mammoth Cave–Proctor Cave connection route.

"It's a bunch of bullshit," he said, looking around. "Is this the river? It's a cinch to get here. You guys move slow." Geary was a different sort of caver. He reminded me of the Cincinnati newspaper reporter with his bravado. But unlike the sport in the new red coveralls, Geary's clothes were shredded. Here was a caver who could handle himself extremely well in a cave while burdened with an antenna resembling an eighteen-pound toilet seat.

Radio antenna used to determine precise surface location of a cave passage

Geary continued: "The politics of this stuff suck. It's totally outrageous to hold off hooking together this river with Roppel Cave."

I said the delay was the desire of the CKKC. They alone could decide when and whether to connect the caves. From the viewpoint of the CRF, we had plenty to do without connecting. There were walking leads in several places heading toward Cave City, Park City, Nashville, Louisville . . .

"I wouldn't be bound by anything as dumb as that. Nobody can dictate where you can go and can't go." Geary's eyes narrowed. "What's to stop me or anyone else from just making the connection?"

"If you were part of the CRF, we'd throw you out."

"I'm not part of anything," said Geary. "I might just walk in here and make that connection myself, if people don't get their asses in gear."

"I think it won't be that easy," I said. "The river sumps upstream at the Roppel end. I've been there a couple of times. There's no easy connection to be walked into. I know."

For all of his abrasiveness, Geary seemed intelligent, capable, and an enjoyable caving companion. We set up and operated the radio successfully at the appointed times and heard Frank Reid's encouraging voice in response.

In later months, the radio survey was performed again, at a better location for the entrance drilling rig in Doyle Valley. However, the memory of Geary Schindel's threat remained alive in my mind: "I might just walk in here and make that connection myself, if people don't get their asses in gear."

Surveying in the breakdown

22

Gasoline on the Fire

The Control Freaks Find Themselves Out of Control

All of us—particularly me—were curious about the Logsdon River upstream siphon and the upper-level leads there. Would those passages connect to Roppel Cave? Pete Crecelius, the CKKC's president, was willing to run trips there, and we egged him on. If one of those trips found a connection with Roppel Cave, the blame would fall on the CKKC's own man. If we found miles of additional cave passageway but no connection, so much the better. On one of these trips, in February 1983, Norm Pace found a new dome complex near the domes my son Tom and I had discovered. There seemed to be leads everywhere.

In January 1983, Lynn Weller and I were married. We spent our honeymoon in the Mammoth Cave area at a much more leisurely pace. Most weekends, we had led separate parties on grueling river trips in wetsuits. Occasionally we would cave together, which I much preferred.

In mid-August 1983, Pete Crecelius led Billy Matlin, Don Paquette, and me back to the area of Norms Dome. We entered through the new Ferguson Entrance to Mammoth Cave. The water was low through the U-Tube, and it was extremely hot wading through the shallow pools of the river in our wetsuits. The neck-deep cooling pools of the past were mostly gone.

We explored several side leads off the river on the way to the domes at the end, most of them promising. We surveyed more than thirteen hundred feet through more domes and canyons and were sure that the area would yield additional discoveries through the leads we found.

The most interesting discovery was that the upstream siphon was gone. Now there was a one-inch airspace with a steady breeze issuing from it. I had not expected this! When I first saw the siphon in 1979, and again on the "Secret Trip" in April 1981, the river welled up in a pool at survey station Z129. Deep down we could see the underwater passage that fed the pool, and it gave every indication of being a permanent siphon. The streambed at the pool's exit was floored with cemented rocks. There was no gravel, nothing loose to dig to drain the pool. But now, during low water, the pool level had lowered and the siphon was breached. But one inch of airspace did not exactly thrill us.

John Branstetter, who had been on the trip with us to survey the connection between Mammoth Cave and Logsdon River in Proctor-Morrison Cave, puzzled about the mystery of where all the water came from. He later discussed the question with Don Coons, Sheri Engler, and Jim Quinlan. The problem was the same one Tom Brucker, Diana Daunt, and I had used as a cover story for the Secret Trip. About half the volume of water just appeared somewhere between the Mammoth Cave connection point and the upstream sump. "We need to search for the source when the water is low," John told Quinlan.

Quinlan left no river unsurveyed so long as he had money in his hydrology project budget. His crew owed him many hours of caving for pay advances made to them; now, armed with John Branstetter's news, he called them in. The water level had been dropping all summer, so he ordered a thorough search for the missing water input. John agreed to lead the search for water coming in from Cave City.

John contacted Pete Crecelius, the active leader of exploration in this part of the cave. They selected 27 August 1983 for a large trip in two parts. The first group would enter the Morrison Cave Entrance, where some of the party had stashed their wetsuits. John Branstetter, Don Coons, Sheri Engler, and Chris Kerr changed into their wetsuits in Thrill Shaft for their plunge into Logsdon River.

Pete Crecelius, Darlene Anthony, Lynn Brucker, and I entered the Ferguson Entrance to Mammoth Cave. Darlene was new to the Ferguson Entrance, but she was reputed to be a strong caver. She was about five feet, seven inches tall and stocky with an animated face, engaging smile, and southern accent. She was part of Quinlan's summer hydrology crew of cave explorers and mappers. She carried her wetsuit in a pack and maneuvered through the tight Rat Scratch passage with yells and protests.

At the end of this passage, a canyon opened in the floor. We would have to chimney down the wide opening for thirty feet; the descent was difficult for most cavers. Darlene stopped.

"I'm not going down this without a belay," she said.

Nobody had a hand line.

"You can make it," I said. "Everyone else has." I lied.

"Give me a line or I'm not going."

I climbed down partway and braced between the walls. "I'll spot you and tell you where to put your hands and feet." Her request was reasonable, but we had no line, so the safest approach was not possible. Lynn spoke calmly, persuading her to start down the climb.

"If I slip, there's no way you can stop me," Darlene yelled down.

"You climb. I'll stop you. Money-back guarantee." She made it, but her lecture about safety continued for some minutes.

We reached the changing room where Pete Crecelius, Lynn, and I had hung our wetsuits on rock projections at the end of the last trip. We wiggled into the clammy sponge-rubber suits. Darlene asked if we were concerned about the sanitary aspects of leaving wetsuits hanging in the cave between trips. We told her we had done it for years.

Pete explained why explorations in the area between Mammoth Cave and Roppel Cave were sensitive. Now here he was, on CRF urging, leading CRF trips to prevent a connection. He thought it was a reasonable solution. And in truth, he was as curious and as excited as anyone about that part of the cave.

"We're going to look at the end of the river, but other things too," Pete said. "We want to find the Oasis River that Quinlan dye-traced from the Cave City

Oasis Motel into this river. There are also several cutarounds to survey. And there's plenty to survey around the domes above the end of the Z Survey."

We walked and waded upstream in the river. The water was over our heads in only one short stretch of pool. The ceiling began to lower, a sign that we were near the end of the river.

As we covered the last few feet of the passage, we were astonished to see that the one inch of air space at the former siphon had opened to eighteen inches! Apparently the stream had cut down its bed or somehow had reduced the bar at the exit of the pool. We ate a meal, waiting for the other party to reach us.

Pete led onward through the formerly flooded siphon. We were in virgin passage, twenty feet wide, filled with water to a depth of two feet with eighteen to twenty-four inches of headroom above the water. The ceiling rose to four feet. We alternated between crouching and walking in the flowing river, our excitement mounting. After 120 feet, the passage had grown to twenty feet wide by ten feet high. We passed a crawlway on the right. I fancied this would be the other end of the drain I had found from the passages above on the Secret Trip.

"I'm counting the distance," said Pete. "We'll go one thousand feet or until we're blocked."

He guessed that the distance between Roppel Cave's breakdown and the former end of the river was more than two thousand feet. This agreed with my own estimate, but I was skeptical that Pete would stop as long as the passage continued. My heart pounded at the thought of going right under the upper-level pits and domes we had seen on earlier trips. There was no stopping this passage, I thought.

"Right here!" Pete was standing at a constriction of black boulders that filled most of the airspace. The main ceiling slab had fallen and formed an open book, spine down, at the left side of the passage. "This is a good place to stop the exploration, so we'll build a cairn." We stacked up a teetery pile of muddy rocks that would be obvious to anyone coming from either direction. Then we retreated to Z129, expecting to find the other party there.

They were not there, so we started a survey. We carried the survey almost due east, more than one hundred feet though the low ceiling reaches to where we could stand in thirty-foot-wide passage. At Station Z134, we noted a small waterfall entering from a too-small crack in the ceiling. Fifty feet farther, a major tributary entered from the south wall. The entering stream passage varied from two to four feet high, and it broadened from a few feet to fifteen feet for three hundred feet. A strong breeze blew toward us from the continuing passage.

This one would go, but we merely noted that fact, then returned to our main survey in time to meet John Branstetter's party.

Sheri Engler said, "We got started late. It's so far out here. I tried to get these guys to slow down a little. My legs aren't as long as theirs."

Chris Kerr and Don Coons were ready to go to work.

Pete Crecelius told them to continue walking upstream until they came to the cairn. "Start a survey back toward us. When our surveys join, we'll decide what to do next."

As the other party slogged past, Pete continued, "Don't go beyond the cairn. We don't want to connect the caves now."

They didn't answer. John's party soon had waded out of sight around the next corner.

We surveyed for several hours in passageway twenty-five feet wide by four to five feet high. After about a thousand feet, we set Station Z152 at the beginning of some breakdown. There were thick piles of twigs and other organic fragments in the pool bottom and fifty or more blind white crayfish. This was where the large passage became constricted, but Pete found a small crawlway high to the right that led one hundred feet to a plug of rocks with wind blowing through it. We checked the breakdown in the center of the passage. After two hundred feet, it became too low to follow. Based on their tracks they had left, the other party had climbed through the open book route on the left-hand wall. The cairn was sitting right where we had built it, obstructing the natural path. But where was John Branstetter's party?

I smelled a rat. Pete had told them to survey from the cairn back to us. Lynn wondered if they had just decided to connect the caves and worry about the consequences later.

Pete didn't speculate. "Let's go find them."

Within two hundred feet, we encountered John's survey party calling off a bearing, heading toward us. Pete asked if they had missed the cairn. They said they had not seen a cairn but thought they might be going a little far, so they had turned around in virgin cave and started an R Survey back toward our Z Survey. There was some low conversation I did not hear, except for a defiant proclamation by Chris Kerr.

"No way!"

Chris took off into virgin cave. Don Coons was right behind him. John Branstetter and Pete Crecelius raced after Don and Chris.

In ten minutes, Pete and John caught up with Chris and Don. Chris's lamp was nearly out. Ahead, the route narrowed and was tighter.

"We really ought to leave this," Pete said.

Chris just looked at Pete. He continued to try to nurse a few more minutes out of his failing light.

"I'm out of light anyway," Chris said.

Chris got up and headed back toward the R Survey.

Fifteen minutes later, the party was back together. Pete explained what they had found.

"We went back to R1, Branstetter's first survey station," he said. "It's marked with orange flagging tape. We went 250 to 300 feet farther into virgin cave where there's more breakdown, but plenty of air blowing. We stopped there and talked over the situation. We agreed to turn back."

The situation, I thought, is out of control. Pete somehow had the trust of the CKKC, and yet he seemed to be parleying with Don Coons and John Branstetter about when and how to make the connection. And with Chris Kerr? Wasn't Chris friends with Geary Schindel? I thought that in about two minutes, Pete was about to get run over in a connection stampede! Cold wind blew at us from what I knew was Roppel Cave.

Back at Z152, we discussed what to do for the remainder of the trip. John's party would survey cutarounds downstream and search for the Cave City lead. We might not see each other after our parties split, so we made no plans to get together again. Our party would check other leads at higher levels. Lynn and I set off to find the drain from the pits above us to see whether or not it bypassed the sump. Pete and Darlene went off to retrieve a hand line left behind on an earlier trip.

I could not find the pits or the drain. The area was more complex than I recalled from two previous trips. I felt foolish, not being able to show Lynn the master drain that had been at the center of my connection dreams for the past two years and four months.

We found Pete and Darlene and ate a meal. I told Pete I was worried about his ability to keep a lid on explorers like Don Coons and John Branstetter. Not only had they kept their own counsel on secret trips, they had also unsuccessfully tried to stand in the way of the Proctor-Morrison Cave connection. They were bulldozed aside there and now were pissed. Scooping this connection might be their form of vengeance.

Despite the patching of relations between Don and the CRF, he and Sheri continued to refer to themselves as "nonaligned." I once questioned this, and Don replied that he worked for Quinlan and thus represented neither the CRF nor the CKKC. This was bullshit, given the full and open participation in the two organizations that Don had enjoyed over the years. No, "nonaligned" here meant simply he could screw anybody or any organization if he wanted to.

"Pete, I don't think you have much time to sit on this connection possibility," I said. "There's too much cold air blowing."

"It's pretty hot, you mean? Yeah, I've been thinking about that," he replied.

It was getting late. We surveyed a few hundred feet of cutaround and left plenty for future parties.

We traveled out of the cave in near silence. We struggled out of our wetsuits and pulled on our damp clothes in the changing room beyond the U-Tube. Darlene punctuated the long silence by asking if we kept the wetsuits here so we could skinny in and make the connection between Mammoth Cave and Roppel Cave!

"Why else would anyone think of stashing those nasty, smelly wetsuits?"

Lynn explained again that there were plenty of passages to explore without going for a connection.

"Don't lie to me," said Darlene. "You guys and Pete don't fool me for a minute!"

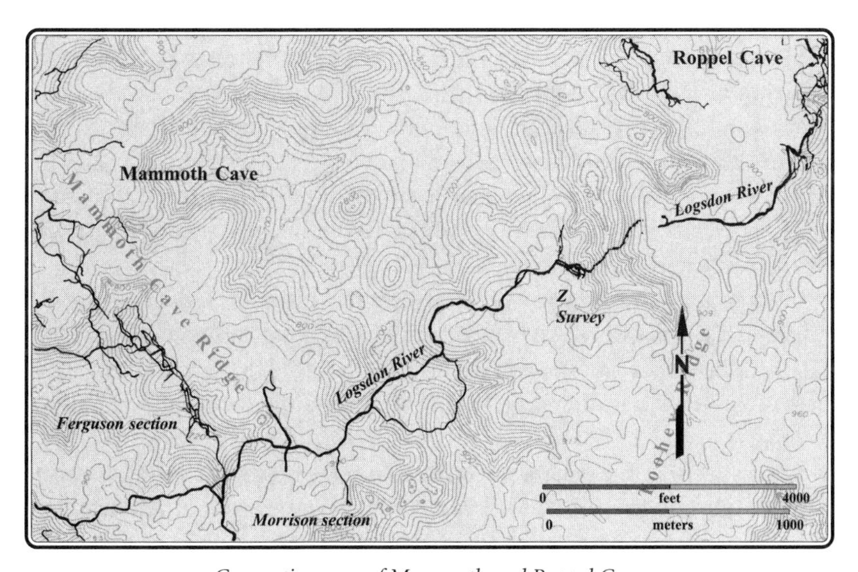

Connection area of Mammoth and Roppel Caves

Darlene had to be talked up the chimney approaching the Rat Scratch Passage, a few hundred feet from the entrance, but it was easier than talking her down had been. She was motivated to get out of the cave.

Near the beginning of the Rat Scratch, I pushed my pack ahead of me as I squirmed through the tight belly crawl.

Boom! My eyes were blasted with a ball of flame and sand. My carbide lamp flame had ignited a cloud of acetylene gas escaping from my spent carbide bag inside my pack. My face stung, my eyes were centers of excruciating pain, my ears rang. In the distance I heard Lynn say, "Are you okay?"

"I blew up. I'm hurt!" I could feel and smell my singed eyebrows and eyelashes. The hair on my wrist had crinkled. The blast had blinded me. I tried to force my eyes open, but it hurt too much. Lynn and Pete were at my side in a few seconds.

Pete was an M.D., professionally calm but reassuringly concerned. "Let's get this sand out of your eyes. I'll use some water from your water bottle. It's not sterile, but that's not the main thing right now." He squirted the water into both of my eyes to irrigate the tissues and wash away the sand. I was glad to be able to see the light from his headlamp, but I could not focus on anything, and it was painful to sneak a blurred peek. How would I get out of the cave?

"I'll take your pack," Pete said. "Think you can move okay?" He slipped a sourball candy in my mouth.

Lynn said, "I'll go in front of you and talk you through."

I started crawling. I remembered the route as a succession of tight squeezes requiring wedging in the top of a canyon. Once, when I had just started caving, a friend and I challenged each other to find our way out of Carpenter Cave in Pennsylvania in pitch dark. Instead of an ordeal, it turned out to be fun. The trick was to close your eyes, not keep them open. In a few minutes we began to get clues about the route from sound echoes as we moved. Knowing the route helped, of course.

Now, it was not fun. I hurt. But the sound clues did help, and I made steady progress. It was reassuring to touch Lynn's ankle now and then. I'd squeeze it to show my appreciation and affection. Lynn rigged my ascending harness on the forty-foot standing rope at the entrance. Going up blind was almost automatic.

"On your way home you may want to stop at the hospital emergency room and get your eyes looked at," said Pete. "There's some trauma. They can put antibiotic ointment in to prevent infection and lubricate it."

"What do I owe you, Doc?" I said as we moved along the path to our parked car.

"Not a dime," he said. "I don't have malpractice insurance in this state, so I didn't treat you."

We telephoned the Caverna Hospital near Horse Cave. I told them my carbide lamp had exploded, blowing sand into my eyes, and that we were on our way to the emergency room. When Lynn and I arrived, they were all set to neutralize battery acid in my eyes. I explained the chemistry and physics of carbide lamps and acetylene gas versus unsafe electric lights that used acid batteries while the EMT washed out a few more grains of sand and applied soothing ointment.

I was thankful for Lynn's coolness and Pete's cave-side manner. We wondered, however, whether Pete could control the nonaligned cavers who now knew that the Mammoth Cave–Roppel Cave connection was no longer a matter of mere politics. An open shot at making it seemed there for the taker. Would Pete's order be sufficient to stay a "Let's do it!" attitude?

Jim Borden and Roberta Swicegood discussing connection plans in a quiet corner at the Old-Timers Reunion

23

Stark Reality

The CKKC Surrenders to the Odds

Pete Crecelius called me—Jim Borden—following the trip that had surged through the sump towards Roppel Cave. I winced as he told the story. The situation appeared to be desperately out of control. The damage was evident, but we were powerless to stop it now. The Ferguson Entrance was uncontrolled. At least a half dozen individuals—loyal to neither the CRF nor the CKKC—knew the situation and would think nothing of pursuing the virgin passage independently to its ultimate conclusion. These marauders would not care about political sensitivities, fair play, or history. Neither of our groups could do anything. But did the CRF even want to do anything? The CRF could sit back; we could not.

The CKKC had the most to lose by a connection, no matter what the circumstances.

At first, I decided to lie low, delaying a decision until the facts could be better evaluated. A short delay did not seem to be a risk; after all, there was a gate on the Ferguson Entrance. Yet, there were still many questions. Indeed, Roppel Cave and Mammoth Cave were now no more than six hundred feet apart, likely closer since Chris Kerr had followed the breakdown beyond the last survey station. The caves could, in fact, be less than one hundred feet apart. I thought again about Bob Anderson's report of the last trip he took in Roppel Cave: "It does not go." I repeated the words to myself.

Was the breakdown penetrable? Chris Kerr turned around in virgin going cave and the wind was blowing. I did not know Chris well, although I had been acquainted with him back in high school and had gone on caving trips in Virginia with him. Would he hide anything in his description of his trip into the breakdown beyond the sump?

It became clear to me that we would have to mount a trip soon to resolve the questions, or someone else would answer them. It was not a question of if or even when a connection would be made; it was a question of who would make it. I would not permit the connection to be made by a few self-serving cave rapists who cared little about the caves or the CKKC.

The dream was unraveling, a victim of other passionate cavers in Mammoth Cave. So be it. However, I knew one thing for sure: those of us who shared in the dream of Roppel Cave would be the ones to end this dream. Nobody else had that right, and nobody else could steal that from us. Our years of negotiations with the CRF were not going to be for naught.

I had kept my ear to the ground but had heard no scuttlebutt indicating that secret scheming was going on. Things seemed calm, and I wanted the time to think things through completely.

However, I was deluding myself in thinking that no one else had taken an immediate interest in the recent events.

The weekend after Chris Kerr's trip, I attended the Old-Timers Reunion near Elkins, West Virginia. The Old-Timers Reunion was the ultimate caving party. Fifteen hundred beer-drinking and howling party fanatics would descend upon a small campground, set up tents, and indulge themselves in almost every conceivable excess for nearly three straight days. This was a cherished annual rite-of-passage event for many cavers from all over the United States and Canada; some would drive for days to attend. Fortunately, its organizers had a few years back located the Old-Timers Reunion in a remote area, isolated from normal society. In the past, unwitting and innocent weekenders would find themselves surrounded by the melee, their children wide-eyed at the spectacle. Appalled beyond belief, these fleeing families would forever stereotype all cavers as incorrigible hooligans to be avoided at all cost. Now, an entire campground was cordoned off for the enormous blast to spare innocents from the shock. Amazingly, attendance for the weekend shindig was greater than the more-touted annual NSS conventions that lasted a week, were better organized, and were more socially responsible. Some cavers just like to unwind. They never claimed that the reunion was a Sunday school picnic.

It was a fine and sunny Labor Day weekend. Everyone was out socializing, beer in hand, or participating in the remotely cave-related activities that the attending grottoes had organized. The cave was far from my mind, and I was having a great time. I was looking forward to the Speleo-Olympics in the afternoon and the big party later that evening. By chance, I ran into Geary Schindel among the sea of tents. Geary lived in Bowling Green and was doing thesis research at Western Kentucky University. I had known him, though not well, for years and had caved with him a time or two. I had thought of Geary as a fringe caver, not in the mainstream of caving in the Mammoth Cave area, but he seemed to know what was happening. A true gossip, he could cause trouble, but I was totally unprepared for his coming assault.

"I heard all about the trip last weekend. What did you think?" He was baiting me.

"How'd you hear about it?" What was he up to?

Geary smiled. "Oh, I have my ways." Then he set the hook. "I hear Chris Kerr probably made the connection."

"Really?" I asked. "What are you talking about?" I was trying very hard not to play his game.

Several cavers circled us, listening. Geary continued, "Well, he pushed pretty far into the breakdown. He must have gotten into Roppel Cave."

I was already emotionally drained by the recent whirlwind of events—thoughts of a connection, events out of my control. It was a big tornado, and I was worn out. Geary's provocations struck a nerve, and struck it hard.

"You don't know what the hell you're talking about! I heard about what Chris found and I know all about the breakdown at the end of Logsdon River in Roppel."

Geary was itching to argue this. "Oh, I heard all about Anderson's trip to the breakdown, too. Anderson was bullshitting you. He left going cave back there and didn't want to tell you—especially after you guys threw him out for no reason."

The crowd around us was growing. I could feel more anger welling up from within me. What was Geary trying to pull?

"Don't think so. Anderson found nothing—anything else doesn't make sense."

"It doesn't matter." Geary laughed. "If you guys are sure he didn't connect, you guys better do it, and do it quick."

"Why?" I was defiant now.

"If you guys don't connect right away, someone will do it for you."

"Who might that be?" My sarcastic tone was giving away my anger.

Geary kept pressing. "I might just go in there when I get home and connect the caves myself!"

"You're full of shit! You wouldn't dare!" My face was burning.

"Watch me," he snorted.

I turned and walked off. Hot. I heard Geary over my shoulder.

"Remember, if you don't . . . I will."

He had made me furious. I had taken his bait like a stupid bass lunging for a fisherman's lure. As I walked among the hundreds of nylon tents, my indignation increased. I was incensed at this attack—and annoyed at my response. Although I had known immediately after the trip that the situation was out of the CKKC's control, I was not ready for such a blatant and immediate threat as Geary's.

The situation was grave. Whether Geary Schindel was playing some kind of game or not did not matter. He was capable of whipping up a frenzy in other cavers, setting the ball rolling even if he himself was not involved. We could not wait. It was time to move . . . now!

Roberta Swicegood also attended the Old-Timers Reunion, along with her numerous friends who had made the long drive from Alabama and Georgia. It took me fifteen minutes to find her. I told her about my shouting match with Geary. "I think we're going to have to connect. Connect now!"

"Well, James, you're probably right." She looked out across the campground, admiring the multicolored display of tents. "We're dealing with unpredictable people here. Best we act now, while we still have the chance."

"And one chance is all we've got," I added.

What to do next? In the distance, I saw cavers swinging wildly on a rope hanging from a tall tree, hooting before dropping spectacularly into the river. Insanity.

"We can't let on that we're planning this trip. We have to keep it an absolute secret."

This time, someone was swinging naked on the rope, screaming like Tarzan.

"And do it soon."

"Makes sense," Roberta said.

"We'll get a whole bunch of people, some from the CRF and some from Roppel." I frowned as I tried to cobble together the final pieces of the plan. "Since this is a breakdown area, it will probably be hard to connect. We'll have to send a party in each way and meet at the breakdown."

"And bring hammers," Roberta added.

"Of course. We'll shout to each other to find the way through."

Roberta nodded her head in agreement and smiled as she thought about it.

"But who should come?" I asked.

Roberta mused. I knew she wanted to go.

"Well," began Roberta, "we should invite those who are most currently active."

We ironed out the details in the next couple of hours, talking quietly in one corner of the campground. We watched carefully for anyone who might be listening—especially Geary Schindel. The sun had long set and the big parties were cranking up. I was missing it. Loud music throbbed from the central pavilion that was crammed with happy cavers.

We would connect the coming weekend during an already scheduled CKKC expedition. In the next few days, we would assemble two parties, one made up of representatives from the CKKC, the other from the CRF. Also, we would ask some of the more active nonaligned participants. This would ensure that we included Don Coons, John Branstetter, and Sheri Engler, who were as vested in this cave as any of us. One party would use the Ferguson Entrance to Mammoth Cave and the other the Weller Entrance to Roppel. Each party would proceed to the breakdown on a timetable. By hammering, shouting, moving rocks, swimming, or any other necessary means, we would attempt to link the two caves. If the connection was successful, each party would proceed out the other's entrance. This would be a double crossover trip with each party making the complete connection through-trip.

Because of all the controversy, I wondered if it would look like a staged event—not that it mattered. However, to maximize the probability of success, it seemed best to approach the endeavor with this large group of cavers, however ceremonial their selection appeared. All the cavers would be strong and capable. And we did not want to fail. A failure would not only be a bitter disappointment but would also leave the door open for Geary Schindel, or anyone else, to connect the caves. Our failure would be their incentive. We had to throw everything we had at it.

During the next couple of days, I called the people on the list that Roberta and I had compiled and later cleared with Pete Crecelius. The CKKC group was easy to enlist, although I encountered three surprises. Pete declined, for reasons I did not fully understand. He just said, "Let someone else do it." Tommy Brucker refused also: "I don't need to be on the connection; the CKKC expedition needs me." I shrugged, recalling his refusal eleven years ago when offered a chance to participate in the Flint Ridge–Mammoth Cave connection. I was as perplexed now as others had been then.

With mixed feelings, I finally asked Jim Currens. I actually thought he might accept, but he too declined. "No. Not interested at all," he said.

I shook my head, "Unbelievable." I hung up the phone.

The final CKKC list included Roberta Swicegood, Dave Weller, Dave Black, Bill Walter, and me. It was not perfect and would inevitably be second-guessed; certainly, everyone deserving could not go—we had limited our party to five. But it was the best list we could come up with.

For the CRF party, I first asked John Wilcox. Although he had been inactive since the French Connection, I admired him greatly and wanted him along—I owed him one. He refused also. Was he really that modest?

I moved on, eventually arriving at Roger Brucker and his new wife, Lynn. Nobody had invested as much in the caves as Roger, and his emotional investment in the connection was as high as mine. I thought he deserved a shot at this connection, his first as far as I knew. Lynn had been a strong contributor in exploring the difficult passageways in Mammoth Cave that extended towards Doyle Valley, as well as in the river itself.

They said yes. With Don Coons, Sheri Engler, and John Branstetter, we had a complement of ten.

The need for secrecy was the common theme in all my calls. Geary Schindel had shell-shocked me to the point where I feared that any leak before the connection parties were in the cave would have devastating effects. I felt sure that Geary was busy in Bowling Green trying to put together his own renegade connection team. I was counting on him to not know how fast we were moving. We would tell no one outside the circle of participants.

This became more complicated due to the planned CKKC expedition scheduled for the same weekend as our connection trip. Many of us who could usually be counted on as regular participants on CKKC expeditions were now inexplicably busy elsewhere. The plague of dropouts was puzzling to those unaware of the secret plans. Some of the tales we told were quite outrageous: I, for example, said I had to brush up on my flying skills or risk losing my pilot's license—I never missed caving trips if it could be avoided. Dave Black inexplicably decided to take photographs in a nearby cave—he had never before expressed such an interest in this cave. Roberta Swicegood suddenly remembered a previous commitment to lead a Smithsonian field trip to a cave in West Virginia—she never forgot appointments. Pete Crecelius attended the expedition, as planned. He would provide cover for our trip and explanations for our absence. He knew our timetables and could dissemble and divert the Roppel cavers, if necessary.

"What was going on?" everyone asked.

Since the connection was not a certainty, we agreed that we had to extend the deception to well past the start of the trip. Should the trip fail, we would have to try again as soon as possible, or the caving renegades might swoop in before we could regroup. No, we would keep it secret before, during, and also after the trip. The CRF and CKKC had agreed to release news of any connection jointly, and the CRF had an agreement with the National Park Service not to release news of discoveries unilaterally. So, the connection participants agreed to keep the lid on the story until the park announcement could be scheduled. This fact was painful. We had never kept news of discoveries in Roppel a secret from any CKKC members. There would be hurt feelings.

To maintain secrecy, we planned to enter the cave at six o'clock in the morning, long before any other cavers would be awake. We would complete the trip after dark, but early enough so any other Roppel parties would most likely still be deep in the cave. We would hide our cars far from the entrances, slipping in and out undetected.

As we began to implement the plan, I fretted about the implications of our actions. In just a few days, the existence of Roppel Cave as a separate cave might cease forever. My dream of a cave as large as Mammoth Cave would be shattered beyond recall. A connection can never be undone. However, I resolved that Roppel Cave would not simply be swallowed up to become a nameless extension of Mammoth Cave. If I had anything to say about it, the CKKC would also forever retain its identity. I vowed also that the connection event would be momentous, forever etched into the history of caving in the area. Capitulating to the CRF's demands that the Park Service handle this as a press briefing would help ensure that the event would be remembered.

The whole plan was based upon secrecy, right or wrong. Unfortunately, secrecy meant deception, and deception meant lying to our friends. The decision weighed heavily upon me. Why was the entire history of the stormy relationships among all the opposing parties in the Mammoth Cave region's cave connections laced throughout with lies, deception, breaches of trust, and damaged friendships? Regrettably, I could not ensure that it would stop for this one last trip. In any event, the price would be high. Plans went forward with their own momentum.

On Friday, 9 September 1983, Roberta Swicegood and I flew out of Montgomery County Airpark in an old single-engine Cessna airplane. The day was spectacular. Puffy cumulus clouds dotted the skies around us as we flew over the mountains of West Virginia and the site of the Old-Timers Reunion. I looked down with regret at the now empty fields of the party site. A few tents of family campers dotted the peaceful grounds. I thought back over the events that had led us to this point. I was saddened. A grand era was ending, perhaps never to be recaptured. I looked down at the instruments, confirming our altitude and position. We were right on course and ahead of schedule.

As we sped over the Big Sandy River into eastern Kentucky, the sun lazily slipped beneath the razor-sharp horizon. Ahead, the inky blackness lay between us and a three-hundred-mile cave.

Roberta and I headed silently into the deepening gloom.

The connection through the breakdown becoming reality as the two parties meet

24

Swallowed Up

A New Beginning

The bittersweet reality of waking up on Roppel Cave's last day impressed me more by its pain than its pleasure. It was depressing to face the inevitable shattering of my dream. Realistically, there was a low probability of connecting any two caves on a trip deliberately aimed at that objective. Why didn't I just plead sick? Secret reasons for not joining the trip had served others—Pete Crecelius, Tom Brucker, and John Wilcox had removed themselves. Jim Currens, not a likely connection participant anyway, also said he would not attend. His tone of indifference seemed a delicious kind of disdain. Did anyone except me wonder why? Wasn't their refusal an ennobling act of pride and defiance?

My curiosity about whether the cave connection could be found drove me to swing my feet out of my sleeping bag in the Collins House at Flint Ridge. I smarted at the irony that we were ending it where it had all begun—at Floyd Collins' Crystal Cave.

It was just morning's first light, but the Austin House was abuzz with life. Roger Brucker was already up cooking breakfast. The enthusiastic banter from the kitchen reinforced my belief that nobody else saw the serious side of this trip. They were all curious, go-for-broke, to-hell-with-Roppel-Cave, press-on-regardless commandos. Their trip planning was what one would expect from military special forces.

"Let's Simoniz our watches," Roger ordered. He must have made that joke a hundred times in my presence, and I still had no idea what it meant. We set departure times, rehearsed cover stories, decided where to park vehicles, and agreed to operate as independent teams in case we didn't connect. Most of us had packed the night before for the trip, so there was a minimum of scurrying around. We trooped out to the assigned cars. In this veteran group, the experience and skill level was so high that each of us was capable of leading the trip. We were interchangeable in every way—the CKKC and the CRF were as one, and that was a sobering thought.

About two hours into the connection trip, I saw my chance to take the lead. Roger Brucker was taking his sweet time explaining things to those who had not seen this part of the cave. I felt patronized. As he prattled about cave blindfish and the wiping out of the aquatic ecosystem when the Park Service emptied a sewage lagoon into a sinkhole, I charged past him. Was he running a classroom or a connection trip? The pace was too slow.

Hours later, at the so-called sump, I was astonished. There was a two-foot-high opening with cold air pouring out. Was this ever a siphon? Had the previous reports been lies? Dave Weller later speculated that the flood on Kentucky Derby Day, 7 May 1983, must have wiped out the last trace of any dam and scoured the river channel deeper.

"Time." Don Coons was pointing at his watch.

His warning caught me. We were ten minutes late according to the timetable I had set; we should have been well into the breakdown where we would start pounding on the rocks—our signal for the other party coming from Roppel Cave. Roger's lamp troubles had delayed us. Smoothly, Don slipped into the hole following the river upstream. In a few minutes, I realized that he was moving fast and I was no longer leading. I scrambled to catch up. Shortly, we were squeezing along the left wall in a V-shaped crack formed by the breakdown blocks.

A little beyond where someone had hung an orange streamer, we heard a distant cry: "Hello!" It came from ahead of us.

Don sprinted ahead with me struggling to keep up. The rest of the party was behind, nowhere to be seen.

"Hello!" we shouted.

The voice ahead was closer now. I saw faint flashes of light shooting through the cracks among the fallen rocks.

I squeezed through a narrow crack into a room formed between the ceiling and fallen ceiling blocks. John Branstetter, the lead caver on the Roppel party, and Don Coons were sitting on a block in the wide room. John greeted me with an outstretched arm and a broad smile.

"We did it!" he said.

I clasped his hand firmly. Before I could compose a response, cavers from both sides started pouring into the room.

I looked at the growing horde and sighed. The moment was lost.

"How far have you come through the breakdown?" I asked.

"Not real far, but the route was confusing. We arrived at the breakdown early and decided to explore the crawlways under the big slabs. The farther into the breakdown we went, the tighter it got. But, it was wide. Anderson must have turned around because it looked impassable. His scuff marks stopped pretty far back. If you didn't know anything was here, you would think it ends. It wasn't until I heard your shouts that I knew which way to go. I had to squeeze up between two slabs of rock. It looked like it would pinch for sure. But that's where the voices were coming from. When I reached this room, I could hear you clearly. Then I knew we'd connected."

The cavers shouted greetings. It was pandemonium. No time to reflect. I felt this must be like falling into the Niagara River and being swept over the falls. Jubilation took many forms. Roger Brucker organized group photos. He joked and set the camera self-timer, then sprang into the scene with a stupid grin on his face. Several of us built the victory cairn. It was a kids' birthday party without the cake.

After the parties separated, we continued toward Roppel. The route led through a couple of thousand feet of low, miserable river I had not seen from the Roppel side. I thought about how this stretch of river had been "explored to its end" by Bob Anderson and Chris Welsh and how later it had become the focal point of fierce political maneuvering by both the CRF and the CKKC. Now it was just one passage among thousands in the one great cave, nothing more. No one would ever see it as anything else again.

The minutes marched on as we traveled through the mind-numbing succession of cave passages that countless explorers had strung together. Discoveries were made like beads on a string—no one discovery more or less important than the previous, or the one that followed. But, I felt I had shepherded each of them. I thought about the vastness of this cave called Roppel and of my brief glimpse of the vast reaches of Mammoth Cave for the past eight hours or so. It was now obvious: it was not Roppel Cave that was swallowed up by the connection we had found; it was me who was swallowed up in a vastness that I recognized was the beginning, not the end, of what lies beyond Mammoth Cave.

In Tinkle Shaft

Afterword

300 Miles to 365 Miles . . . and More

When the Mammoth Cave–Roppel Cave connection excitement died down at the end of 1983, we added up the mileage of the longest cave and discovered that there were 296 miles instead of the 293 we had thought in the cave. In 1984, after the Memorial Day expedition, the cave mileage rolled over the 300-mile mark. The cave has continued to grow each year. By 1996, it was 365 miles long, making Mammoth Cave more than 3.5 times as long as the second longest cave in the world.

The rate of cave length extensions has slowed a little since the 1970s and 1980s. This is because some of the underground effort is devoted to resurveying passages to higher accuracy standards than the original surveys. The maps are more comprehensive and win prizes in cave cartography contests.

Contrary to some predictions, Roppel Cave was not entirely "swallowed up" by the CRF and the Mammoth Cave effort. The CKKC operates cave trips and recruits volunteers to go through one or more of the new Roppel entrances constructed by Dave Weller. The CRF, of course, continues to run expeditions in the caves of Mammoth Cave National Park.

The big news in 1995 was the Fisher Ridge Cave System, which has been explored and surveyed to a length of eighty miles. The breakthrough areas are miles from an entrance. Deep push trips require camping. Passages in Fisher Ridge and Roppel are less than twelve hundred feet apart. Those Fisher Ridge cavers don't want their cave "swallowed up," so they continue to issue stern warnings to the CKKC to steer clear of a connection—or else!— an ironic reversal of roles.

Some unexpected cave connections were made in 1996. Some cavers, including Jim Currens, gained entrance to the large Jackpot Cave, near Long

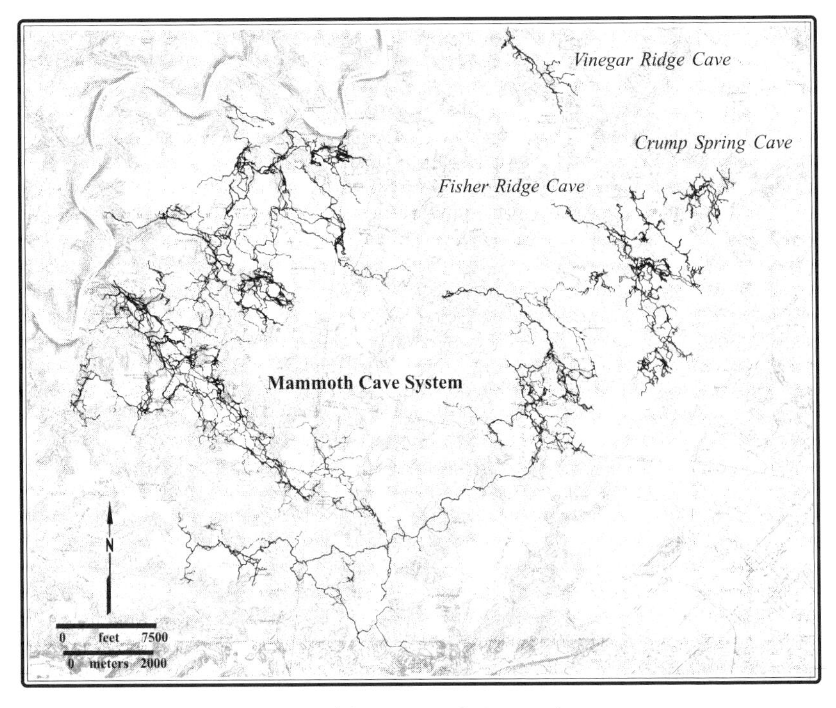

Caves of the Mammoth Cave region

Cave in Mammoth Cave National Park. They went to great lengths to keep their explorations secret. Meanwhile, cavers from Bowling Green had found a second cave, Martin Ridge Cave. This cave was also secret, apparently one better kept than Jackpot Cave. Martin Ridge proved to be the key to several cave connections. First, Martin Ridge was connected to the back of Whigpistle Cave; then, to the chagrin of its explorers, Jackpot Cave was connected to Martin Ridge (by Don Coons and Pat Kambesis). Triple connection! The whole system already exceeds thirty miles in length and will, some day, connect to Mammoth Cave.

Red Watson said we should go out on a limb and predict the five-hundred-mile cave. And we're almost there! Don Coons has compiled a Mammoth Cave area regional map that shows all the surveyed caves. It doesn't take a genius to see that sooner or later, they will all be connected. Jim Quinlan's dye tracing has shown that the passages can go right under the surface drainage basin divides. We'll be dead then, but one day there will be one thousand miles of connected cave under the upland between Munfordville and Bowling Green, Kentucky. Guaranteed!

PARTICIPANTS IN THE MAMMOTH CAVE–ROPPEL CAVE EXPLORATION, 1972–1983

GLOSSARY OF CAVING TERMINOLOGY

INDEX

PARTICIPANTS IN THE MAMMOTH CAVE–ROPPEL CAVE EXPLORATION, 1972–1983

Acree, Ed
Adamson, Greg
Alexander, R.
Alfred, Janet
Alfred, Tom
Allin, Tom
Allured, Dave
Allured, Vie
Anderson, Bob
Anderson, Eric
Anderson, Jennifer
Anderson, Terry
Anthony, Darlene
Arp, Jeff
Artz, Mike
Backstrom, Neil
Baker, Brian
Baker, Linda
Balister, Phil
Ballman, Carolina
Ballman, Hans-ruedi
Banther, Mike
Barnes, John
Baus, Bill
Beard, Bruce
Bennett, Steve
Bennington, Tammy
Berglund, Donna
Berk, Eric
Bersheim, Bill
Binney, Frank
Bishop, Sarah
Bishop, Steve
Bishop, Sue
Bishop, Bill
Black, David
Black, Greg
Black, Tom
Blankenship, Diane
Bonner, Bob
Borden, Jim
Borgerding, Tracy
Bosted, Peter
Bowers, Ken
Bradley, Paul
Bradshaw, Karla
Branstetter, John
Breiding, Mike
Breisch, Richard
Bridge, John

Bridgemon, Ron
Bridges, Hal
Broer, I.
Brotzge, Jeff
Brucker, Ellen
Brucker, Roger
Brucker, Thomas
Bruno, Kevin
Buckner, Joel
Buecher, Debbie
Buecher, Robert
Burns, Denver
Burns, Pam
Butler, Bonnie
Cahall, Donald
Campbell, Frank
Carlton, Keith
Carpenter, Mary
Carter, Jim
Catozzi, Lou
Cavenaugh, Maureen
Ceters, Robert
Chalin, Stu
Chesnick, Tom
Clemmer, Greg
Clifford, Paul
Coates, Fred
Cobb, Bill
Cochran, Alexis
Conner, Steve
Cook, Bob
Cook, Holly
Coons, Don
Cottrell, Tom
Crann, David
Crawford, J.
Crecelius, Pete
Crowther, Pat (later Wilcox)
Crowther, Will
Currens, Deborah
Currens, Jim
Currens, Samantha
Curtis, Sue
Dale-Mule, Renato
Danobich, John
Dasher, Elisa
Dasher, George
Daunt, Diana (later Mergens and Miller)
Davidson, Joe
Davis, Jerry
Davis, Nevin
Davis, Roy
Debevoise, Tom
DePaepe, Duane
DePaepe, Veda
DesMarais, David
DesMarais, Shirley
Dible, Danny
Dickenson, Lou
Dickerson, John
Dickerson, Kathleen (later Womack)
Dickey, Fred
Dillon, Bud
DiTonto, Mike
Divine, Ed
Doerschuk, Dave
Doolin, Dave
Downes, Peter
Downey, David
Downey, Margaret
Drake, Miles
Duchon, Kip
Duft, Andy
Durica, Anne
Dyas, Michael
Dyson, J.
Ecock, Kevin

Eggers, Bob
Eggers, Steve
Eidson, Bill
Eller, P. Gary
Elliot, Mark
Emery, Walt
Engler, Sheri
Estes, Beth
Estes, Gerry
Ewers, Ralph
Farr, Bill
Fausold, Marshall
Fehrmann, Thom
Finger, Ernie
Finkel, Don
Finkel, Liza
Fletcher, Bill
Flint, Sandy
Foeller, Ward
Forbes, Jeff
Forster, Armand
Forsythe, Preston
Frank, Ed
Freeman, Jack
Frushour, Sam
Gariepy, Ron
Garrett, Mike
George, Donald
Gerace, Chris
Glasser, Gwenne
Goatley, Lajuana (later Wilcher)
Goldstein, Harold
Goodbar, James
Goodlet, Collier W.
Goodman, Mark
Gracanin, Tom
Grady, Fred
Graham, Dick
Greenlay, Terry
Grissom, Tim
Grover, Buzz
Grover, Rick
Groves, Darwin
Hacker, Chris
Hall, Brian
Hall, James
Hampel, Chris
Hand, Richard
Hansel, Daryle
Hardison, Richard
Hardy, Jim
Harmon, Russ
Harter, Diana (later Emerson)
Haskin, Dirk
Hatleberg, Eric
Hawes, Bill
Hawkinson, Edward
Hayden, Brian
Heazel, Sue
Helfman, Jon
Helfman, Sheldon
Hensley, Kenneth L.
Herbst, Charlotte
Hess, Letitia
Hiett, John
Hildebolt, Charles (Scooter)
Hildebolt, Louise
Hill, Alan
Hill, Carol
Hobbs III, Horton
Hoechstetter, Rick
Hoey, Kathy
Hogan, Arlie
Holland, Bob
Holland, John
Houchen, Tom
House, Scott
Hughes, Lynne

Hull, Curt
Hummer, Bob
Jackson, Randy
Johnson, Alan
Johnson, Bob
Johnson, Tracy
Kahre, Greg
Kane, Thomas
Kastning, Ernst
Keeler, Ray
Keller, Ben
Kelly, Dave
Kerr, Chris
Keys, Peter
Kiefer, Chuck
Kihara, Deane
Kissling, Randy
Klein, Karen
Knutson, Steve
Koerschner, Bill
Koerschner, Robyn
Komisarcik, Kevin
Korevaar, Nick
Larason, Shari
Lepro, Bill
Levy, Ellen
Lindberg, Ralph
Lindsley, Karen
Lindsley, Pete
Link, Mike
Lipton, Walter
Lisowski, Ed
Lloyd, Robert
Lord, Brenda
Lowery, Joe
Luyster, Bill
Magill, Mike
Mann, Bill
Mann, Charlotte
Maphis, Scott
Marks, Franklin
Martin, Hugh
Martin, Jim
Martini, Jacque
Matlin, Billy
Mavity, John
Mayne, Leland
Mayne, Walter
McAdams, Raymond
McClure, Roger
McCollum, Gerry
McCuddy, Bill
McDonald, Robert
McGill, Sue
McInski, Jeff
McKee, Mike
McMillan, Roger
McNamara, Greg
McThomas, Mike
Melloan, Barry
Mergens, Michael
Merrick, Rob
Metcalfe, Rod
Metcalfe, Wanda
Mezmar, Mike
Michener, John
Miessen, Mike
Miessen, Robin
Miller, Roger
Miller, Tom
Mills, Paul
Mollett, Steve
Molzon, Bob
Moni, Gerald
Moore, Kate
Morgan, Eric
Morgan, Joe
Morgan, Keith

Morley, Tom
Morris, Jon
Muir, Jean
Mulkewich, Jane
Mullett, Frank
Mummery, John
Murray, Lloyd
Neff, Kevin
Netherton, W.
Newton, Geoffrey
OíDell, Phil
Oberlies, Jim
Ogden, Albert
Oliver, Joe
Olson, Rick
Ortiz, Keith
Osinski, Bill
Pace, Norman
Palmer, Art
Palmer, Margaret (Peg)
Paquette, Don
Parker, Judith
Parseley, Andy
Pavey, Andrew
Peloquin, Martin
Perez, Mike
Perry, Luther
Peterson, Butch
Peterson, Charlie
Peterson, Gil
Petranoff, Ted
Picard, Blu
Pickle, John
Pinnix, Cleve
Pollock, Don
Poulson, Tom
Prentice, Dave
Price, Greer
Price, Kelley
Pringle, Lisa
Pugsley, Chris
Radcliffe, Ben
Radford, Mike
Raffle, Mary Ann
Randall, Bru
Rasmus, Kevin
Rausch, John
Reid, Frank
Reynolds, Sam
Rigg, Rick
Robinson, J.
Robnett, Marie
Rogers, J.
Roppel, Jerry
Rumer, Randy
Russell, Ken
Saunders, Joe
Schaffner, Marty
Schafstall, Tim
Scheide, Alfred
Schindel, Geary
Schindel, Sue
Schomer, Barb
Schufeldt, Bob
Scott, Herb
Shifflett, Peter
Shifflett, Tom
Shuster, Evan
Sides, Stan
Simmons, Ron
Simpson, Lou
Sims, Jeff
Skipworth, Joe
Smiley, Larry
Smith, Cindi
Smith, J.
Smith, Marion
Smith, Trey

Smithson, Steve
Snell, David
Snider, Robert
Soukup, Cady
Sperka, Roger
Stecko, Joe
Steele, Bill
Stephens, Bill
Stevens, Lee
Stevens, Paul
Stewart, Steve
Stock, Charlotte
Stock, Mark
Stoessel, Debbie
Storrick, Gary
Sudhoff, Mike
Sumner, Ken
Swicegood, Roberta
Szukalski, Bernie
Taylor, Chris
Taylor, Robert
Thomas, Chuck
Thomas, G.
Thorpe, Dave
Thorsell, Dave
Tinker, Gary
Tobias, Gary
Townsend, Margaret
Trexler, Carol
Ulrich, Jeff
Van der Werf, John
Vansant, Jeff
Veluzat, Phil
Veni, George
Vernot, David
Wagner, Gail
Walter, Bill
Ward, Steve
Warner, Gary
Warner, Mike
Warthman, Bruce
Watson, Anna
Watson, Patty Jo
Watts, Scott
Webb, Beth
Weedman, Claire (later Wood)
Weedman, Curtis
Weimer, James
Welbourn, Cal
Welch, Norbert
Weller, Chip
Weller, David L.
Weller, David W.
Weller, Jean
Weller, Lynn (later Brucker)
Weller, Sheri
Wells, James
Wells, Steve
Welsh, Chris
West, William
Wheeler, Richard
White, John
Wilcox, John
Wilcox, Patricia
Williamson, Suzanne
Wilson, Bill
Wilson, Kent
Wilson, Ron
Womack, Chris
Womack, Ralph
Wood, George
Worthington, Steve
Wright, Michele
Wright, Winfield
Yonge, Chas
Zabrok, Peter
Zidian, John
Zopf, Richard

GLOSSARY OF CAVING TERMINOLOGY

acetylene, n. The flammable gas burned by carbide lamps, produced by the chemical reaction of water with calcium carbide.

anticline, n. An upward arching of rock strata.

arrow, n. A pointed line indicating the way out of a cave. Explorers sometimes mark arrows on walls or rocks with a rock, a finger, or soot from the flame of a carbide lamp.

ascenders, n. Mechanical rope-gripping devices used to climb standing ropes.

base level, n. The elevation of the bed of the largest surface stream in a region. The Green River is the base-level river in the Central Kentucky Karst.

bearing, n. The angular direction measured from one survey station or point to another with reference to magnetic north. In cave surveying, the bearing is read from a compass and is written in a survey notebook.

bedding plane, n. A narrow parting between two layers of rock—in caving, usually a horizontal slot too low for human travel.

belay, v. To secure a climber on a rope for protection in case of a fall.

belly crawl, n. A cave passage so low that one can travel through it only by squirming along in a prone position. Ceiling height is typically twelve to eighteen inches or less.

blowhole, n. An opening in bedrock or between boulders on the surface from which air blows perceptibly from underground.

bolt, n. An anchor, usually for a rope, fastened to a wall with a steel rod hammered into a drilled hole. A hanger is attached to the bolt and serves as means to attach the rope.

borehole, n. See *trunk passage.*

bowline, n. A slip-proof climber's loop knot, usually used with a safety belay.

brake bar, n. An aluminum rod that fastens onto a carabiner or rappel rack to provide friction. The standing rope is threaded past two or more brake bars to enable the climber to control the rate of descent.

breakdown, n. A jumble or pile of rocks in a cave passage produced by ceiling and wall collapse. A *breakdown block* is an angular piece of rock with dimensions ranging from a few inches to tens of feet on a side. *Terminal breakdown* is a pile of breakdown blocks that closes off a cave passage.

breathing cave, n. A cave passage in which air moves perceptibly in one direction and then in the other.

cable ladder, n. Two metal cables—usually made of stainless steel—with rungs of lightweight metal tubing, such as aluminum, forming a ladder six or eight inches wide with rungs spaced about eighteen inches apart. It can be rolled into a compact, lightweight bundle.

cairn, n. A pile of rocks constructed for use as a marker, ranging in height from a few inches to several feet.

canyon, n. A cave passage in which height exceeds width, usually by two or more times.

carabiner, n. Metal "snap-rings" used in climbing. Sometimes used with brake bars for short descents.

carbide, n. The chemical compound calcium carbide, which produces acetylene gas when it reacts with water. To *change carbide* is to recharge a carbide lamp by removing the spent carbide and putting in fresh carbide; with the carbide lamps used in Mammoth Cave, this must be done every three or four hours.

carbide lamp, n. A two-part container—usually made of brass—for generating acetylene gas. Water drips from the top part onto carbide in the bottom part to generate gas that escapes through a tiny hole in a tip or nozzle directed horizontally from the front of the container. The gas is ignited by sparks from a small steel wheel spun over flint, mounted on a reflector 2.5 inches in diameter. The lamp is mounted on the front of a hard hat, so that light shines in the direction the caver's head is turned.

cartography, n. With reference to caves, exploring and surveying with attention to passage detail, recording data, plotting, and drafting to produce a map showing the contours and features of the cave passages, names, topographic contours, and other details.

cave, n., v. A natural cavity beneath the earth's surface that is long enough and large enough to permit human entry, extending into total darkness. Caves in the Central Kentucky Karst are produced by solutional action of groundwater draining through natural openings and flowing to the base-level Green River. Also called caverns. *To cave* or *to go caving* is to explore a cave. *Big cave* usually means walking passages. *To look for cave* or *to look for more cave* is to look for new cave passages.

cave conservation, n. A policy of managing or using caves to protect and preserve their natural appearance by minimizing the adverse environmental effects of human activities.

cave meal, n. A small can of meat, a small can of fruit, and some candy bars.

Cave Research Foundation, n. A nonprofit organization founded in 1957 for the purpose of supporting cave science, interpretation and education, and conservation. The CRF is not an open membership organization.

cave river, n. Any stream of water in a cave is called a river. A cave river may be a few inches (Black River) or tens of feet (Logsdon River) wide and deep.

caver, n. A cave explorer.

Central Kentucky Karst, n. An area of karst terrain centered about one hundred miles south of Louisville. It is bounded on the north by the Hilly Country and on the west by Barren River and includes the Mammoth Cave Plateau and the Sinkhole Plain. It contains about two hundred square miles.

Central Kentucky Karst Coalition, n. A nonprofit organization founded in 1976 to explore and describe the caves of the Mammoth Cave Plateau outside Mammoth Cave National Park. The CKKC is an open membership organization with annual dues.

chain, n. A metal or fiberglass tape measure used in cave surveying, usually fifty or one hundred feet long. Also called a tape.

chert, n. A very hard, flintlike rock that occurs in beds or nodules in limestone.

chest compressor, n. A low, horizontal belly-crawl passage that one can get through only by squeezing—and often only by exhaling—to reduce the size of one's chest.

chimney, v. To climb the walls of a narrow passage or vertical cleft in a cave wall by bridging the opening with back and hands against one wall and feet against the other.

chock, n. A natural or artificial obstruction in a crack used as a hold for climbing or for securing ropes in caves.

collapse, n. A rockfall that usually closes off a passage with breakdown.

column, n. A pillar-like deposit extending from ceiling to floor of a cave passage, formed by the natural joining of a stalactite and a stalagmite.

commercial cave, n. A cave containing trails and lights that is exhibited to the public for an entrance fee. Kentucky Caverns in the central Kentucky cave country is a classic commercial cave.

compass, n. A pocket-sized surveying instrument or hand-held transit consisting of a magnetic needle or dial and one or more sights for reading bearings.

connect, v. To find and survey through a natural cave passage that joins what were previously known as two independent caves.

connection, n. In caves, the naturally formed passage between two or more caves previously known through separate entrances.

crack, n. A narrow opening in the wall, floor, or ceiling of a cave passage.

crawlway, n. A cave passage so low that one can get through it comfortably only on hands and knees. Ceiling height is typically two to three feet.

crouchway, n. A cave passage so low that one can get through it only in a stooped or duck-walk position. Ceiling height is typically three to five feet.

cutaround, n. A section of secondary cave passage that departs from a main passage and returns to it after a short distance.

dome, n. A vertical, circular, or oval opening in the ceiling of a cave passage produced by the solutional activity of descending water or by collapse. Vertical shafts viewed from below are often called domes; viewed from the middle, they are called domepits.

dripstone, n. A secondary mineral deposit within a cave. Common forms include stalagmites, stalactites, and rimstone.

exfoliate, v. To come off in sheets, flakes, or layers parallel with the surface; for example, pieces from shaley rock walls and ceiling of a cave passage.

fill, n. Any indigenous loose material in the cave, such as sand or mud. Or, as in the *passage ends in fill:* a passage that terminates in a floor to ceiling fill. Or, as in the *passage filled:* a passage that was found to end in fill.

Flint Ridge, n. The northernmost karst ridge on the Mammoth Cave Plateau in Mammoth Cave National Park, containing Floyd Collins' Crystal Cave, Unknown Cave, Great Onyx Cave, Salts Cave, and Colossal Cave. The four connected caves (all except Great Onyx Cave) were known as the Flint Ridge Cave System (FRCS) from 1961 to 1972. In 1972, the FRCS was connected to Mammoth Cave to form the Flint Mammoth Cave System.

Flint Ridge con, n. A pattern of speech and body language calculated to win the confidence of the listener and to promote expectations of wonderful underground discoveries. As perfected by Jim Dyer, it fascinates and disarms the listener and creates eagerness to explore the place described, even if aware that he or she is being manipulated by an artist.

flowstone, n. A secondary deposit in caves of calcium carbonate, usually in the form of the mineral calcite, precipitated from groundwater. It occurs in the form of sheets, drapery, dams, lily pads, and the like. (This is a local usage for what is more generally called travertine.)

fluorescein dye, n. A concentrated, nontoxic chemical that colors water a vivid green. It is used to trace the course of underground streams. It can be detected in concentrations of a few parts per million when intercepted by an activated charcoal dye trap or "bug."

flutes, fluting, n. Vertical striations, grooves, or alcoves in cave passage walls or vertical shaft walls produced by the solutional activity of concentrated streams of descending water.

free-climb, n. A climb that does not require any equipment to ascend or descend.

frost action, frost-wedging, n. Mechanical weathering of rock caused by the freezing of water and expansion of ice in fractures of exposed rocks.

go, v. Cave passages that go are passages open for exploration, or *going cave.* A lead that looks passable is said *to go.* One also says of an explored passage that *it went.*

grape, n. A secondary deposit of calcium carbonate on cave walls, with surface projections consisting of globules ranging from pea size to grape size.

groundwater, n. Subsurface water that lies at or below the water table.

gypsum, n. A white to colorless mineral, calcium sulfate, deposited in caves in a variety of crystalline forms resembling needles, flowers, cotton, feathers, and wood shavings and as faceted crystals. Gypsum is sometimes colored yellow, orange, or brown by impurities.

hanger, n. A metal fastener, attached to a bolt, with a hole to attach or clip in a carabiner.

hanging survey, n. Cave passages that are surveyed, or the data of that survey, that are not connected to the main grid of surveyed passages in a cave.

hard hat, n. A protective helmet or head covering of fiberglass or strong plastic held on the head by a chin strap and containing an attachment on the front for the placement of a carbide lamp.

hydrology, n. The study of groundwater.

hydrologist, n. A scientist who studies groundwater.

Joppa Ridge, n. The southwesternmost karst ridge of the Mammoth Cave Plateau in Mammoth Cave National Park, containing Proctor Cave and Lee Cave.

Jumar, n. A mechanical clamping and camming device used to ascend ropes. Used in combinations of two or three, they allow the climber to ascend step-like up a rope.

karst, n. A characteristic landscape produced by solution and underground drainage in areas of soluble bedrock such as limestone and dolomite. Karst landforms include sinkholes and sinking streams, irregular ridges such as Flint Ridge, and blind or closed valleys such as Doyle Valley.

karst ridge, n. A cavernous upland, bounded by karst valleys, base-level rivers, or sinkhole plains. Mammoth Cave Ridge is an example.

karst valley, n. A cavernous lowland, usually formed by solutional activity of water along the course of a former surface stream, the waters of which have been captured underground. Sometimes called a sink valley. Doyle Valley is an example.

knee-crawlers, n. Molded rubber pads, each with two straps for fastening on one's knee. They protect knees from bruises and abrasions when one crawls on rocks in caves. Also called knee pads.

layback, v. A method of rock climbing in which the climber's fingers pull one way and the feet push the opposite way in a crack in a wall, with the climber's body out in space parallel to the wall like that of an inchworm. The vector of opposing forces effectively neutralizes gravity to permit the climber to move hands and feet alternately to climb vertically.

lead, n. A cave passage that looks big enough to explore.

limestone, n. A sedimentary rock composed principally of calcium carbonate. Mammoth Cave is developed primarily in the Girkin Limestone, the Ste. Genevieve Limestone, and St. Louis Limestone of Carboniferous age.

line plot, n. A map made by drawing straight lines between survey points.

longbar, n. A steel crowbar three or four feet long.

Mammoth Cave National Park, n. Established in 1941 as a part of the National Park System. For information, write to Mammoth Cave National Park, Mammoth Cave, Kentucky, 42259.

Mammoth Cave Plateau, n. An upland in Mammoth Cave National Park in the Central Kentucky Karst containing major karst ridges separated by major karst valleys. Flint Ridge (10 square miles) is separated by Houchins Valley (2 square miles) from Mammoth Cave Ridge (3.5 square miles), which in turn is separated by Doyle Valley (2.7 square miles) from Joppa Ridge (4 square miles). Toohey Ridge (3.5 square miles) is a spur of Mammoth Cave Ridge. These individual ridges join each other.

Mammoth Cave Ridge, n. The central karst ridge of the Mammoth Cave Plateau in Mammoth Cave National Park containing Mammoth Cave, which has been known since 1798 and which became a part of the Flint Mammoth Cave System in 1972.

mirabilite, n. A water-soluble colorless mineral, hydrated sodium sulfate, occurring in a variety of crystalline forms, often resembling needles, coconut, or powder. Useful as a salt substitute and as a cathartic.

National Park Service, n. Founded in 1916 as a branch of the U.S. Department of the Interior. For information, write to National Park Service, Washington, D.C., 20240

National Speleological Society, n. A membership organization founded in 1941 and affiliated with the American Association for the Advancement of Science. The purpose of the NSS is to encourage scientific study and conservation of caves, and it maintains extensive conservation and publication programs. Local branches of the NSS are called grottoes and are located in all areas of the U.S.A. For further information, address queries to the National Speleological Society, 2813 Cave Avenue, Huntsville, Alabama, 35810-4431.

passage, passageway, n. A horizontal opening in a cave large enough for one to enter.

pinch, pinch-down, n., v. Where a passage gets so low or narrow that one cannot penetrate it; at that time, the passage has *pinched out.*

pit, n. A vertical opening in the floor of a cave passage produced by collapse of rock, slumping of breakdown, or the solutional activity of descending water. Vertical shafts viewed from above are often called pits.

pitch, n. An uninterrupted vertical drop.

piton, n. A nail-like device used for climbing or securing ropes in caves. Pitons are hammered into cracks with a rock hammer.

plotting, v. Use of a protractor and scale to draft a cave map. Sometimes a computer plotter is used to draft the map automatically from data supplied in digital form.

point, survey point, n. A marked, numbered location in a cave passage on which the compass is placed for measuring the bearing and on which the chain is placed for measuring the distance to the next point. (This is a local usage for what is more generally called station or survey station.)

point person, n. The first person in a cave survey party who selects and marks each survey point.

popcorn, n. A secondary deposit in cave passages, characterized by rough, globular surface projections.

pothole, n. A round, bowl-like pocket in the floor of a cave.

prussik knot, n. A special sliding hitch made of rope used for climbing standing ropes. The hitch clamps on the standing rope when loaded. Also called a prusik.

rack, n. A mechanical device containing multiple brake bars that is attached to a climber's seat harness to enable a controlled descent. The standing rope is threaded around the bars to provide friction to control the speed.

rappel, n., v. The act or method of descending a vertical shaft or cliff by means of a rope passed around the body through any of a variety of devices that produce friction, permitting a controlled rate of descent.

rebelay, n. A secondary anchor for a standing rope.

rinky-dink, n. A cave tour calculated to confuse, entertain, and fascinate the victims by taking them on a circuitous route over obstacles and at a dizzying pace. Sometimes called a run-around.

short bar, n. A steel crowbar twelve to eighteen inches long with a curved end, using for digging and prying at breakdown rocks in caves.

sinkhole, n. A closed, often oval basin-like depression on a karst land surface through which water drains underground. A sinkhole entrance is an opening into a cave from a sinkhole.

sinkhole plain, n. A geographic unit of the Central Kentucky Karst consisting of a gently rolling land surface with drainage carried underground through sinkholes and sinking streams. Water drains underground to base-level Green River and Barren River. The sinkhole plain is bounded by the Mammoth Cave Plateau on the north, the Green River on the east, the Warsaw Limestone on the south, and the western divide of the Barren River on the west.

sinking stream, n. A stream that flows in a valley that terminates in a headwall beneath which the stream plunges underground.

siphon, n., v. The place at which a cave passage is drowned by the ceiling extending underwater. A passage that is closed off by water in this way is said *to siphon.* (This is a local usage for what is more generally called a sump.)

speleologist, n. A scientist who specializes in speleology.

speleology, n. The science of the origin and nature of physical and biological features and processes of the karst and cave environment.

spelunker, n. A derogatory term for caver. "Who is that spelunker with new red coveralls and flashlight?"

stalactite, n. A pendulant, conical, or icicle-like deposit. Stalactites are usually of calcite and hang from the ceilings of cave passages. They are precipitated from mineral-bearing solutions dripping from cave ceilings.

stalagmite, n. A cylindrical, conical, or mound-shaped mineral deposit. Stalagmites are usually of calcite and project from floors of cave passages. They are precipitated from mineral-bearing solutions dripping from the ceiling and often form under stalactites.

station, survey station, n. See *point.*

sump, n. See *siphon.*

survey, n., v. In caves, the systematic process of, or the data from, measuring and recording the bearings and distances of passages, using a compass and chain for the purpose of making an accurate map. Often a cave passage is referred to by the name of the survey in the passage, such as the Q Survey. Thus "the Q Survey" can refer either to the survey data or to the passage from which that data is taken.

survey party, survey team, n. Usually consists of three or four cavers, one of whom reads the compass or other surveying instrument, one or two of whom set points and read the chain, and one who takes notes in a survey book.

survey shot, shot, n. In cave surveying, one of a series of bearing and distance measurements between points.

tape, n. See *chain.*

terminal breakdown, n. Rocks that have fallen from ceiling and wall to form a barrier that closes a cave passage.

Toohey Ridge, n. A karst ridge immediately to the east of Mammoth Cave National Park, containing Roppel Cave. In 1983, the forty-nine-mile-long Roppel Cave was connected to Mammoth Cave.

transit, n. An instrument for measuring horizontal and vertical angles of bearings in surveying.

traverse, n., v. To move laterally across the face of a wall, such as a wall of a pit, using climbing techniques, or a section of cave passage requiring traverse movements. A *compass traverse* is a survey.

trunk, trunk passage, trunk stream, n. A major conduit, usually of walking height. Logsdon River, Elysian Way, and Arlie Way are considered trunk passages.

tube, n. A horizontal cave passage roughly cylindrical or elliptical in shape.

tyrolean, n., adj. A traverse using a horizontal, standing line, usually to cross pits. In a tyrolean traverse, the caver moves hanging like a sloth beneath the line.

vertical shaft, n. A vertical, well-like, usually oval cave opening underground, produced by the solutional activity of water seeping downward at the intersection of joints or fractures in the limestone or dolomite. Shafts range in diameter from a few inches to more than forty feet and in height from a few feet to more than a hundred feet. In the Central Kentucky Karst, vertical shafts are the underground heads of drainage, and their drains are tributaries to underground trunk streams.

virgin cave, virgin passage, n. A cave or passage that has not been explored.

walking cave, n. A cave passage high enough to walk in.

water table, n. The level below which the ground is saturated with water.

wetsuit, n. A rubber suit used to insulate against cold water.

wild cave, n. A cave that has not been developed with trails and lighting.

INDEX

The abbreviations CKKC and CRF stand for Central Kentucky Karst Coalition and Cave Research Foundation respectively. Abbreviations locate passages within their respective caves: Mammoth Cave (MC), Morrison Cave (MO), Proctor Cave (PC), and Roppel Cave (RC). Page numbers in **boldface** type refer to illustrations or maps; **cp** refers to color plates.

James D. Borden is a program manager at the IBM Corporation in Poughkeepsie, New York. He has authored a number of technical papers on computer system performance and taught classes in mathematics and in the technical support of computer subsystems and software. For over twenty-five years, he has directed exploration work in Roppel Cave, and he has written a number of articles documenting this work. Borden is a fellow of the National Speleological Society, a past director of the Cave Research Foundation, and the managing director of the Central Kentucky Karst Coalition.

Roger W. Brucker is the former president of OIA Marketing Communications in Dayton, Ohio, an adjunct professor of marketing at Wright State University and at the School of Advertising Art in Dayton, and the author of many articles on marketing and advertising. A cave explorer and speleologist, he has coauthored three books on cave exploring in the Mammoth Cave area of Kentucky: *The Caves Beyond, The Longest Cave,* and *Trapped! The Story of Floyd Collins.* In addition, he taught speleology for twenty years at Western Kentucky University. Brucker is an honorary life fellow of the National Speleological Society and a past president and founding director of the Cave Research Foundation.